Biotreatment Systems

Volume I

Editor

Donald L. Wise, Ph.D., P.E.

Cambridge Scientific, Inc.
Belmont, Massachusetts
and
Cabot Professor of Chemical Engineering
Northeastern University
Boston, Massachusetts

CRC Press, Inc.
Boca Raton, Florida

Library of Congress Cataloging-in-Publication Data

Biotreatment systems.
  Bibliography: p.
  Includes index.
  1. Sewage—Purification—Biological treatment.
I.Wise, Donald L. (Donald Lee), 1937- .
TD755.B49 1988   628.3'51   87-24234
ISBN 0-8493-4848-X (v. 1)
ISBN 0-8493-4849-8 (v. 2)
ISBN 0-8493-4850-1 (v. 3)

Direct all inquiries to CRC Press, Inc., 2000 Corporate Blvd., N.W., Boca Raton, Florida, 33431.

© 1988 by CRC Press, Inc.

International Standard Book Number 0-8493-4848-X (v. 1)
International Standard Book Number 0-8493-4849-8 (v. 2)
International Standard Book Number 0-8493-4850-1 (v. 3)

Library of Congress Number 87-24234
Printed in the United States

PREFACE

Biotreatment Systems is a uniquely valuable reference text consisting of contributed chapters in which are described the most insightful research and development programs around the world. The authors of these contributed chapters are those very conscientious and thoughtful technologists who are investigating pragmatic solutions to environmental problems. This important text has as the major theme the biotreatment of organic residues. This major theme primarily encompasses the field of anaerobic methane fermentation, with an emphasis on treatment of complex wastes. The text is intended to present a comprehensive overview of the most practical research programs that are being carried out in this emerging field of international significance. Due to the fact that both research and development have been carried out at major centers around the world, great care has been taken to include chapters from an international perspective. Further, as a perusal of the chapter titles will indicate, a special emphasis has been made to address both the important research aspects and the practical aspects of the work on biotreatment systems. It is to be noted that each chapter included in this text is the work of a particular individual or group. There are no multiple chapters by more than one author or group. Thus, each of the included chapters most often reflects the dedicated career efforts of these workers. Further, each contributed chapter is presented on a stand-alone basis so that the reader will find it helpful to consider only the theme of each chapter. On the other hand, there is the unifying theme with all chapters of addressing biotreatment systems research and development. A reader of this text, just entering the field, will find this text provides an excellent state-of-the-art presentation of the international import of work on biotreatment systems, with an emphasis on methane fermentation. A reader of this text, who has experience in this field, will find the text to be essential for assessment and referral of this increasingly valuable area of technology.

THE EDITOR

Donald L. Wise, Ph.D., P.E., is Founder and President of Cambridge Scientific, Inc., Belmont, Massachusetts. Dr. Wise also holds the Cabot Chair of Chemical Engineering at Northeastern University, Boston, Massachusetts. Dr. Wise received his B.S. (magna cum laude), M.S., and Ph.D. degrees in chemical engineering at the University of Pittsburgh. Dr. Wise is a specialist in process and biochemical engineering as well as advanced biomaterials development. During his career he has managed a series of programs to develop processes for production of fuel gas, liquid fuels, and organic chemicals from municipal solid waste, an array of agricultural residues, and a wide variety of crop-grown biomass, especially aquatic biomass. Dr. Wise has also been primarily responsible for the initiation of development work on fossil fuels such as peat and lignite to gaseous fuel, liquid fuels, and organic chemicals, and he also originated work on the bioconversion of coal gasifier product gases to these products. Dr. Wise initiated a program to establish the engineering feasibility of converting large-scale combined agricultural residues to fuel gas by the action of microorganisms, a project ultimately involving joint effort with research workers in fifteen countries around the world.

Dr. Wise has worked in the area of biotechnology research and development for 2 decades, has approximately 50 publications in the field, and has edited a number of reference texts. As Associate Editor of *Solar Energy,* the journal of the International Solar Energy Society, he is responsible for the review of manuscripts in the biomass/bioconversion area. Dr. Wise is also on the Editorial Board of *Resources and Conservation,* an international journal published by Elsevier, Amsterdam. He has served as an international consultant in bioconversion for the United Nations and for the U.S. Agency for International Development (AID).

A meaningful portion of these programs that Dr. Wise initiated, and has been carrying out, is his meeting with experts across the U.S. and around the world, to become familiar with both current and practical aspects of bioconversion systems.

CONTRIBUTORS

Ulrich Behrens
Institute of Biotechnology
Academy of Sciences
Leipzig, German Democratic Republic

Günter Bürger
Institute of Biotechnology
Academy of Sciences
Leipzig, German Democratic Republic

J. F. de Kreuk
Division of Technology for Society
TNO
Delft, The Netherlands

Phillip M. Fedorak
Department of Microbiology
University of Alberta
Edmonton, Alberta, Canada

Michael J. Hanchak
Ecology and Environment, Inc.
Buffalo, New York

A. O. Hanstveit
Division of Technology for Society
TNO
Delft, The Netherlands

Philip A. Herzbrun
CECOS International
Buffalo, New York

Steve E. Hrudey
Department of Civil Engineering
University of Alberta
Edmonton, Alberta, Canada

Robert L. Irvine
Center for Bioengineering and
Pollution Control
and
Department of Civil Engineering
University of Notre Dame
Notre Dame, Indiana

D. B. Janssen
Groningen Biotechnology Center
Groningen, The Netherlands

Iman W. Koster
University Lecturer
Water Pollution Control Department
Wageningen Agricultural University
Wageningen, The Netherlands

P. Kumaran
Scientist
Environmental Microbiology Division
National Environmental Engineering
Research Institute
Nagpur, India

Kenneth C. Malinowski
CECOS International
Buffalo, New York

Günther Martius
Institute of Biotechnology
Academy of Sciences
Leipzig, German Democratic Republic

Manfred Ringpfeil
Institute of Biotechnology
Academy of Sciences
Leipzig, German Democratic Republic

W. H. Rulkens
Division of Technology for Society
TNO
Apeldoorn, The Netherlands

N. Shivaraman
Scientist
Environmental Microbiology Division
National Environmental Engineering
Research Institute
Nagpur, India

Ulrich Stottmeister
Institute of Biotechnology
Academy of Sciences
Leipzig, German Democratic Republic

W. J. Th. van Gemert
N.V. Nederlandse Gasunie
Groningen, The Netherlands

H. J. van Veen
Division of Technology for Society
TNO
Apeldoorn, The Netherlands

Lutz Wenige
Deceased
Dipl. Ing.
VEB Chemieanlagenbaukombinat
Leipzig, German Democratic Republic

SERIES TABLE OF CONTENTS

Volume I

Aerobic Treatment of Sewage from Lignite (Brown Coal) Processing
Literature Study on the Feasibility of Microbiological Decontamination of Polluted Soils
Treatment of Hazardous Wastes in a Sequencing Batch Reactor
Anaerobic Degradation of Phenolic Compounds with Applications to Treatment of Industrial Waste Waters
Biological Treatment of Toxic Industrial Wastes
Microbial, Chemical, and Technological Aspects of the Anaerobic Degradation of Organic Pollutants

Volume II

Toxicity of Heavy Metals to Thermophilic Anaerobic Digestion
In Situ Biological Treatment of Hazardous Waste-Contaminated Soils
The Role of Phenolic and Humic Compounds in Anaerobic Digestion Processes
Bacterial Leaching of Heavy Metals from Anaerobically Digested Sludge
Biodegradation of Recalcitrant Industrial Wastes

Volume III

Anaerobic Biological Processes for the Prevention of Noxious Odors in Pulp Manufacturing
Potential for Treatment of Hazardous Organic Chemicals with Biological Processes
Anaerobic Treatment of Sulfate-Containing Waste Water
Enhanced Biological Phosphorus Removal from Waste Waters

TABLE OF CONTENTS

Chapter 1
Aerobic Treatment of Sewage from Lignite (Brown Coal) Processing................... 1
**Manfred Ringpfeil, Ulrich Stottmeister, Ulrich Behrens, Günther Martius,
Günther Bürger, and Lutz Wenige**

Chapter 2
Literature Study on the Feasibility of Microbiological Decontamination of
Polluted Soils.. 63
**J. F. de Kreuk, A. O. Hanstveit, W. J. Th. van Gemert, D. B. Janssen,
Wilhelmus Henricus Rulkens, and H. J. van Veen**

Chapter 3
Treatment of Hazardous Wastes in a Sequencing Batch Reactor 157
**Philip A. Herzbrun, Robert L. Irvine, Kenneth C. Malinowski, and
Michael J. Hanchak**

Chapter 4
Anaerobic Degradation of Phenolic Compounds with Applications to Treatment of
Industrial Waste Waters 169
Phillip M. Fedorak and Steve E. Hrudey

Chapter 5
Biological Treatment of Toxic Industrial Wastes 227
P. Kumaran and N. Shivaraman

Chapter 6
Microbial, Chemical, and Technological Aspects of the Anaerobic Degradation
of Organic Pollutants 285
Iman W. Koster

Index 317

Chapter 1

AEROBIC TREATMENT OF SEWAGE FROM LIGNITE (BROWN COAL) PROCESSING

Manfred Ringpfeil, Ulrich Stottmeister, Ulrich Behrens, Günter Martius, Günter Bürger, and Lutz Wenige

TABLE OF CONTENTS

I.   Introduction.. 2

II.   Brown Coal Formation and Processing 3
      A.   Brown Coal Formation 3
      B.   Brown Coal Reserves of the World 5
      C.   Refinement of Brown Coal by Thermal Methods........................ 5
            1.   Carbonization 5
            2.   Coal Gasification 6
            3.   Liquefaction of Brown Coal 8
            4.   Summary of Processes of Brown Coal Refinement and
                  Composition of Water 8

III. Treatment of the Waste Waters from Brown Coal Refinement................. 10
      A.   Physical and Chemical Methods....................................... 10
            1.   Physical Methods 12
                  a.   Preliminary Purification 12
                  b.   Extraction Methods for Valuable Substances 12
                  c.   Koppers Recirculation Method 14
                  d.   Adsorption Method....................................... 14
            2.   Chemical Methods 14
      B.   Microbial Methods of Purifying Brown Coal Processory
            Waste Waters.. 17
            1.   Detailed Composition of Brown Coal Processory
                  Waste Waters 17
                  a.   Profile of the Fatty Acids 18
                  b.   Profile of Polyhydroxy Benzenes........................ 18
                  c.   Profile of Further Compounds 19
            2.   Air Oxidation and Biological Oxygen Demand after 5 Days...... 19
            3.   Microbiological Fundamentals of the Purification of Brown
                  Coal Processing Waste Waters................................. 21
            4.   Biochemistry of the Degradation of the Substances Contained
                  in Brown Coal Processory Waste Waters 23
            5.   Kinetics of the Aerobic Degradation of Brown Coal
                  Processory Waste Waters 30
            6.   Nitrification and Denitrification............................ 35
                  a.   Nitrification 35
                  b.   Denitrification 36
      C.   Possibilities for the Utilization and Disposal of Sludge from
            Plants for the Biological Treatment of Brown Coal Processory
            Waste Waters.. 37

IV.   Technology of Biological Waste Water Treatment of Brown Coal
       Processory Waste Waters 40
       A.   Substrate-Specific Demands on the Technology 40
       B.   Processes of Biological Purification 41
             1.   The Magdeburg P-Process 41
             2.   The Tower Trickling Filter Method 41
             3.   Methods of High-Performance Biological Treatment 42
             4.   Final Purification Stage 46
             5.   Advanced Biological Purification of Sewage 46
       C.   Technical Plants Realized in the German Democratic Republic for
             the Biological Purification of Brown Coal Processory
             Waste Waters 48
             1.   High-Performance Biological Purification by the PKM
                   Process with Aeration by a Centrifugal Aeration with
                   Automatic Foam-Drawing Device 48
             2.   High-Performance Biological Purification with CLG Biotank
                   Reactors 49
             3.   The PKM Process for Advanced Biological Purification 49

V.   Discussion and Prospects 53

Acknowledgments 55

List of Abbreviations 56

References 56

I. INTRODUCTION

On an international scale the utilization of brown coal is only beginning. A few countries are beyond the simple use for heat production, being the lowest stage of utilization. The pressure for utilization of indigenous raw material sources in the Middle European countries led, in the first half of our century, to a highly developed industry aimed at the refinement of the products to be obtained from brown coal processing. Following this way directive and technical developments were attained. However, due to the numerous technical and economical advantages of the petrochemistry, carbochemistry based on brown coal was pushed back in these countries.

In the German Democratic Republic (G.D.R.) and Czechoslovakia carbochemistry has always been important. Newly erected, large plants in the G.D.R. were designed in the 1950s and 1960s with a fully biological waste water purification system. On the basis of some experiences of old, brown coal processing plants in the field of biological waste water purification, the necessity of an intensive research work concerning these new plants arose. Appropriate to the interest at that time the results have been published only in the German-speaking area. The mentioned turn of interest toward a more extended utilization of brown coal also in other countries challenges a summarized representation of the biological waste water purification. In this way parallel developments may possibly be avoided and new suggestions for further investigations may be stimulated.

The experiences gained from decades of operation of biological purification plants in the brown coal industry have led to industrial plants taking into account the latest knowledge of biotechnology. To recognize the features of waste waters yielding from brown coal processing and to understand the treatment technologies applied to it, a short representation of the brown coal formation as well as a compendium of processing technologies seem to be necessary.

II. BROWN COAL FORMATION AND PROCESSING

A. Brown Coal Formation

Formation of peat, brown coal, and hard coal from vegetable material presumed special conditions in the course of earth history. The local redox condition in connection with surface subsidence and water covering as well as the prevailing climate were fundamental for humification of the plants. The plurality of compounds included in plants can be confined to the groups celluloses, hemicelluloses, sugar, lignin, albumins, waxes, and resins. Cellulose and lignin are of importance for humus formation. Albumins are completely decomposed in the humification process. Phosphorus, nitrogen, and sulfur of the coal result from albumin compounds. Waxes and resins go scarcely altered and they provide the protobitumen.

In the process of coal formation the organic components of the plants are converted to humic acid in the primary oxidation zone with differentiated ozygen access. By means of condensation and polymerization reactions compounds with an aromatic nucleus and aliphatic side-groups are formed; for example, fulvic acid, humolignin acid, and various humic acids have been identified.

In deeper zones from the geological point of view a reduction of the humus substances takes place by microbial processes. Since no oxygen is available sulfate, nitrate, and organic compounds are used as electron acceptors (biochemical phase of coalification).

In the course of coalification the local geological conditions are leading to products of different composition. Elementary analyses of different products of the coalification process are illustrated in Figure 1 (A and B).

An elementary analysis, however, is not sufficient for the characterization and the comparison of coals summarized under the term "brown coal". The microscopically detectable single-constituents (macerals) are used for a classification of brown coals (e.g., humite, liptinite, and intertinite) indicating the chemical, physical, and technological properties of brown coals. To enable a more complete evaluation of coals from different fields concerning the technological properties of coals, classification characteristics have been amplified. Along with a coefficient for the coalification degree, one or two numbers have been introduced for the characterization of the technological behavior of the brown coals named ISO classification (ISO 2950 — 1974 [E]). It differentiates into six classes of freshly mined coals concerning the water content and within these five classes concerning the tar yield.[1] The technological properties of the coal determine the establishment of specific processing technologies in the vicinity of certain coal fields (e.g., in the G.D.R., brown coal coking to the west and gasification to the east of the Elbe river).

Within a deposit the different seams may show great differences with regard to composition and, regularly, only one seam is worth mining. Extraction of brown coal takes place exclusively by open-cast mining. Temporarily large influences on the cultural landscape are inevitable. An open-cast mine causes a serious engagement in the water conservation of the landscape concerned. The pit waters have to be pumped out and are partly used as industrial water. In case of a high salt load as is found in salt coal deposits, additional problems with regard to water treatment come into being which are not taken into account in the following considerations.

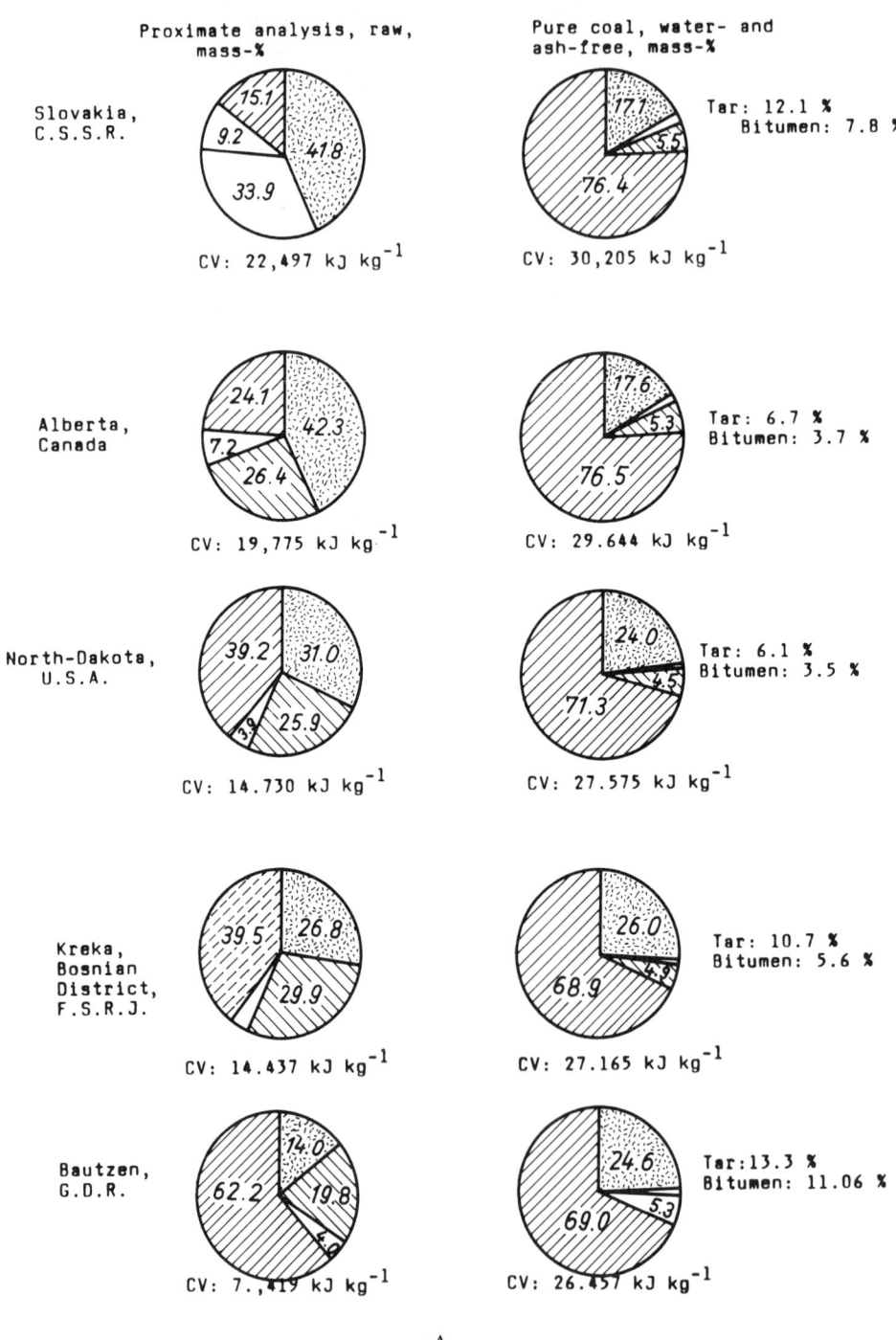

Proximate analysis, raw, mass-%

Pure coal, water- and ash-free, mass-%

Slovakia, C.S.S.R.

15.1, 9.2, 33.9, 41.8
CV: 22,497 kJ kg^{-1}

17.7, 5.5, 76.4
Tar: 12.1 %
Bitumen: 7.8 %
CV: 30,205 kJ kg^{-1}

Alberta, Canada

24.1, 7.2, 26.4, 42.3
CV: 19,775 kJ kg^{-1}

17.6, 5.3, 76.5
Tar: 6.7 %
Bitumen: 3.7 %
CV: 29.644 kJ kg^{-1}

North-Dakota, U.S.A.

39.2, 31.0, 25.9
CV: 14.730 kJ kg^{-1}

24.0, 4.5, 71.3
Tar: 6.1 %
Bitumen: 3.5 %
CV: 27.575 kJ kg^{-1}

Kreka, Bosnian District, F.S.R.J.

39.5, 26.8, 29.9
CV: 14.437 kJ kg^{-1}

26.0, 68.9
Tar: 10.7 %
Bitumen: 5.6 %
CV: 27.165 kJ kg^{-1}

Bautzen, G.D.R.

14.0, 62.2, 19.8
CV: 7.,419 kJ kg^{-1}

24.6, 5.3, 69.0
Tar: 13.3 %
Bitumen: 11.06 %
CV: 26.457 kJ kg^{-1}

A

FIGURE 1 (A and B).   Composition of brown coal from different fields. (According to Kurtz, R., *Ullmanns Enzyklopädie der technischen Chemie,* Vol. 14, 4th ed., Bartholomé, E., Biekert, E., Hellmann, H., Ley, H., Weigert, W. M., and Weise, E., Eds., Verlag Chemie, Weinheim, West Germany, 1977, 491.)

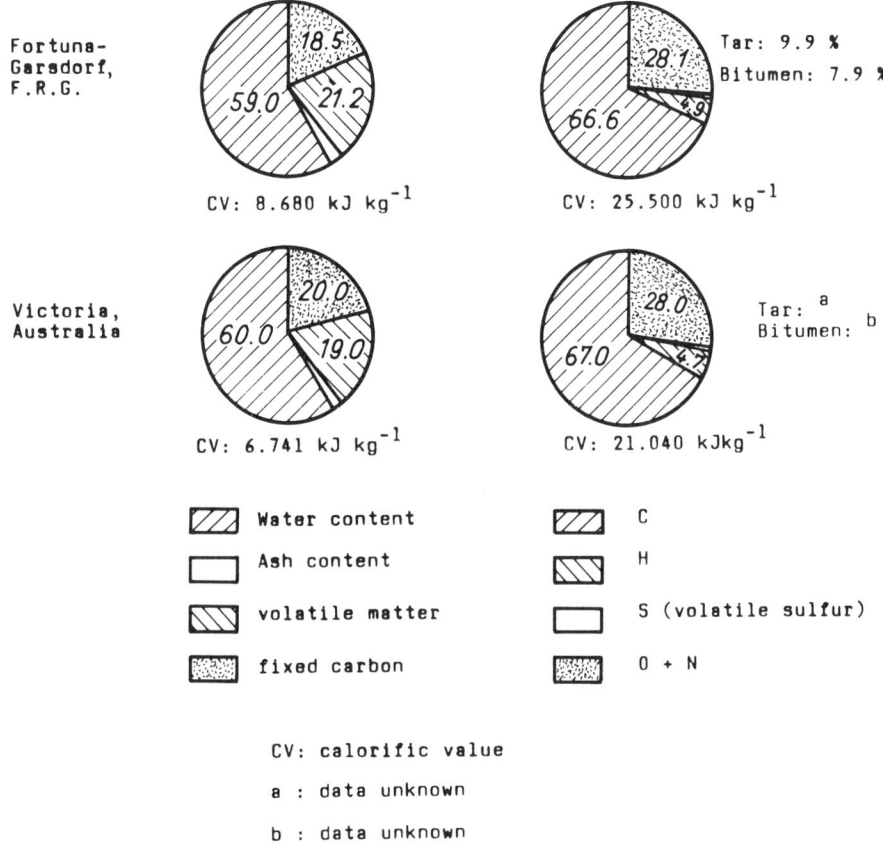

Fortuna-
Garsdorf,
F.R.G.

Victoria,
Australia

Tar: 9.9 %
Bitumen: 7.9 %

CV: 8.680 kJ kg^{-1}

CV: 25.500 kJ kg^{-1}

Tar: [a]
Bitumen: [b]

CV: 6.741 kJ kg^{-1}

CV: 21.040 kJkg^{-1}

Water content

Ash content

volatile matter

fixed carbon

C

H

S (volatile sulfur)

O + N

CV: calorific value

a : data unknown

b : data unknown

FIGURE 1B.

B. Brown Coal Reserves of the World

In Table 1 the known brown coal reserves of the world are illustrated. Reserves and production of some countries are also shown. About 70% of the world production is realized by the G.D.R. (27.5%), West Germany (13.4%), the U.S.S.R. (17.6%), and Czechoslovakia (10.2%).

C. Refinement of Brown Coal by Thermal Methods

Thermal refinement methods are leading to products which are energetically more favorable to use than brown coal with regard to utilization in combustion processes. Products of these refinement processes seem to be usable more favorably, e.g., as high-temperature coke, than comparable products from hard coal. For that reason brown coal refinement methods have found again an increasing interest in the last years.

1. Carbonization

Low-temperature carbonization (LTC) is the heating of brown coal to a temperature of 500 to 600°C in the absence of air. High-temperature carbonization (HTC or coking) is the heating of brown coal to a temperature of about 1200°C in the absence of air. Solid bitumen-free residues as well as liquid and gaseous hydrocarbons result from these pyrolyses. LTC aims at the production of tar whose constituents are the basis of a comprehensive branch of carbochemistry.[2] Coke is the main product of HTC and the available portions of gas and liquid products are adequately modified. In principle there is no other difference between the two methods than the different heating temperatures. Heating-up velocity and size of

Table 1

BROWN COAL RESERVES OF THE WORLD AS WELL AS RESERVES AND PRODUCTION OF SOME COUNTRIES[6]

Country/area	Year	Reserves (metric tons, millions)				Production (metric tons, thousands)		
		Known economic reserves						
		Total	Total known recoverable	Additional resources	Total resources	1970	1976	1978
World		342,070	213,865	2,472,916	2,186,874	792,652	891,481	922,169
Australia	1977	68,058	39,000	54,628	122,686	24,175	30,939	32,868
Bulgaria	1972	4,356	4,356	840	5,196	28,854	25,184	25,531
Czechoslovakia	1966	8,234	3,870	1,623	9,857	81,783	89,468	94,879
G.D.R.	1966	30,000	252,000	—	30,000	261,482	246,879	253,264
F.R.G.	1977	55,000	28,890	6,100	61,100	108,437	134,535	123,587
Hungary	1966	2,900	1,450	2,779	5,679	23,679	22,323	22,716
India	1977	1,868	934	231	2,099	3,545	3,900	3,606
Indonesia	1974	1,960	980	—	1,960	—	1	—
Poland	1976	6,449	4,840	8,413	14,862	32,767	39,305	40,985
Romania	1966	1,367	1,100	2,533	3,900	14,129	18,731	21,845
Turkey	1977	2,968	3,575	1,875	4,843	4,437	8,252	9,326
U.S.S.R.	1971	107,402	53,700	1,612,922	1,720,324	144,745	160,031	162,871
U.S.	1977	30,497	15,249	764,664	795,161	5,409	23,101	32,318
Yugoslavia	1971	17,894	16,800	3,753	21,642	27,779	36,259	39,238

lumps of the feedstock have a decisive influence on the products yielding. Realizing a heating-up velocity of, for example, 2°C/min in the coke oven the vaporization of the inherent water will begin at 70°C. Within the range of 105 to 220°C the inherent moisture vaporizes almost completely, after that the formation of decomposition water begins and above 270°C the tar formation phase will follow.

In the case of carbonization involving gas recirculation, hot flue gases are used for coal heating. Therefore, using this method in addition to inherent water and decomposition water, a portion of combustion water is in the mixture of condensed waters from carbonization process. The mass portions of the waters formed in carbonization process are illustrated in Figure 3.

2. Coal Gasification

In principle coal gasification consists of a joint heating of coal with a gasification medium. The objective of the conversion is the production of combustible and reactive gases, respectively. In the case of autothermal gasification of brown coal, water vapor and oxygen or air are used. The heat necessary for the conversion is obtained by partial combustion of brown coal.

In the case of allothermal gasification the heat necessary for the conversion is generated outside of the reactor and transmitted to the reaction chamber.

A plurality of reactor types and process variants have been elaborated[3] for the technical realization of autothermal gasification. Technological innovations such as, for example, pressure gasification of pulverized coal, have been developed for the utilization of brown coals which are not suitable for processing with known technologies due to the composition of the coals (e.g., salt coal).

Relating to allothermal gasification methods, efforts can be recognized to combine nuclear energy and coal gasification in an energetically profitable way.[4] With the employment of a

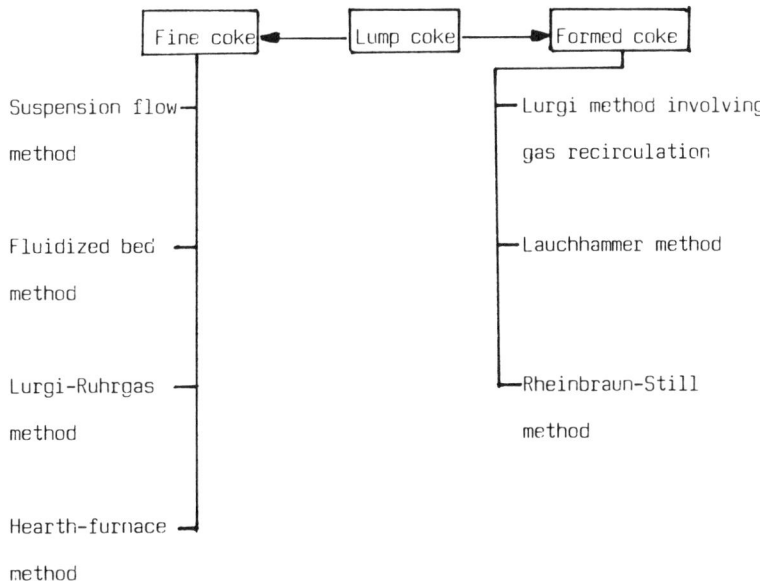

FIGURE 2.   Methods of coke production from brown coal.

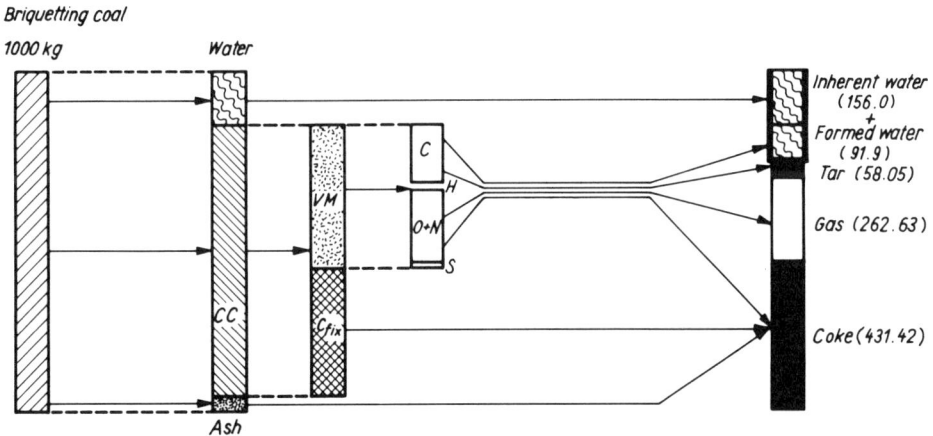

CC : clean coal
VM : volatile matter

FIGURE 3.   Matter and elementary balance for coking of brown coal briquette. (According to Kurtz, R., *Ullmanns Enzyklopädie der technischen Chemie*, Vol. 14, 4th ed., Bartholomé, E., Biekert, E., Hellmann, H., Ley, H., Weigert, W. M., and Weise, E., Eds., Verlag Chemie, Weinheim, West Germany, 1977, 491.)

gas-cooled high-temperature nuclear reactor, an improved utilization of brown coal connected with reduced contaminant emission as well as lower production costs are expected. Hydrogen and water vapor are used as gasification mediums.

Theoretically, a complete conversion of the carbonaceous substances of coal is possible leaving only ash as the solid phase. In practice the reactions are carried out in the way that carbonization and gasification are parallel. Reactive gases are formed containing hydrogen as well as carbon monoxide. The gases can be arranged into the following groups concerning caloric value.

1.   Lean gas and water gas: 4600 to 12,500 kJ/m³ (101.3 kPa; 0°C)
2.   Synthesis gas and reduction gas: 3000 to 12,500 kJ/m³ (101.3 kPa; 0°C)
3.   Town gas and sewer gas 16,700 to 20,000 kJ/m³ (101.3 kPa; 0°C)
4.   Rich gas and substitute natural gas 25,000 to 37,000 kJ/m³ (101.3 kPa; 0°C)

Independent of origin all these gases contain water vapor originating from coal water content and nonconverted reaction water vapor. Further, a portion of formed water originates from the partial oxidation of hydrogen in raw material. By condensation the gas waters are separated out of the gases. The composition of condensation waters may vary widely. Some examples are given in Table 2.

The composition of the condensation waters has to be regarded when technology for the coal-converting process is chosen. Particularly, the acidic waters of the east of the Elbe River (eE) coals cause corrosion problems which are harnessed by the application of suitable materials.

Contrary to hard coal condensation waters the content of cyanide and rhodanide is no object with regard to the treatment of brown coal waters.

3. Liquefaction of Brown Coal

In addition to the above-mentioned methods of coal carbonization and gasification yielding mainly solid and gasified products, the processes of coal hydrogenation mainly produce liquid products as benzines and oils of different boiling ranges. At reaction temperatures of 300 to 500°C added hydrogen becomes attached under pressure to the sites of fracture of pyrolysis. As products high- and low-boiling oils and benzines are recovered.

Large coal hydrogenation plants run on a brown coal basis had been changed over to a petroleum-run basis and were put out of operation, respectively. In the case of the Bergius process pulverized coal was pasted with high-boiling oils and the suspension was pressed into reaction chambers. The pressure required for the conversion with hydrogen depends on the type of coal and is in the range of 20 to 80 mPa. The waters resulting from this process are loaded to a lower degree than the waters from coal carbonization or gasification.

The new synthesis of hydrocarbons out of synthesis gas is not often practiced due to the high energy demanded (Fischer-Tropsch synthesis). In South Africa plants are in operation based on the Sasol-process scheme (Gasol I and II). The treatment of tars and tar oils also leads to condensation waters burdened with phenolic compounds.

4. Summary of Processes of Brown Coal Refinement and Composition of Water

In Table 3 the possibilities of secondary product processing with an output of phenolic waters are summarized. An example for the composition of such waters is illustrated in Table 4.

All processes of brown coal refinement are characterized by producing condensed waters which contain several volatile organic and inorganic constituents. Because of the phenolic content of these waters they are sometimes called phenolic effluents or phenolic waters. Table 3 shows a systematic overview concerning phenolic water-bearing processes and the main products of these processes. Table 4 shows some examples of typical phenolic waters from several brown coal refining processes. The values of the content of different groups change with the specific conditions of the mentioned processes as well as with the applied coal.

Table 2
COMPOSITION OF WASTE WATERS FROM BROWN COAL PROCESSING BY DIFFERENT ORIGINS OF BROWN COAL[3,7,8]

Process	Plant (East Germany)	Origin of coal	Water content (m³/t)	pH value	Monohydric phenols (water vapor volatiles) (g/ℓ)	Polyhydric phenols (g/ℓ)	Fatty acids	CO_2	H_2S	NH_3	COD_{MN}	BOD_5
Low temperature carbonization	Böhlen	wE	0.2	8.5	8.0	9.0	8.8	5.6	0.8	5.7	0.5	31
	Espenhain	wE	0.2	8.0—8.4	8.3—8.6	11.0—12.6	12.4—13.0	3.7—4.6	1.2—2.1	6.5—7.6	105—120	20—30
Fixed-bed pressure gasification	Böhlen	wE	1.0	8.5	4.4	3.9	2.0	5.4	0.7	3.5	66	19
	Schwarze Pumpe	eE	1.0	8.8	5.2	1.9	2.3	4.6	0.3	4.0	49	25
Coking	Schwarze Pumpe	eE	0.2	5.6	12.1	6.0	10.4	0.6	0.1	3.5	140	49
	Lauch-hammer	eE	0.2	5.0—6.0	12.0—14.0	6.0—8.0	13.0—18.0	0.5—1.0	0.1	3.0—4.0	150	50—60
Hard coal coking			0.2	9.0	2.0	1.0	0.3	—	2.0	8.0	10	5

Note: wE — brown coal from west of the Elbe River and eE — brown coal from east of the Elbe River.

Table 3
SUMMARY OF PROCESSES OF BROWN COAL REFINEMENT PRODUCING PHENOL-BEARING WATERS

Process	Feedstock(s)	Products
Coking	Briquette	Gas, formed coke, and light and middle oil
Pressure gasification	Raw brown coal and briquette	Town gas
Low-temperature carbonization involving gas recirculation	Briquette	Tar, light oil, and middle oil
Gasification	Low-temperature coke	Synthesis gas, hydrogen, and "Winkler" gas
Hydrogenation	Low-temperature tar and tar oil	Gasoline and heavy oils
Delayed coking	Tar	Paraffin and electrode coke

Table 4
COMPOSITION OF PHENOLIC WATERS FROM BROWN COAL CARBONIZATION, GASIFICATION, AND TAR HYDROGENATION[3]

	Gas liquor (g/ℓ)	Carbonization liquor (g/ℓ)	Tar distillation (g/ℓ)	Washing oil distillation water (g/ℓ)	Gasoline distillation water (g/ℓ)	Phenosolvan weak gas liquor (g/ℓ)
Monohydric phenols	4,2	9,9	18,7	2,0	4,7	See later
Polyhydric phenols	4,6	8,3	4,0	7,5	6,5	2,2
Fatty acids	2,0	9,9	3,8	3,5	4,4	5,6
N-compounds (ketones, nitriles, etc.)	3,3	3,7	2,0	—	2,5	0,7
H_2S	0,3	0,5	0,2	8,5	6,0	See later
NH_3	4,6	5,1	1,8	19,8	13,5	4,9
Total	19,0	37,4	30,5	41,3	37,6	13,5

Note: Compositions from brown coal west of the Elbe River.

III. TREATMENT OF THE WASTE WATERS FROM BROWN COAL REFINEMENT

A. Physical and Chemical Methods

A number of methods for the treatment of the condensation waters of brown coal as well as hard coal refinement have been developed to reduce the organic load and to recover valuable substances such as ammonia, hydrogen sulfide, phenols, pyridine, and ketones. In Figure 4 a summarizing view is given. Some of these methods has been applied successfully in the case of condensation water from brown coal. Others have been proven less suitable or are only of historical interest or provide suggestions for the future.

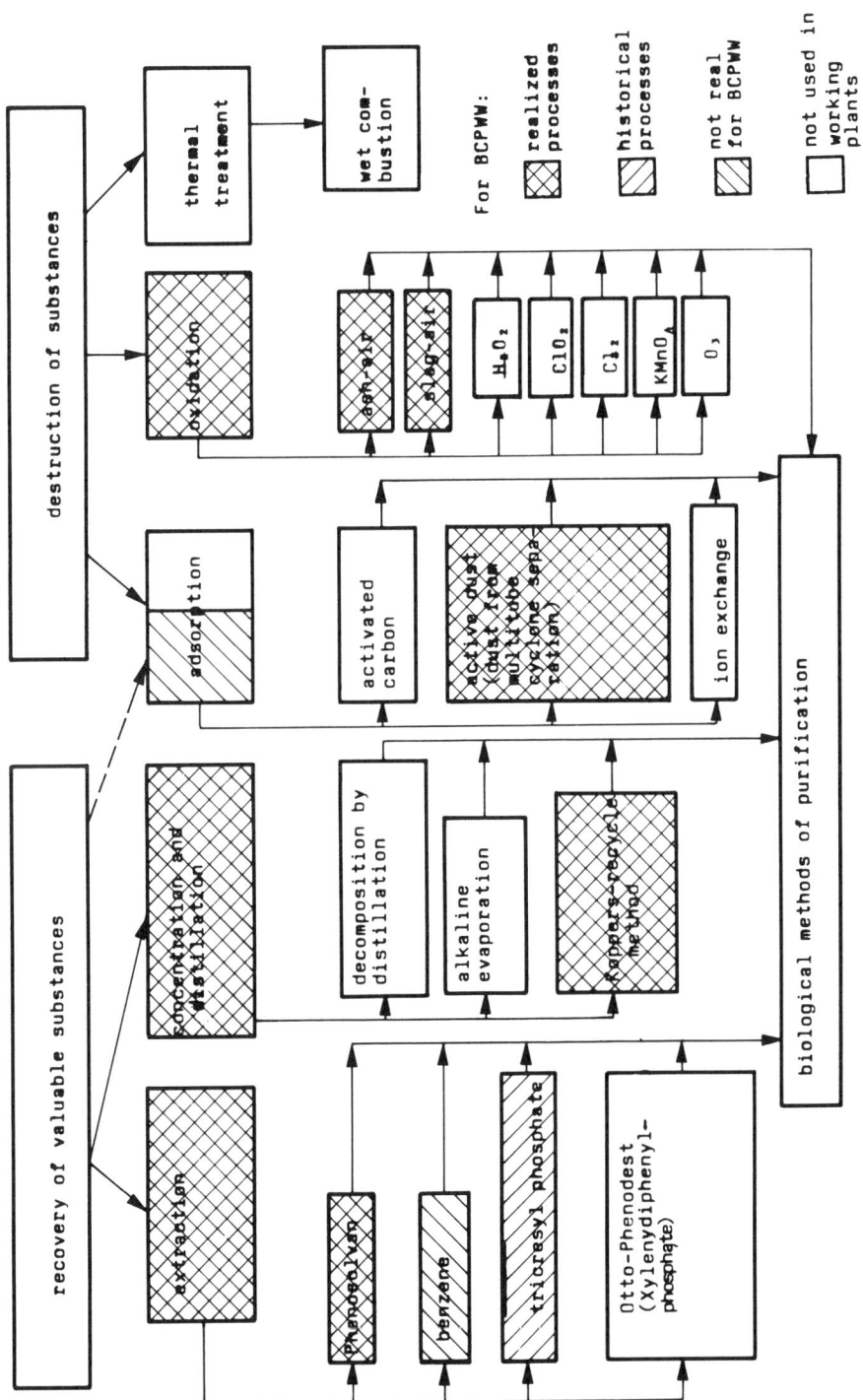

FIGURE 4.   Physical and chemical methods for treatment of BCPWW.

1. Physical Methods

As generally can be seen in sewage treatment, preliminary steps of purification contain mainly physical methods. Therefore, in condensation water treatment, purification processes start with physical operations.

a. Preliminary Purification (Figure 5)

Mechanical purification and deoiling — Tar sludge, tar oil, and neutral oil are separated from phenolic waste waters in gravity oil separators (1) which are equipped with an automatically working scraper to remove the deposed oil and tar sludge. The separation of nonemulsified oil — the so-called "free oil" — is achieved additionally. The oil and tar sludge are conveyed to a sludge thickener (2) and the oil separated from tar sludge is led to the oil separating tank (3).

Demulsification — The breakdown of tar-oil-water emulsions takes place in a specially designed separating tank (4) by heating up and simultaneously injecting CO_2-containing flue gas which also promotes a salting-out effect. The removal of remaining fine solids is achieved by means of gravel filters (5). The waste water is led via a storage tank (6) for equalization of quantity and concentration of substances to the extractive dephenolization process.

Waste oil dewatering — Waste oil from the individual purification stages is centrally dewatered in the heated oil separating tank (3) and then recycled via oil receiver (7) as secondary raw material for further processing.

b. Extraction Methods for Valuable Substances

Extractive dephenolizing — For the recovery of phenols predominantly the phenosolvan process is used nowadays. Solvents for phenols must have a low water solubility, may not form emulsions with water, and must be easily regenerable. Out of the great number of solvents for phenols isobutyl acetate and diisopropyl ether have been proven. For both of the solvents the name "phenosolvan" has been introduced (Table 5).

In spite of some disadvantages of isobutyl acetate (high price and saponification in alkaline environment) at present plants using this extraction medium are under operation. The most important process stages in extractive dephenolizing are (Figure 6) extraction, solvent and NH_3 stripping, and NH_3 rectification.

In the deacidification column (1) high-volatile neutral oils occur in the form of ketone oil and pyridine bases.

Sour gases (H_2S and CO_2) are stripped by indirect thermal distillation. The crude ketone oil obtained on top of the column is led to a H_2SO_4 treatment via cooler (2) and vessel (3).

The extraction of phenolic substances is carried out in the rotary-disk extractor (4). The extracting agent dissolved in the dephenolized waste water as well as the free NH_3 are stripped by steam distillation (5 and 6). The solvent is recycled and the NH_3 desorbed is processed to NH_3-strong liquor in a rectification column (7). The phenol extract is drawn off from the solvent recovery column (8).

Recovery of ketone and pyridine bases (Figure 7) — The treatment of crude ketone oil (composition: 50% methanol, 20 to 30% acetone, 10 to 20% acetonitrile, and 1 to 10% esters and other compounds) from the deacidification stage with H_2SO_4 in the agitator (1) results in a raffinate in the form of pyridine sulfate. After distillation (2) in order to concentrate the pyridine sulfate it is treated with concentrated sodium hydroxide solution in the following agitator (3). The distilled ketone oil being free from pyridine is treated with sodium hydroxide solution at a temperature of 80°C in order to improve the odor due to saponification of the nitriles. The waste water separated from the pyridine base is a harmless effluent and may, therefore, be discharged into the environment after dilution with the bulk of other treated waste waters. Crude pyridine and crude ketone oil can be purified by further distillation.

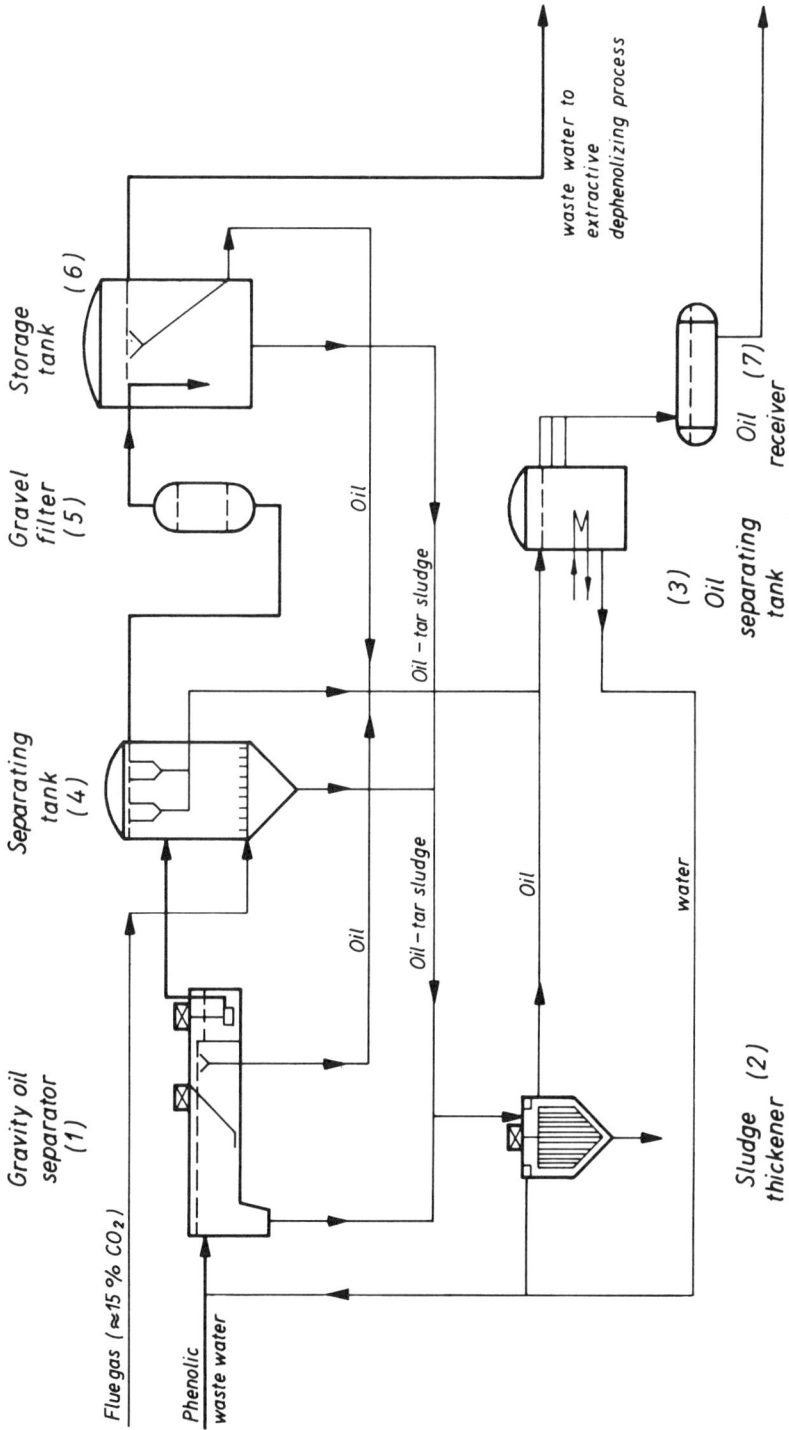

FIGURE 5.   Process flow diagram: preliminary purification of waste water (CLG process).

Table 5
COMPARISON OF ISOBUTYL ACETATE
WITH ISOPROPYL ETHER[9]

	Isobutyl acetate	Isopropyl ether
Distribution coeff against 2% phenol solution	49 (22°C)	20 (20°C)
Boiling point (°C)	124.5	67.5
Density (kg/dm³)	0.88	0.73
Water solubility (%)	1	0.8
Emulsification with water	None	None
Normal dilution with condensation water	1:10	Not known
Regeneration	By distillation	By distillation
Saponification in alkaline surroundings	Noticeable	None
Biodegradable	Yes	Yes

Another extraction process, the Otto-Phenodest-method, uses xylenyl diphenylphosphate. This solvent has a distribution coefficient of 40 for phenol extraction. The separation of the phenol extracted is accomplished by vacuum distillation.[10] The effect of temperature on the extraction of phenolic solution by octanol and mixtures of benzenes with butylacetate and diisopropyl ether, respectively, is described by Gravelle and Panayiotou.[11]

c. Koppers Recirculation Method

This process developed in the 1930s uses stripping of phenols by using water vapor. The process includes a preliminary stripping of hydrogen sulfide and CO_2 in a souring column. The steam-volatile fatty acids are kept in solution by saturation of the recirculation water vapor with ammonia. The volatile phenols carried over are washed out from the recirculation water vapor and recovered by carbonization and causticitation.

d. Adsorption Method

The use of activated carbon for phenol adsorption from brown coal condensation waters is not possible. Because of the high content of polyphenols polymerization reactions take place and the pores as well as the surface of the activated carbon are covered.

The use of ashes and dusts from Winkler generators is more suitable by far. This method reduces the phenol content down to 5 mg/ℓ (e.g., carbon content of the dust from a multitube cyclone separator is 50%, utilization of 100 kg dry dust for 1 m³ of phenolic water). An oxidative adsorbing purification with ashes[12] or slags leads to humic acids by catalytic oxidation of the phenols. The method of slag adsorption is especially important in Czechoslovakia.[13]

2. Chemical Methods

There is a series of proposals for processes regarding the removal of phenol by chemical oxidation. Although they are highly efficient, they are not technically important up to now. In the presence of iron (II) ions H_2O_2 oxidizes the phenols to hydroquinone and catechol. These in turn are oxidized to the corresponding quinone and further on to carboxylic acids and finally to CO_2.[14]

In the presence of phosphate the catalytic action of iron is reduced and the oxidation cannot be accomplished. If peroxidase (horseradish peroxidase) is added to phenolic waste

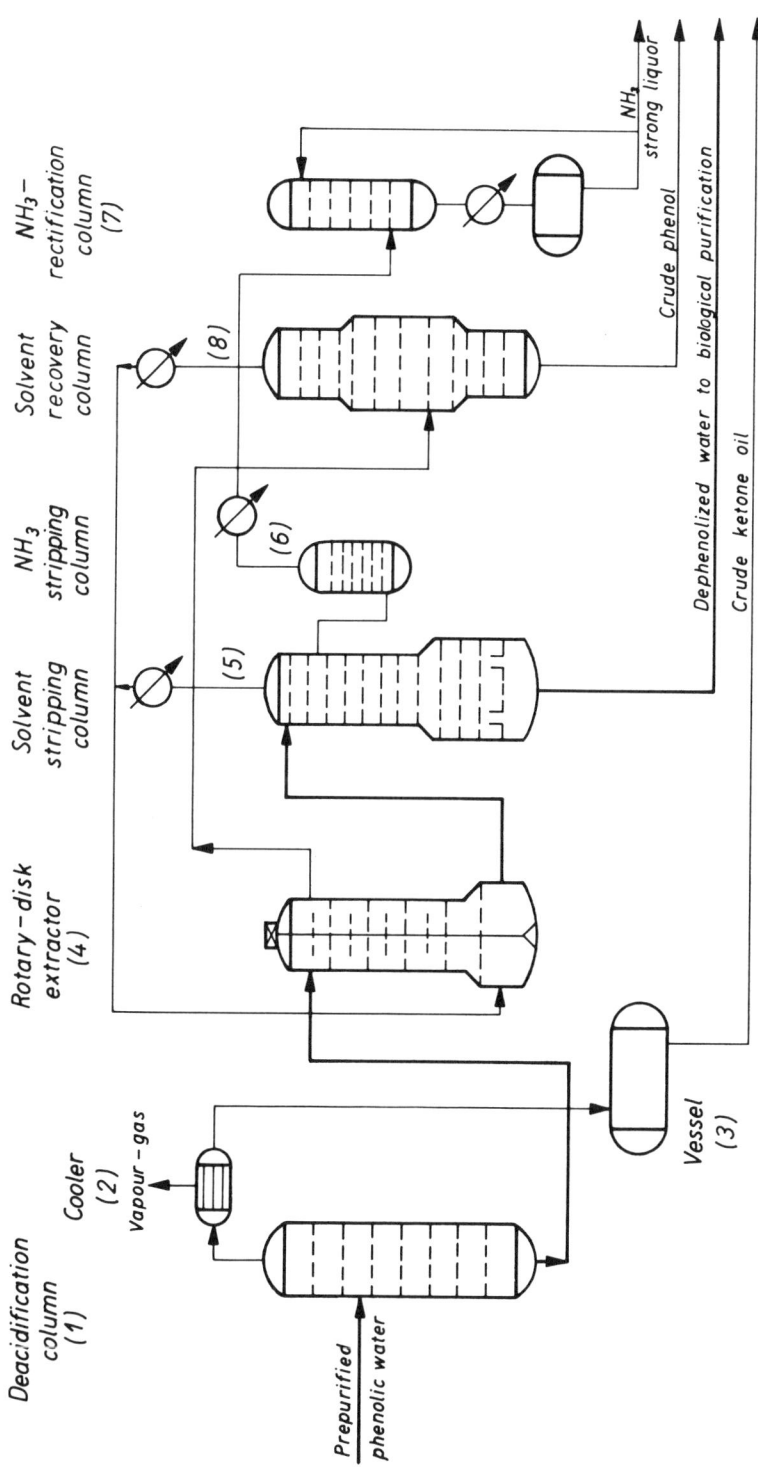

FIGURE 6.   Process flow diagram: extractive dephenolizing (CLG process).

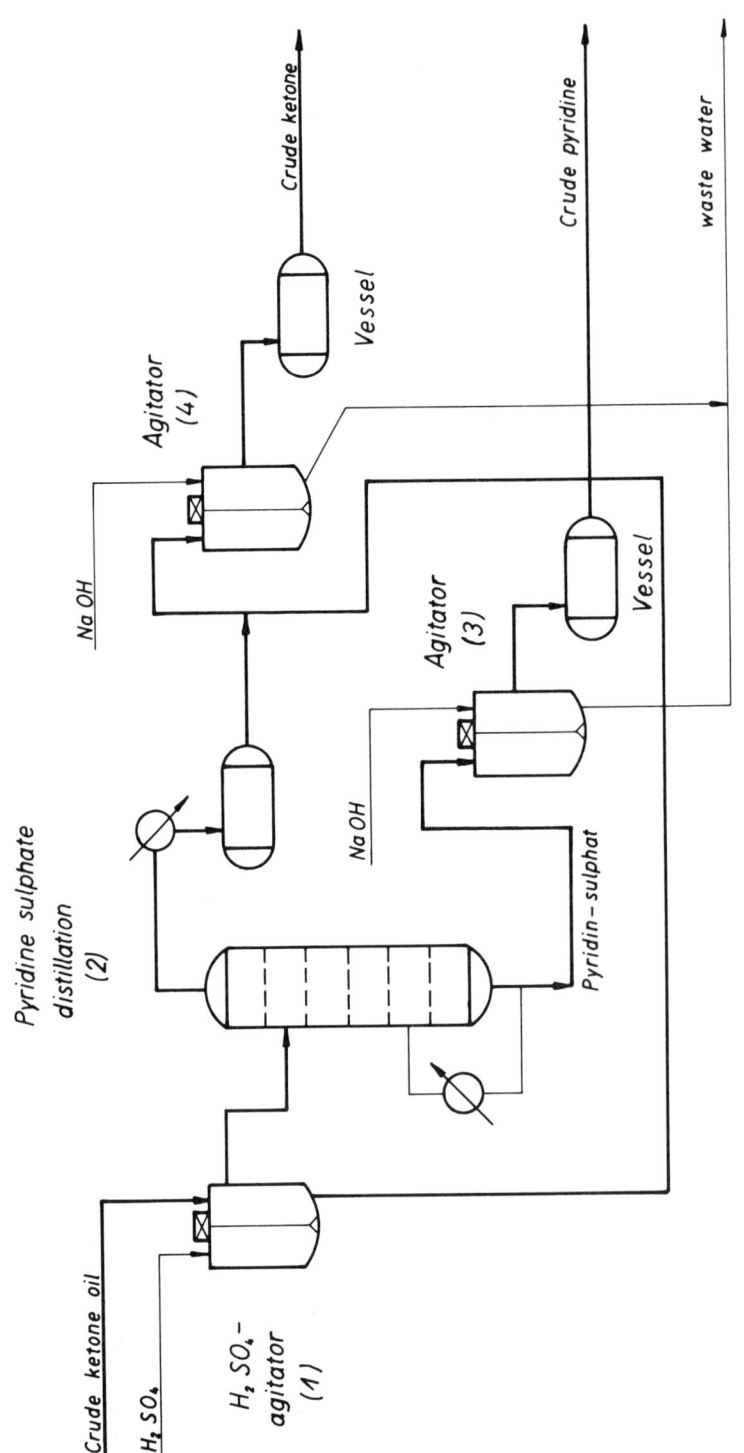

FIGURE 7. Process flow diagram: recovery of ketone bases and pyridine bases (CLG process).

Table 6
CONTENTS OF A CONDENSATION WATER OF A SOFT COAL TAR
DISTILLATION (mgℓ^{-1})[31]

	Tar dist effluent	Phenosolvan extraction effluent	Adsorption (C powder) effluent	Mixed plant effluent	Activated sludge treatment effluent	Receiving water (pond)
pH	8	10	9	8—8.5	7—7.5	7—7.5
BOD₅	22,000	9,000	3,000	120—400	30	15
TOC	15,000	6,000	1,500	100—170	30	10
H₂S	0	0	17—51	0—6	0	0
Ammonium	5,000	4,500	1,200	50—200	50—200	5
Ketones	563	487	58—244	0—12	0	0
Nitriles	0	0	21—82	0—40	0	0
Ethylenehydrocarbons	171	158	0	0	0	0
Alcohols	612	228	12—120	6—36	1	0
N-bases	810	680	40—70	15—40	5	0
Volatile organic acids	6,000	5,900	1,200	60—180	0	1
Polyhydroxy benzenes	11,000	3,600	400—780	60—140	25	5
Aminophenolics	517	435	40—100	0	0	0

waters polymerization of the phenols occurs. The settled products can be removed by simple filtration (see References 15 and 16 and Section III.B.4). At present a technical significance of this biochemical method cannot be recognized.

If the production of cheap enzymes by microorganisms will succeed new aspects of a waste water treatment could arise. An oxidation of the phenols by chlorine, chlorine dioxide, potassium permanganate, and ozone is possible. According to Besselievre,[17] residual contents of 3 μg phenol per liter can be attained.

B. Microbiological Methods of Purifying Brown Coal Processory Waste Waters

The large-sized plants for the production of brown coal, high-temperature coke, and coal gas, put into operation in the G.D.R. during the 1950s and 1960s, have been equipped with physical and biological waste water purifying units (VEB Gaskombinat Schwarze Pumpe, 1958 and VEB Braunkohlenveredlungswerk Lauchhammer, 1964). Extensive investigations have been performed in order to know the chemical composition of the waste waters before projecting the plants. In the previous sections the data about the contents represent only substance classes. After starting the first biological waste water purifying units, knowledge was enriched about the microbial degradation of the substances in the waste waters.

1. Detailed Composition of Brown Coal Processory Waste Waters

After the introduction of chromatographic techniques, the fate of particular compounds of the most important substance classes, volatile fatty acids and hydroxy benzenes, has been investigated. Derivatives of neutral substances and nitrogenous bases are investigated only sporadically. Analytic methods for the determination of substance classes have been developed by Meissner.[18-21]

Table 6 shows the results obtained in the investigation of the composition of condensation water and the decrease of concentrations of the determined substance classes after general treatments of the waste water. Polyhydroxy benzenes, volatile fatty acids, and monohydroxy benzenes have the highest concentrations. As mentioned above, the extraction with butyl acetate results especially in the removal of monohydroxy benzenes. Along with the volatile fatty acids the polyhydroxy benzenes are those organic substances the waste waters were loaded with most. Because an effective absorbing with dust from the factory is impossible

Table 7
**INFLUENCE OF PROVENANCE OF SOFT COAL AND PROCESSING
ON THE COMPOSITION OF VOLATILE ORGANIC ACIDS IN
CONDENSATION WATERS**

| | n + iso-acids | | | | | | | |
	C_1	C_2	C_3	C_4	C_5	C_6	C_7	Total
Older soft coal								
Gas water (g ℓ^{-1})	—	2.5	1.0	0.5	0.5	1.0	0	5.5
Relation (%)	—	46	18	9	9	18		100
Younger soft coal								
Gas water (g ℓ^{-1})	—	10.3	1.6	0.6	0.6	0	0	13.1
Relation (%)	—	78	12	5	5			100
Older soft coal								
Effluent from brown								
coal hydrogenation (g ℓ^{-1})	—	1.3	0.6	0.6	0.6	0.4	0.4	3.9
Relation (%)	—	35	15	15	15	10	10	100

Adapted from Leibnitz, E., Behrens, U., and Ringpfeil, M., *Wasserwirtsch. Wassertech.*, 7, 266, 1957.

in most cases, the polyhydroxy benzenes are very interesting in regard to the reactions of the benzenes to the microorganisms in a biological reactor.

a. Profile of the Fatty Acids

The influence of the brown coal provenance and the used method of brown coal processing on the composition of the waste waters has been discussed. As shown in Table 7, the content of volatile fatty acids (predominantly of acetic acid) is more conspicuous in waste waters from younger lignitic brown coals. It was also shown that different brown coal processing had a minor distinct effect on the composition of the fatty acids. Gas chromatographic analysis showed a small amount of formic acid (0.5 g ℓ^{-1}) and C_4- and C_5-carboxylic acids with an n C_4/i C_4 ratio of 2.5:1 and the n C_5/i C_5 ratio of 1.7:1, respectively.[22] The use of gas chromatography/mass spectrometry for the determination of organic substances in high BTU coal gasification quench waters and the quench water pretreatment has been described by Stamoudis and Luthy[23] in the case of hard coal. Yet there are no investigations about the brown coal processory waste waters (BCPWW).

b. Profile of the Polyhydroxy Benzenes

The evaluation of polyhydroxy benzenes from condensate water was successfully accomplished by Ringpfeil[24] in an adaption from paper chromatographic methods for the structural analysis of chemical homologous substances such as the structure determination of di- and trihydroxy benzene methyl derivates (Table 8) on the basis of the retention factor.[25,26] Using the method of Blackburn et al.[27] after distillative separation, separate distillates were investigated by column or paper chromatography and the eluates by IR and UV spectroscopy in combination with the preparative synthesis of some hydroxy benzenes. Similarly, good correlation could be obtained with a method developed by Ringpfeil[24] between the structures calculated by the authors and the paper chromatographically determined retention factors.[28] For quantitative analysis of dihdyroxy benzenes in condensate water the correlation of Fisher et al.[29] was checked and deemed as sufficiently faithful.[26]

The Fisher correlation means that the area of a substance spot on the paper chromatogram is in conformity with the common logarithm of concentration. The detailed results from the quantitative analysis of a waste water before and after the phenosolvan extraction is shown

Table 8
METHYLDERIVATIVES OF DIHYDROXY
BENZENES, FOUND IN CONDENSATION
WATERS OF THE SOFT COAL PYROLYSIS
AND HYDROGENATION BY THE STRUCTURE
DETERMINATION OVER THE RETENTION
FACTOR, PARTLY OVER REFERENCE
SUBSTANCES

	Number	Position of the OH groups	Position of the CH_3 groups
Catechols	1	1.2	3
	2	1.2	3
	3	1.2	4
	4	1.2	3.6
	5	1.2	4.5
	6	1.2	4.6
	7	1.2	?
Resorcinols	8	1.3	—
	9	1.3	4
	10	1.3	5
	11	1.3	4.5
	12	1.3	4.6
	13	1.3	?
	14	1.3	?
	15	1.3	?
	16	1.3	?
	17	1.3	?
	18	1.3	?
Hydroquinones	19	1.4	—
	20	1.4	2
	21	1.4	2.3/2.5/2.6
Pyrogallol	22	1.2.3	—

Adapted from Leibnitz, E., Behrens, U., and Ringpfeil, M., *Wasser-wirtsch. Wassertech.*, 6, 299, 1956.

in Table 9. Table 10 shows the biological degradation of the dihydroxy benzenes most essential in quantity in dilute gas water as a function of aeration. Catechol was used as an indicator of the rate of treatment of the BCPWW also in the receiving water. Yet in a distance of several kilometers of the inlet tube were found catechol (0.1 g ℓ^{-1}) and derivatives of the methyl catechol (1 to 10 mg ℓ^{-1} in each case).

c. Profile of Further Compounds

A 1:10 dilution with activated sludge-inoculated ammoniacal water was fermented in a discontinuous way and then used to determine the volatile fatty acids and the neutral compounds (methyl alcohol, butyl alcohol, butyl acetate, pyridine, and the water-vapor volatile phenoles, especially monohydroxy benzene) (Table 11). The high concentration of butyl acetate and of the saponification products of butyl acetate shows that this extraction medium is a poor one in the phenosolvan process.

Considerable amounts of formalin could be detected. The formalin amounts found depend on the brown coal and the process used.

2. Air Oxidation and Biological Oxygen Demand after 5 Days

One part of the substances in the BCPWW is oxidizable by air oxygen. This gives a markable failure in the determination of the biological oxygen demand after 5 days (BOD_5).

Table 9
DIHYDROXY BENZENE
CONTENT OF A BCPWW (FROM
GASIFICATION AND
HYDROGENATION) BEFORE
AND AFTER PHENOSOLVAN
EXTRACTION

Position OH	Position CH$_3$	Effluent (g ℓ^{-1})	
		Plant	Phenosolvan
1.2	—	1.70	0.77
1.2	4	0.76	0.13
1.2	3	0.76	0.13
1.2	4.5	0.48	0
1.2	4.6	0.16	0
1.2	3.6	0.19	0
1.3	?	0.16	0.04
1.3	4	0.10	0.03
1.3	5	0.08	0.04
1.3	4.5	0.03	0.01
1.4	—	0.10	0.07
1.4	2	0.06	0

Adapted from Leibnitz, E., Behrens, U., and Ringpfeil, M., *Wasserwirtsch. Wassertech.*, 7, 266, 1957.

Table 10
DEGRADATION OF DIHYDROXY
BENZENES IN AN ACTIVATED SLUDGE
BASIN AS A FUNCTION OF AERATION

Phenolic substances		Influent (g ℓ^{-1})	Effluent (g ℓ^{-1})	
-OH	-CH$_3$		Low aeration	High aeration
1.2	—	0.99	0.9	0
1.2	4	0.57	0.07	0
1.2	3	0.51	0	0
1.2	4.6	0.36	0	0
1.3	—	0.16	0.05	0.04
1.3	5	0.14	0.01	0.01
1.4	—	0.1	0.01	0.02

Adapted from Leibnitz, E., Behrens, U., and Ringpfeil, M., *Wasserwirtsch. Wassertech.*, 7, 266, 1957.

About 10% of the overall oxygen consumption is used by chemical oxygenation at the beginning of the discontinuous fermentation or at the start of a manometric BOD$_5$ determination. The biological oxygenation starts with a clear time contrast after the completion of the air oxygenation. In ammoniacal condensation waters after the phenosolvan extraction the BOD$_5$ is, for example, 10 to 10,000 mg O$_2$ ℓ^{-1}. Accordingly, the share of oxygen needed for the air oxygenation is about 1000 mg ℓ^{-1}.[30]

Table 11
DISCONTINUOUS FERMENTATION OF A 1:10 DILUTED GAS WATER

Time (hr)	pH	Temp (°C)	Volatile acids[a] (mg ℓ^{-1})			Neutral substances[a] (g ℓ^{-1})			Pyridine[a] (g ℓ^{-1})	Volatile[b] phenols (g ℓ^{-1})
			C_2	C_3	C_4	Methyl alcohol	Butyl alcohol	Butyl acetate		
0	8.3	18	518	127	38	42	11	19	3.0	17.2
5	8.4	28	173	40	9	31	8	13	2.6	9.1
8.5	8.6	29	106	5	0	9	4	7	2.0	5.1
24	8.5	27	—	0	0	0	0	0	0	1.2

[a] Determination by gas chromatography.
[b] Determination by chemical standards.[33]

Adapted from Thielemann, H., Doctoral thesis, Karl Marx University, Leipzig, German Democratic Republic, 1962.

3. Microbiological Fundamentals of the Purification of Brown Coal Processing Waste Waters

The condensation water which is largely liberated from monohydroxy benzenes by phenosolvan extraction is a waste water characterized by a high content of volatile fatty acids — especially acetate — and an alkaline pH resulting from the high content of ammonia.

Among the aromatic constituents, the dihydroxy benzenes (polyphenols) are the dominant substances. Due to the biological degradation of the fatty acids, the pH continues to increase. The said conditions as well as the content of other difficulty degradable substances may have a selective effect on the composition of the microbial community of a biological treatment plant. There are, however, very few reports in the literature, so the information given about the essential microbial species, secondary species, and qualitative as well as quantitative changes are limited.

As early as 1910 it had been possible to isolate phenol-utilizing bacteria from trickling filters.[34] Not long after that, Wagner[35] reported on "benzene bacteria" he had been able to isolate from soil, manure, saliva, various foods, etc. As a consequence of the development of carbochemistry on the basis of brown coal in the 1930s, it was necessary to develop methods for the biological purification of phenol-containing waste waters. At the same time basic research work was carried out by three working teams: (1) Kalabina and Rogovskaya,[36] U.S.S.R., 1934; (2) Sierp and Fränsemeier,[37] Germany (Ruhr district), 1934; and (3) Nolte, Meyer, and Franke,[38] Germany (Magdeburg), 1934. The fundamental findings can be summarized as follows:

- Phenol-degrading bacteria are ubiquitous.
- Phenol degradation requires that a sufficient amount of oxygen is present.
- The mineral salts (especially of phosphorus) necessary for the microbial growth must be available.

On the basis of these findings, the so-called Magdeburg P method was developed and introduced in several carbochemical plants and gas works (see below).

Soviet working groups attempted to utilize the heat of the condensation waters by treatment with thermophilic or thermotolerant bacteria in order to achieve a higher reaction rate.[39] By running the process at 50 to 60°C it was possible to isolate a number of species such as *Pseudomonas, Zoogloea, Achromobacter, Alcaligenes, Flavobacterium, Micrococcus,* and *Vibrio* as well as *Escherichia coli.* Bringmann[40] was able to adapt *Nocardia* sp. to high phenol concentrations (phenols 1.6 g ℓ^{-1} and cresols 1 g ℓ^{-1}). A "Nocardia method" based on these results is the first, although not very successful, attempt to accomplish a biological purification by means of a "high-performance strain of bacteria".

Similar attempts to adapt bacteria and mixed bacterial populations to high phenol concentrations were carried out in the 1950s, because the toxicity of the phenols was still considered to be a decisive criterion. Although the plants for biological water treatment were operated in a continuous process mode from the beginning, the principles of continuous cultivation, which had been developed by Monod[41] and Herbert,[42] were still widely unknown.

Moreover, the activated sludge processes were operated in oblong basins with the inlet at the narrow side, so that here as well as, necessarily, in the case of the trickling filters, the feed concentration in the first sections of a basin or trickling filter inevitably affected the biological activity. Depending on the constitution of the hydroxy benzenes, the cultivation conditions, and the strains or mixed populations of the activated sludge employed, the results obtained were similar to those of the Nocardia process. The question of whether bacteria adapted to high phenol concentrations can have a favorable effect on the process of a biological treatment was discussed controversially. According to a recent paper,[43] bacteria adapted to high phenol concentrations are claimed to bring no appreciable advantage over nonadapted ones, even in the case of a sudden increase of the phenol concentration in the purification plant (phenol surge). It is only the start-up phase which can be shortened by the addition of such high-performance strains.

It was not until lately that the degradation of hydroxy benzenes by yeasts and fungi was recognized. *Oospora lactis* was frequently found in plants for the purification of brown coal condensation waters, and was also used for the degradation of monohydroxy benzenes.[44] Rieche et al.[45] reported the utilization of phenols by *Torula utilis* in the presence of sugars. Zimmermann[46] was the first to show that both mineral and cultivated yeasts utilize phenols.

Basic investigations of the alteration in the bacterial population as a function of external conditions (aeration and temperature) were carried out by Zülke[47] (see Table 12). In this case aerobic, Gram-negative asporogenic rod-bacteria dominate. *Achromobacter, Arthrobacter, Bacillus,* and *Pseudomonas* seem to be the prevailing genera. Thus, the content of dihydroxy benzenes should hardly have any influence on the composition of the biological community. However, it turned out clearly that the maintenance of thermophilic conditions and good aeration led to a decrease in the number of genera. In experiments with high-performance fermenters and a slight (1:1) dilution of the waste water from a brown coal gasification plant (COD = 15,000 mg ℓ^{-1} O_2), activated sludge flocculation was no longer observed. This could be attributed to an extreme depletion and to the destruction of the flocs by an increased turbulence. In this case the population largely consisted of Gram-negative asporogenic rods with polar flagella (*Pseudomonas*).[47a]

In the growth of an adapted mixed culture (starting culture: activated gas liquor sludge) in a medium with phenol and acetate as the only carbon sources, a diauxic course of the growth phase was clearly observed.[47] The diauxic course of the growth phase is equally observed in the degradation of a phenosolvane-treated gas liquor.[48] Thus, the readily degradable volatile fatty acids can be clearly distinguished from the slower-utilizable hydroxy benzenes (see below). From the knowledge of the biocoenosis of the activated gas liquor sludge floc and of the alkaline pH of the condensation water, it was not to be expected that yeasts contribute to phenol degradation. It turned out, however, that numerous yeasts could be isolated from a pilot trickling filter charged with gas liquor[49] (see Table 13).

Similar investigations were carried out on a three-stage pilot plant consisting of two trickling filters and an activated sludge basin for final purification[47] (Table 14).

It can be supposed that the presence of yeasts is due to ecological niches in the trickling filters, which allows the utilization of the fatty acids contained at a pH favorable for growth. Yeasts that occurred in the activated sludge basin of the above-mentioned three-stage pilot plant should have been flushed out of the trickling filters, because no yeasts were detected in the activated sludge from BCPWW purification plants without trickling filters inserted before them.

Table 12
**CHANGES IN THE BACTERIAL POPULATION AS A FUNCTION OF
TEMPERATURE AND AERATION**

	Waste water I: brown coal high-temperature coke			Waste water II: brown coal gasification	
Microorganisms	Fermenter mesophilic	Basin thermophilic	Fermenter thermophilic	Fermenter thermophilic	Basin thermophilic
Aerobacter	−	+	−	−	−
Acetobacter	+	−	−	−	−
Achromobacter	+	+	+	+	+
Actinomyces	−	+	−	−	−
Alcaligenes	+	−	−	−	+
Arthrobacter	+	+	+	+	−
Bacillus	+	+	+	+	+
Bacterium	−	+	−	−	−
Brevibacterium	−	−	−	−	+
Corynebacterium	+	−	−	−	+
Escherichia	−	+	−	−	−
Flavobacterium	−	−	−	−	+
Micrococcus	−	+	−	+	−
Mycobacterium	−	−	−	−	+
Nocardia	+	−	−	−	+
Paracolobacterium	+	−	−	+	+
Pseudomonas	+	+	+	+	+
Sarcina	−	−	+	+	−
Staphylococcus	+	−	−	+	+
Xanthomonas	−	+	−	−	−

Note: Mesophilics: 28 to 42°C, thermophilics: 43 to 72°C, laboratory fermenter: optimum aeration, and activated sludge basin: oxygen limitation.

Adapted from Zülke, H.-J., Doctoral work, Ernst Moritz Arndt University, Greifswald, German Democratic Republic, 1971.

Experiments investigating the degradation of defined fatty acids and hydroxy benzenes have been carried out with some of the yeasts listed in Table 13.[49] Here *Trichosporon cutaneum* proved to be a very good utilizer of acetate, phenol, resorcin (1,3-dihydroxybenzene), and orcin (1,3-dihydroxy-5-methylbenzene), while *Endomyces oventitis* utilized acetate, phenol, n-butyrate, i-butyrate, and phenol with a good efficiency. Methylated mono- and dihydroxy benzenes were only slowly degraded.

In the past few years, successful attempts at a biological purification of gas liquors from hard coal gasification plants under anaerobic conditions have been reported. There are no reports of an application of this technique to BCPWW. It should, however, be possible to degrade the main substances contained in BCPWW (volatile fatty acids and phenols) without any difficulties by a methanogenic population. The methanogenesis from fatty acids, especially with acetate as a methanogenic substrate, is a long-known fact. There have been several papers published on the anaerobic degradation of hydroxy benzenes by methanogenic bacterial populations (e.g., Neufeld et al.[50]).

4. Biochemistry of the Degradation of the Substances Contained in Brown Coal Processory Waste Waters

The degradation of volatile fatty acids up to C_3 is initiated by an acyl-CoA synthetase reaction:

Table 13
**YEASTS ISOLATED FROM TRICKLING
FILTERS FOR THE TREATMENT OF
PHENOSOLVAN-EXTRACTED
CONDENSATION WATER (A) AND FOR
THE TREATMENT OF MUNICIPAL
SEWAGE (B)**

Yeast	A	B
Candida guilliermondii	+ +	−
C. lipolytica	+ + +	−
C. parapsilosis var. *intermedia*	+	−
Endomycopsis oventitis	+	−
Geotrichum candidum	+ + +	+
Species 1	+ + +	−
Species 2	+ +	−
Species 3	+ +	−
Hansenula cultiformica	+ + +	−
Rhodotorula glutinis	+	−
R. mucigalosa	+	+
Trichosporon cutaneum	+ + +	−
T. cutaneum var. *multisporum*	+ + +	−
T. fermentans	+	−

Adapted from Waller, M., Doctoral work, Martin Luther University, Halle, German Democratic Republic, 1962.

Table 14
**CHANGES IN THE OCCURRENCE OF YEASTS
IN A BIOLOGICAL THREE-STAGE PLANT FOR
THE TREATMENT OF PHENOSOLVAN-
EXTRACTED GAS WATER**

Yeast	Trickling filter 1	Trickling filter 2	Activated sludge
Candida curvata	x	x	x
C. scoftii	x	—	—
Cospora sp.	x	x	x
Rhodotorula glutinis	—	x	—
Torulopsis candida	x	—	x
T. ernobii	—	—	x
T. famata	x	x	x
T. inconspicna	x	x	x
Trichosporon cutaneum	x	—	—

Adapted from Zülke, H.-J., Doctoral work, Ernst Moritz Arndt University, Greifswald, German Democratic Republic, 1971.

$$R\ CH_2\ COOH + CoA + ATP \rightarrow RCH_2CoA + ADP$$

This reaction probably also enables the active uptake of acetate by microorganism cells.[51] The metabolism of propionate and of propionyl-CoA (as a final product of the β-oxidation of odd fatty acids) is still a subject of discussion. As with acetate, propionate might be converted to propionyl-CoA, and taken up actively. Propionyl-CoA might then be included in the metabolism by two possible reaction schemes. On the one hand, it can be isomerized

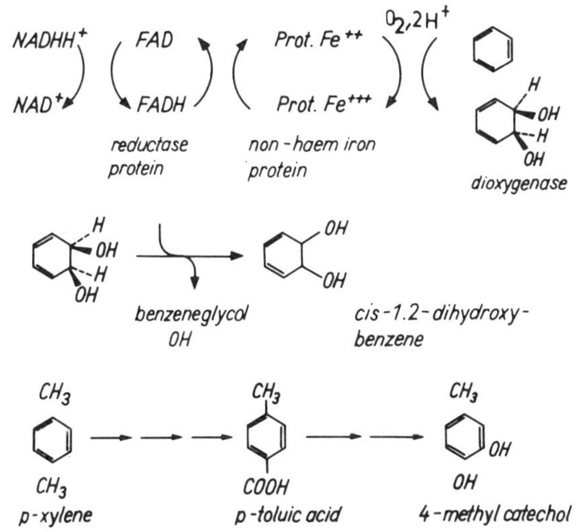

FIGURE 8.   Mechanism of dioxygenation of the benzene nucleus.

to methylmalonyl-CoA in a way analogous to propionic acid fermentation, with a subsequent isomerization to succinyl-CoA. This methyl group shift requires coenzyme vitamin B_{12}, and thus might be one of the causes of the relatively high content of corrinoids in the aerobic activated sludge. Another possibility was proposed by Tabuchi and Serizawa.[52] By analogy with the tricarboxylic acid cycle (TCC) cycle, oxalacetate condenses with propionyl-CoA to give methyl citrate. From methyl citrate, pyruvate is separated, so that the TCC of succinyl-CoA can proceed up to the formation of oxalacetate.

The pH of the medium has an essential influence on the uptake of the water-soluble fatty acids.[53,54] Undissociated acetic acid, which is present in the medium at low pH, is toxic in contrast to acetate (blocking of the uptake of succinate and phosphate). In contrast to this, however, the acetate ion is taken up less easily than acetic acid. This might account for the frequently observed slow and incomplete uptake of acetate by bacteria in the alkaline BCPWW (see Table 11).

The clarification of the biochemical mechanism of the degradation of hydroxy benzenes was highly stimulated by Stanier's[55] investigations on sequential induction. At present there are largely confirmed results on the degradation of benzene derivatives available. Single- and fused-ring systems are prepared by a doubly vicinal hydroxylation of the benzene ring, so that catechin or a catechin derivative is formed. The hydroxylating enzyme is a relatively unspecific (benzene) dioxygenase. The characterization of this enzyme was very difficult because of the instability of the enzyme. It was only Axell and Geary[56] who succeeded in isolating a strain of *Pseudomonas putida* whose dioxygenase system was stable. It could be shown that the dioxygenase consists of three protein fractions and that the hydroxylation system of the dioxygenase is very similar to that of the monooxygenase for aliphatic hydrocarbons (Figure 8).

By an NAD-dependent *cis*-benzeneglycol dehydrogenase the reaction is terminated to give catechin.

For monohydroxy benzenes the second hydroxyl group is introduced by a (phenolic) monooxygenase.

In monoalkyl benzenes the alkyl groups are preferably maintained, so that, for example, 3-methyl catechol is produced from methyl benzene (toluene).[57] Longer-chain alkyl groups also remain unaffected, unless they are so long that the aliphatic chain becomes the determinant group, so that aryl-substituted fatty acids are formed. In dialkyl-substituted aromatics such as xylenes, a methyl group is oxidized to a carboxyl group. Thus, for example, *p*-toluic acid is formed from *p*-xylene. The carboxylic group is involved in dioxygenase reaction, so that methyl catechol is formed.

Cooxidations may lead to incomplete oxidation products. Prerequisite to cooxidation is another carbon source which is, however, suitable for the cell growth. Under these conditions the dihydrogenase may be inhibited, so that the side chains can be oxidized or dehydrogenated. Propyl benzene may give rise to the formation of phenylacrylic acid and dodecyl and ethyl benzene, or of phenylacetic acid (see the summaries given by Raymond and Jamison[59] and Perry[60]).

However, ring fissions are also possible under cooxidative conditions, so that, with the use of *Nocardia corallina,* a dimethyl muconic acid can be formed.[61] Coxidation and, in a wider sense, the cometabolism play an important role in the degradation of xenobiotic substances under natural ecological conditions, e.g., in biological waste water treatment. A diversified microbiological consortium can also degrade persistent compounds, where frequently the majority of the species in a consortium realize cooxidative or cometabolic reactions and subsequently produce a degradable substance. In biological waste water treatment such systems are of advantage for a great variety of cometabolic reactions that allow cell agglomeration, such as those occurring in the activated sludge flocs and in fixed film reactors.

Oxygen deficiency may be another exogenic condition for the concentration of intermediate products. Using a strain of *Pseudomonas* sp., the methyl group in the para position to the hydroxyl in 1-hydroxy-2,5-dimethyl benzene or 1-hydroxy-3,4-dimethyl benzene was oxidized under oxygen limitation, where the resulting derivatives of the benzoic acids accumulated in the medium.[62]

The cleavage of the aromatic ring may take place between the two hydroxyl groups bearing carbon atoms (ortho pathway or intradiol cleavage), or between a hydroxyl-bearing carbon and an adjacent one (meta pathway or extradiol cleavage). Intradiol cleavage forms muconolactone or muconic acid as a cleavage product which is further reacted to β-ketoadipic acid (therefore, this pathway is sometimes also called ketoadipic pathway) and to acetyl-CoA and succinyl-CoA. The intradiol pathway is apparently also found for catechol itself.

For the yeast *Trichosporon cutaneum,* it was found[63] that methyl-substituted and some other monosubstituted phenols are metabolized via intracleavage. The enzymes of the ring fission are dihydrogenases (catechol-1,2 or 2,3-dihydrogenases) which react with a relatively broad range of substrates.

For the purification of waste waters containing hydroxy benzenes, an enzymatic reaction has also been proposed, which was already referred to in Section III. Peroxydases dehydrogenate many organic compounds in the presence of H_2O_2 according to the scheme:

$$AH_2 + H_2O_2 \rightarrow A + 2H_2O$$

The radicals occurring as intermediates of this reaction may separate from the active center of the enzyme and diffuse into the aqueous medium, where they polymerize chemically to substances insoluble in water, which can then easily be removed from the waste water[58] (see Table 15).

The anaerobic treatment of phenolic waste waters gains an increasing interest. For the sake of completeness let us refer to this subject in brief here. The published articles concerning the anaerobic treatment of coal conversion waste wasters are summarized in Table 15a.

The biological methane formation from substances contained in BCPWW is a complex process which has not been fully clarified so far. For the degradation of low fatty acids and aromatics, a group of bacteria (syntrophic bacteria) is required which separate these sub-

Table 15
REMOVAL OF
MONOHYDROXYBENZENES AND
AROMATIC AMINES BY HORSERADISH
PEROXIDASE AND HYDROGEN
PEROXIDE

Substrate	Optimum pH	Removal coeff
Phenol	3.5	85.3
m-Cresol	4.0	95.3
o-Chlorophenol	7.0	99.8
m-Chlorophenol	7.0	66.9
p-Chlorophenol	5.0	98.7
5-Methyl resorcinol	3.5	90.8
2.3-Dimethyl phenol	4.0	99.7
Aniline	7.0	72.5
4-Chloraniline	5.5	62.5
4-Bromaniline	5.5	84.5

Note: Conditions were 0.1 g ℓ^{-1} substrate and one unit peroxidase 1 mM H_2O_2 treatment for 3 hr at room temperature.

Adapted from Klibanov, A. M., Alberti, B. N., Morris, E. D., and Felshin, L. M., *J. Appl. Biochem.,* 2, 414, 1980.

stances into methanogenic substrates, since the methane bacteria can only utilize CO_2/H_2, formiate, acetate, methanol, and methylamines as substrates. Under standard conditions this cleavage reaction is endergonic:

$$C_6H_5COOH + 6H_2O \rightarrow 3CH_3COOH + 3H_2 + CO_2$$

$$G° = +89.7 \text{ kJ mol}^{-1}$$

$$CH_3CH_2COOH + 2H_2O \rightarrow CH_3COOH + CO_2 + 3H_2$$

$$G° = +76.1 \text{ kJ mol}^{-1}$$

$$CH_3CH_2CH_2COOH + 2H_2O \rightarrow 2CH_3COOH + 2H_2$$

$$G° = +48.1 \text{ kJ mol}^{-1}$$

These reactions can take place only if hydrogen is continuously removed from the system by other microorganisms, in this case by the methane bacteria. The methane bacteria, because they are highly affinitive to hydrogen, maintain a low partial pressure of the latter. The presence of methane bacteria in the group of syntrophic bacteria is a prerequisite to the degradation of the volatile fatty acids and phenols. While, until recently, acetate was the methanogenic substrate only the methyl group was considered a supplier of the methane, the cleavage following the reaction scheme:

$$CH_3COOH \rightarrow CH_4 + CO_2 \qquad G° = -31.0 \text{ kJ mol}^{-1}$$

It was shown recently that the carboxyl group is also involved in this process. It was possible to detect a syntrophic bacterium which, in competition with the direct acetate utilization by methane bacteria, cleaves acetate into CO_2 and H_2 according to the scheme:

Table 15a

ANAEROBIC TREATMENT OF COAL CONVERSION WASTE WATER

Coal/process	Reactor type	Remarks	Ref.
Indianhead lignite/carbonization effluent	Two-stage anaerobic activated carbon reactor	$V = 14.4\ \ell$, filled with Raschig® rings/activated carbon	132—135
	Two-stage anaerobic activated carbon reactor	$V = 11\ \ell$, filled with bed-saddle bodies/activated carbon, water 1:10 diluted, $D = 0.04\ h^{-1}$, partial adsorption effects	
H-coal effluent	Serum bottles	$V = 50\ m\ell$, 57 hr shaking	136
Coking effluent	Biofilter	60—70% degradation of soluble organic carbon	137
H-coal effluent	Serum bottles	$V = 50\ m\ell$, semicontinuously 155 days	138
Brown coal/pressure gasification	UASB-reactor	$V = 300\ m\ell$, water 1:1 diluted 0.1 hr^{-1}, 1000 hr continuously, 60% degradation, biogas 80% methane	139

$$CH_3COOH + 2H_2O \rightarrow 2CO_2 + 4H_2 \qquad G° = +104.6\ kJ\ mol^{-1}$$

$$4H_2 + CO_2 \rightarrow CH_4 + 2H_2O \qquad G° = -135.6\ kJ\ mol^{-1}$$

thus forming here a syntrophic system with H_2-utilizing methane bacteria.

Apart from the degradation of aromatics by methanogenic syntrophic cocultures, other anaerobic bacteria which can realize such a degradation are also known (nitrate- and sulfate-reducing microorganisms as well as some photosynthesizing microorganisms). Unlike methane fermentation, here the aromatics are direct substrates. Investigations have in most cases been carried out on benzoic acid, but for other aromatics the results should be basically similar to the degradation mechanism according to the pathway shown for the syntrophic bacteria of the methanogenic coculture (Figure 9).

As a qualification, let us point out that for sulfate reducers no results have been published so far. In other respects it has been possible to detect a complete degradation of the aromatic ring, followed by an ortho cleavage of the reduced ring, for all microorganisms investigated. The discovery of extrachromosomally coded enzymes in the late 1960s[64] and the organization of genes for enzymes of the salicylate degradation in a plasmid (SAL) in *Pseudomonas putida*[65] has led to the discovery of a great number of plasmid-coded enzymes of degradative pathways, where *Pseudomonas* strains have been preferred subjects of investigation, although plasmids could be detected in many Gram-negative bacteria. The enzymes of a degradative pathway are necessarily not completely plasmid-coded. Thus, for the degradation of alkanes (C_5 to C_{10}) the octane degradating plasmids (OCT) contains only two key enzymes, the alkane hydroxylase proteins and alkane dehydrogenase. The other enzymes are chromosomally coded.

On the other hand, in the degradation of toluenes as well as of *m*- and *p*-xylenes most of the enzymes in the TOL plasmid are plasmid-coded. The ring cleavage is effected here by *m*-cleavage. Phenol (monohydroxy benzene) is likewise degradable by TOL strains, provided that the latter can form a chromosomally coded phenol monooxygenase. On the other hand, strains containing phenol degradating plasmids (PHGs) have also been reported.[66]

Of special interest for the degradation of industrially produced xenobiotics was the detection of plasmids for the degradation of alkyl benzene sulfonates (ASL)[67] and chlorobenzoates.[68] In a strain of *Flavobacterium* sp., a plasmid was found which codes two highly specific enzymes that allow the degradation of the ring-shaped as well as of the straight-chain nylon dimer.[69]

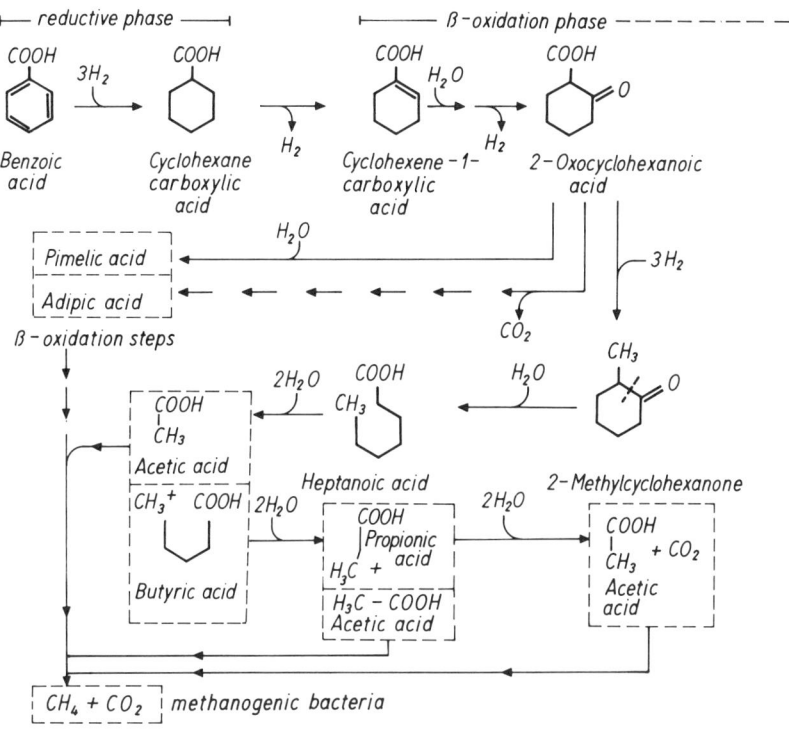

FIGURE 9.   Degradation mechanism of aromatics for syntrophic bacteria of the methanogenic coculture.

The conjugative transfer of most of the degradative plasmids from a donor cell to a recipient cell, i.e., the possibility of transfer without any manipulation, the interplay of the enzymes coded by them and of the chromosomally coded ones, the low specifity of many plasmid-coded enzymes, and the fact that they are inductive, impart a special importance to the plasmids for the adaptation of a strain of a microbial community to environmental changes, and account for the enzymatic versatility that is found in the pseudomonads.

Moreover, the simple realization of conjunctive transfer of plasmids makes it possible to construct multiplasmid strains for the degradation of a broad range of substrates. In the first patented strain, (*P. putida*),[70] which contained the xylene degradating plasmid (XYL) (XYL corresponds to the toluene degradating plasmid (TOL) but is not conjugatively transferable), naphthalene degradating plasmid (NAH), and camphor plasmid (CAM)-octane degradating plasmid (OCT), were additionally inserted.

Usually CAM and OCT are incompatible, that is, there exists a mutual inhibition between the two plasmids; incompatibilities between two different plasmids were frequently found. After a transfer of OCT into a CAM+ strain it has been possible to isolate cells with OCT+ and CAM+ by selection. It cannot yet be estimated to what degree multiplasmid strains with a broad spectrum of substrates will be introduced into waste water treatment. The barriers still to be overcome are in effect given by

●   The incompatibility of many plasmid pairs
●   The instability of plasmids and of some plasmid-coded enzymes
●   The increased instability of a plasmid in a multiplasmid cell

According to the present state of knowledge, a mixed population, an adapted, dynamically responding microbial community, which contains single-plasmid bacterial cells, should be superior to a multiplasmid strain. The construction of strains which make it possible to degrade certain groups of synthetically produced organic substances should be more promising, for example, for the degradation of differently substituted chlorobenzenes.[71,72]

5. Kinetics of the Aerobic Degradation of Brown Coal Processory Waste Waters

For pure cultures, Monod[41] has been able to show the validity of the following relation, which is analogous to the Michaelis-Menten equation:

$$\mu = \mu_{max} \frac{S}{K_s + S} \tag{1}$$

where μ = specific growth rate (T^{-1}), μ_{max} = maximum specific growth rate (T^{-1}), S = substrate concentration $(M_i L^{-3})$, and K_s = half-rate constant, substrate concentration at

$$\mu = \frac{\mu_{max}}{2} (M_i L^{-3})$$

Moreover, Monod derived the relation

$$\frac{dX}{dt} = -Y \frac{dS}{dt} \tag{2}$$

which describes the ratio of the growth rate to the substrate concentration, where X = concentration of microorganisms $(M_i L^{-3})$ and Y = growth yield coefficient $(M_x M_i^{-1})$. A substitution of Equation 2 into Equation 1 gives

$$\frac{dX}{dt} = Y \frac{dS}{dt} = \mu_{max} \frac{S}{K_s + S} \cdot X \tag{3}$$

or

$$\frac{dS}{dt} = \frac{\mu_{max}}{Y} \frac{S}{K_s + S} \cdot X \tag{4}$$

By insertion of

$$\frac{\mu_{max}}{Y} = r_{x_{imax}} \tag{5}$$

into Equation 4 one obtains

$$\frac{dS}{dt} = r_{x_{imax}} \frac{S}{K_s + S} \cdot X \tag{6}$$

or

$$\frac{dX}{dt} = r_{x_{imax}} Y \frac{S}{K_s + S} \cdot X \tag{7}$$

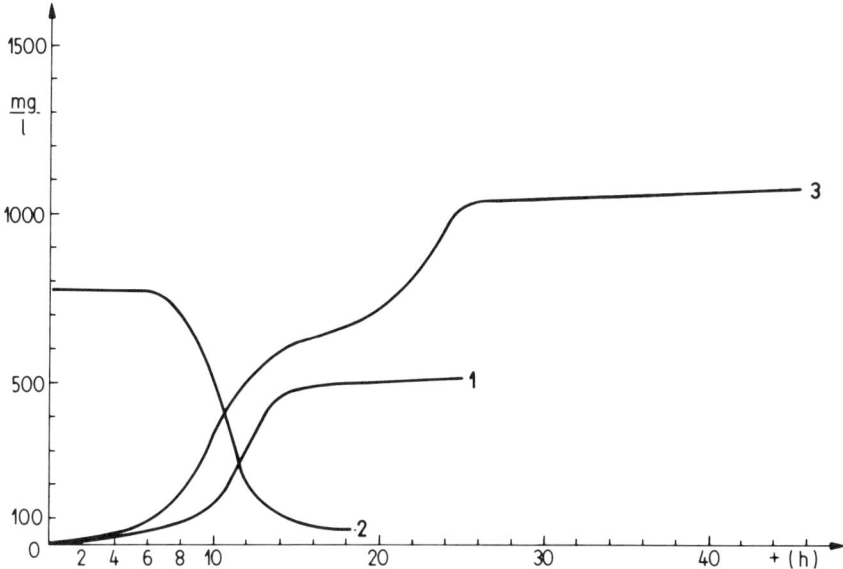

FIGURE 10. Polyauxic course of the degradation of the substances contained in a BCPWW. 1: Biomass (DW), 2: volatile fatty acids, and 3: oxygen consumption.

If the continuous cultivation customary in fermentation engineering is used, then in the stationary state the growth rate μ is equal to the dilution rate D (T^{-1}), which is arbitrarily adjustable below μ_{max}.

The relations established by Monod[41] for pure cultures and a single substrate can also be applied to the mixed populations of the sewage treatment plants with a substrate mixture. The organic compounds contained in the sewage can be utilized by stages (polyauxic effect). By measuring the oxygen absorption it is possible to detect the polyauxic course of the degradation (e.g., see Reference 48). An example of the polyauxic degradation of the organic substances contained in sewage is shown in Figure 10.

From Equation 7 it is possible to derive two limiting cases. If the substrate concentration is very high, so that K_s can be neglected, which is the case for the brown coal waste waters with high organic loads, then Equation 7 reduces to

$$\frac{dX}{dt} = r_{x_{imax}} \, Y \cdot X = \mu_{max} \cdot X \tag{8}$$

This is a direct proportionality between the growth rate of the biomass and the actual biomass concentration. If there are no constraints, then the specific growth rate μ corresponds to μ_{max}. The substrate concentration does not influence the growth rate.

For low substrate concentrations Monod's relations hold in the nonreduced form, reflecting the dependence on the substrate concentration and the growth rate. From these considerations it is possible to draw fundamental conclusions for the biological treatment of BCPWW.[73]

In the first stage, the high-performance biological treatment, there is a rapid substrate degradation combined with a high growth rate of the biomass. This first stage is designed as a stirred reactor in order to enable a high oxygen inclusion, because an oxygen limitation would reduce the degradation performance.

The effect of the actual dissolved oxygen concentration on the degradation performance in condensation waters has been investigated by Martius et al.[74] using a respiration fermenter.[75] For dissolved oxygen concentrations above 10% of air saturation there was no effect on the rate of substrate concentration observable. Below this concentration there occurs a slowing-down of the degradation. The resulting half-saturation kinetics is shown in Figure 11.

An influence of the variation of the dissolved oxygen concentration on the specificity of the degradation performance for groups of substances could not be detected. Differences in the degradation of dihydroxy benzenes between low and high degrees of aeration have already been pointed out (Table 10).[26] In the high-performance stage, a full exhaustion of the substrate is not reached because of the polyauxic course of the substrate degradation. The degradation of this residual load takes place in a tubular reactor, where Monod's relation holds in the nonreduced form (limiting case 2). In this reactor back mixing is completely absent, so that every volume element must be considered separately. In this system the degradation of the organic load is a function of the path traveled. According to Monod, the length of the tubular reactor required for total degradation can be calculated as follows (where the biomass concentration in the reactor is assumed to be constant)[76] (Figure 12 and Equation 9):

$$\int_{S_2}^{S_1} \frac{K_s + S}{S}\, dS = X_{r_{imax}} \int_{t_2}^{t_1 = 0} dt \tag{9}$$

where S_1 = feed substrate concentration $(M_i L^{-3})$, S_2 = discharge substrate concentration $(M_i L^{-3})$, $t_2 - t_1$ = reactor retention time (T), t_1 = starting time, and t_2 = reactor-flow-time.

$$= -\frac{1}{X_{r_{imax}}} K_s\, X\, n \frac{S_1}{S_2} + S_1 - S_2 \tag{10}$$

In the tubular reactor, the discharge biomass concentration reaches a maximum, while the substrate concentration tends to a minimum.

The two-stage sewage treatment process has the advantage that surge loads, which may occur especially in BCPWW, can be taken up without difficulty. However, it turns out that in the high-performance stage a mixed population accumulates which is adapted to the specific conditions in this stage. It exhibits a poor flocculation behavior, which makes it difficult to separate the biomass. Moreover, this population does not give rise to any further substance degradation in the tubular reactor. Therefore, it is of importance that a separate mixed population is built up in the second degradation stage.

To achieve a stable operation of the activated sludge stage, it is of advantage to recycle part of the discharged sludge. This not only effects a continuous inoculation with the adapted culture, but also increases the degradation performance with as short retention times as possible by means of increasing the sludge concentration. Moreover, sludge recycling will compensate a lack of nutrients in the waste water medium, which again stimulates the growth of the waste water culture.

The two-stage biological waste water treatment reduces the load of biologically degradable substances by 85 to 90% in the first stage. The second, i.e., the activated sludge, stage reaches another degradation efficiency of 80%.

On the basis of Pirt's relations, Koné and Behrens[77] extended Monod's equation to include the maintenance metabolism:

$$\mu = \mu_{max} \frac{S}{K_s + S} - m\, Y_{max} \tag{11}$$

where m = maintenance coefficient (T^{-1}).

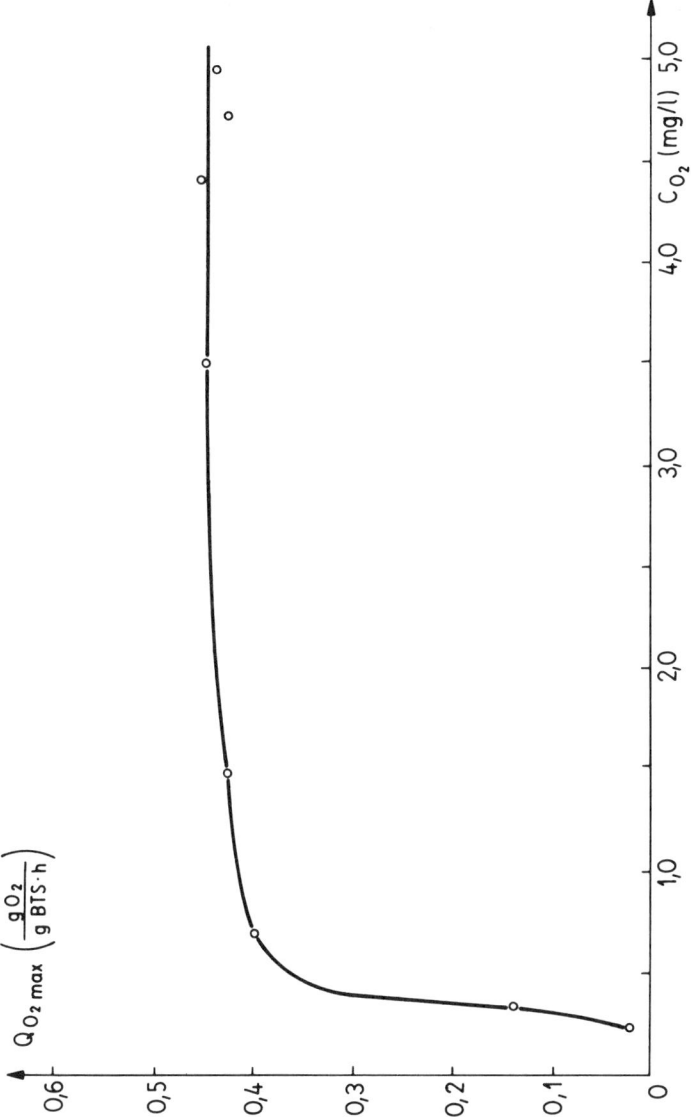

FIGURE 11.   Half-saturation kinetics for a variation of oxygen concentration.

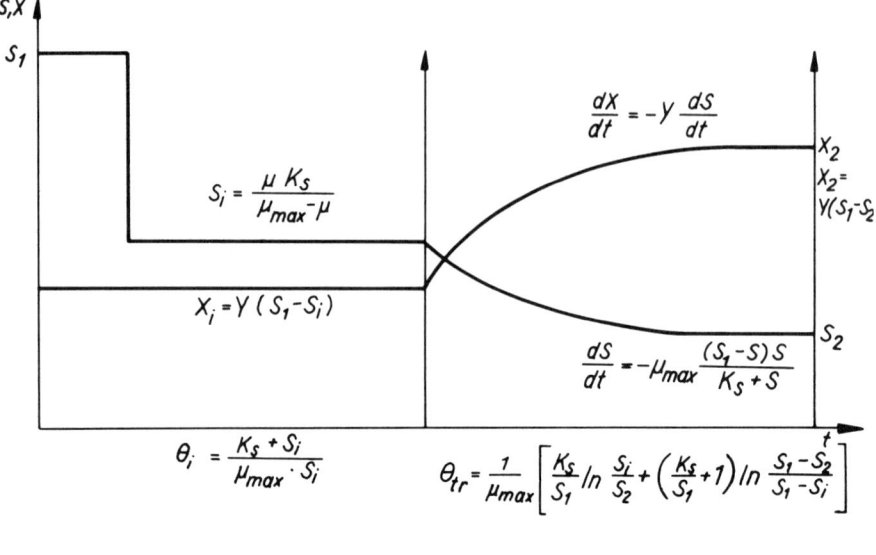

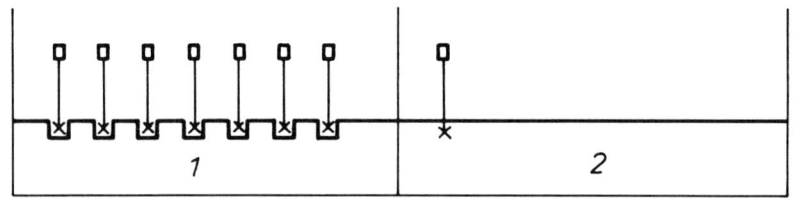

1. high-performance basin   2. low-performance basin

   (stirred reactor)   (tubular reactor)

FIGURE 12.   Reaction-kinetic system of the two-stage waste water treatment according to the PKM process. S_1 = substrate concentration at the inlet of the total plant ($M_i L^{-3}$), S_i = substrate concentration in the high-performance stage ($M_i L^{-3}$), S = substrate concentration at an arbitrary point in the tubular reactor ($M_i L^{-3}$), S_2 = substrate concentration at the outlet of the total plant ($M_i L^{-3}$), X_i = biomass concentration in the high-performance stage ($M_i L^{-3}$), X_2 = biomass concentration at the outlet of the total plant ($M_i L^{-3}$), θ_i = retention time in the high-performance state (T), and θ_{tr} = retention time in the tubular reactor (T).

In another paper,[78] the authors report their results of investigations on the denitrification of brown coal condensation waters.

It was found that the level of the maintenance metabolism in the waste water medium, as compared with the model medium, was clearly increased (0.073 g $N_{NO_3^-} \cdot$ g $X^{-1} \cdot$ hr^{-1} as compared to 0.01 g $N_{NO_3^-} \cdot$ g $X^{-1} \cdot$ hr^{-1}). The higher expenditure for the maintenance metabolism in the waste water medium was attributed to the influence of toxic substances contained in it, where energy-consuming processes due to the change of the energy efficiency of respiration by toxic substances, disturbances of the transport processes, and by lytic processes are included.

Lytic processes occur to an increasing degree in the population of microorganisms as the substrate approaches complete exhaustion, and likewise reserve substances are consumed by endogenic respiration. These processes are of importance in the conditioning of activated sludge (see below). For the description of these processes, a number of authors (e.g., Van Uden[79]) introduced the decay coefficient into Monod's relation:

$$\mu = \mu_{max} \frac{S}{K_s + S} - b \qquad (12)$$

where b = decay coefficient (T^{-1}). Note the analogy between Equations 11 and 12. One obtains

$$b = \mu Y_{max} \qquad (13)$$

By endogenic respiration it is possible to utilize 80% of the cellular mass. The rest is not biodegradable.

6. Nitrification and Denitrification

BCPWW not only exhibits a high organic load, but also has a high content of soluble nitrogen compounds, the major part of which is constituted by free ammonia and ammonium compounds (see Table 4). Small quantities of other nitrogen compounds can be neglected here. In a nonbiological way, the nitrogen compounds can be eliminated by thermal ammonia stripping, ion exchange, or chlorination. However, these techniques have disadvantages, which are attributable partly to technological, partly to economic, causes.

An elimination of the nitrogen content by a conventional aerobic waste water treatment is not realizable, because the nitrogen to carbon ratio (as referred to fatty acids) is shifted in favor of nitrogen far beyond the ratio of 1:0.23 that can be calculated from an empirical biomass formula of $C_5H_7NO_2$.

The excess of NH_3-NH_4^+ can be eliminated by the well-known biological processes of nitrification and denitrification. For these two technological stages there are, however, particular features resulting from the composition of the BCPWW. Nitrification generally requires a high degree of exhaustion of the organic carbon sources. The brown coal waste waters, however, also contain considerable amounts of slowly utilizable and persistent compounds, which in part also have toxic effects.

a. Nitrification

The oxidation of ammonia or ammonium ions takes place in two partial reactions. These can be formulated as follows:

$$NH_4^+ + 1.5\ O_2 + 2HCO_3^- \rightarrow NO_2^- + 2H_2CO_3 + H_2O$$

$$(\textit{Nitrosomonas})$$

$$NO_2^- + 0.5\ O_2 \rightarrow NO_3^- \qquad (\textit{Nitrobacter})$$

It is seen that up to the nitrification stage the oxidation effects a liberation of protons, that is, it leads to a pH shift. If the shift of the pH into the acidic range is not compensated, this will affect the nitrification, because the optimum of the pH ranges is between pH 7 and 8. Nitrification requires the supply of considerable amounts of oxygen (theoretically 4.57 mg O_2 per milligram of NH_3-N).[80]

The nitrifying bacteria are chemolithotrophic microorganisms, that is, they obtain the carbon required for cell synthesis by fixation of carbon dioxide, and the energy by oxidation of NH_4^+ to NO_2^-, or of NO_2^- to NO_3^-. The growth rate of the nitrifying bacteria is extremely low. Sharma and Ahlert[80] indicate specific growth rates of μ = 0.46 to 2.2 day^{-1} for *Nitrosomonas* and 0.28 to 1.44 day^{-1} for *Nitrobacter*. On the other hand, the heterotrophic microorganisms in activated sludge (with glucose as a substrate) exhibit growth rates of μ = 7.2 to 17 day^{-1}.

The growth of the nitrifiers is influenced by a great variety of substances. Thus, it has been found that, for instance, phenols and heterocyclic compounds have an inhibiting effect.[81] Such investigations must, however, be interpreted with reserve, because adaptation processes, the selection of certain species, etc. in practice frequently do not correspond with results obtained in the laboratory. Earlier investigations carried out on BCPWW showed that, for example, in an adapted sessile culture considerable phenol concentrations are tolerated.[81a] Thus, the particularities of the nitrification process enable a number of conclusions to be drawn for an efficient nitrifying treatment of BCPWW:

1.   The waters to be nitrified must exhibit a BOD close to zero; if not, then extremely long retention times must be maintained.
2.   Due to the slow growth of the nitrifiers special emphasis must be placed on the sludge retention.
3.   It is of great importance to maintain the optimum pH.
4.   A sufficient aeration has to be ensured. Heinrich,[82] for example, considered an oxygen concentration of 2 mg/ℓ to be most favorable from the economic point of view for the oxidation up to the nitrite stage in the fixed-bed reactor. A further increase of the oxygen concentration led to a further increase in the nitrification rate.

There is only a small number of publications on the nitrification of BCPWW. Luthy[83] reported a single-stage treatment of waste waters from the processing of bituminous coal and brown coal. Here the degradation of the organic load and the nitrification were run in parallel. Good nitrification efficiencies were obtained with retention times of 10 to 40 days. Nitrificants can be stored for some months at room temperature without loss of the nitrification activity, if they are separated, resuspended together with $CaCO_3$ and $(NH_4)HCO_3$, and dried in at the air.[131] It should, however, be pointed out here that the set of problems connected with BCPWW is closely related to that of hard coal waste waters, and that there is a large number of publications that deal with the latter.

b. Denitrification

Denitrification represents the process of actual nitrogen elimination. Here nitrate is reduced to molecular nitrogen via several intermediate stages. Focht and Chang[84] proposed the following path:

$$NO_3^- \rightarrow NO_2^- \rightarrow NO \rightarrow N_2O \rightarrow N_2$$

This reaction path cannot be completely realized under all conditions and by all microorganisms. Denitrification is a process that takes place under anaerobic conditions, where the links of the above-mentioned reaction chain act as electron acceptors. Moreover, denitrification requires organic compounds as hydrogen donors. Methanol is frequently used under technical conditions. It is economically desirable to utilize substances contained in sewage for that purpose. The following equation shows denitrification with the use of acetate, the typical component of the BCPWW:[77]

$$8.5\ CH_3COO^- + 7.94\ NO_3^- + 1.42\ NH_3 \rightarrow 1.42\ C_5H_7O_2N + 1.7\ H_2O$$

$$+ 3.97\ N_2 + 9.93\ CO_2 + 16.44\ OH^-$$

As the equation shows, the denitrification leads to the liberation of hydroxyl ions. There is a great variety of facultative anaerobes that are capable of denitrification. Beccari et al.[85] listed the following genera:

Acinetobacter	*Halobacterium*	*Pseudomonas*
Alcaligenes	*Hyphomicrobium*	*Rhodopseudomonas*
Bacillus	*Micrococcus*	*Spirillum*
Gluconobacter	*Moraxella*	*Xanthomonas*

Which microorganisms predominate is decisively determined by the electron donor.[86] Acetate is most suitable as a hydrogen donor.

Regarding the role of oxygen, there are discrepant data. Since under technical conditions denitrification may take place in a neighborhood close to nitrification at unaerated points in the same basin, lower oxygen concentrations are obviously tolerated. Denitrification proceeds in an optimum way in the weakly alkaline pH range under mesophilic temperature conditions.

Reports about inhibitory effects on denitrification are rare. Lewandowski[87] investigated the inhibitory influence of cyanide on denitrification. Unlike the nitrifying bacteria, which grow slowly, the denitrifying bacteria reach considerable growth rates. Koné and Behrens[88] carried out investigations to determine kinetic parameters of the denitrification of a BCPWW. In this case a maximum specific growth rate μ_{max} of 0.55 hr^{-1} was determined for a dilute gas liquor with a dilution ratio of 1:1, as compared with 0.65 hr^{-1} in the model medium. For a continuous process management, the yield coefficient Y was found to depend on the dilution ratio. The value of the maintenance coefficient increased from 0.01 g NO_3-N/g BDW·hr in the model medium to 0.073 g NO_3-N/g BDW·hr in dilute gas liquor. This significant increase was attributed to the influence of toxic substances in the waste water. The different maximum specific growth rates in the waste water and in the model medium are explained by different substrate inputs for the maintenance metabolism.[78] Koné and Behrens[77] considered the relation between the maintenance metabolism and the maximum yield, taking BCPWW as an example. They arrived at general conclusions for the microbial growth. The high maximum specific growth rates found by Koné and Behrens[88] for a denitrifying mixed culture could not, however, be realized in investigations with the pH auxostat.[89]

To summarize, the following conclusions can be drawn for the denitrifying treatment of BCPWW:

1.   The denitrifying mixed cultures are relatively insensitive to inhibitory effects.
2.   The denitrifying bacteria exhibit high growth rates; nevertheless, sludge retention or recycling is also frequently applied in order to enable a stable operation.
3.   Setting the anaerobic conditions required for denitrification is unproblematic, because the denitrifying mixed cultures tolerate small amounts of dissolved oxygen.

The incorporation of nitrification and denitrification stages into complex plants for the purification of BCPWW is dealt with in Section IV.

C. Possibilities for the Utilization and Disposal of Sludge from Plants for the Biological Treatment of Brown Coal Processory Waste Waters

As in all biological waste water treatment processes, the disposal of the sludge produced represents a problem. It can be expected that in the steps of microbial purification every ton of BOD_5 load produces 20 to 30 m^3 of fresh sludge, with a water content of 98 to 99%.[76] A disposal by spraying on heaps for recultivation purposes or by flushing with ashes obtained is not in any case possible, but is practiced frequently.

Here we shall make some remarks on the attempts to utilize ammoniacal phenolic waters by direct application to land for fertilizing purposes.[90]

Favorable effects on plant growth have been described by Schultz.[91] According to these results, the following concentration limits are considered tolerable: 50 mg ℓ^{-1} of monophenols, 500 mg ℓ^{-1} of polyphenols, and 3300 mg ℓ^{-1} of steam-volatile fatty acids. The pretreated BCPWW must be diluted to a BOD_5 of 3200 mg ℓ^{-1}.

The biosludges obtained from the complex biological treatment of BCPWW can be classified into fresh sludges and stabilized sludges. Problems encountered in a further treatment of the sludges are based on the colloidal structure or on the presence of stabilized suspensions in which the major part of moisture is fixed in capillaries or in a gel-like form. A subdivision of the sludge treatment methods is shown in Figure 13.

The simple concentration method of flotation by inclusion of air is made very difficult by the negative charge of the hydrophilic colloid, because the small air bubbles are negatively charged, thus counteracting the flotation effect. If waters are acidified, the isoelectric point of the microorganisms is reached in the pH range between 3.5 and 3.8, then by finely dispersed air a complete separation of the microorganisms is achieved.[92] On the other hand, this way is prohibited from practical application by economic reasons and considerations of water resources management.

The addition of cation-active substances in the most favorable cases led to a concentration of the microorganisms to 85%, where the best effect was achieved by application of *n*-diethylamino oxypropyl ether of dextran and hexacetal pyridine bromide in concentrations of 0.01 and 0.05%, respectively.[92] In a wider sense, anaerobic sludge treatment of BCPWW also represents a flotation, because the CO_2 bubbles produced attached to the changed floc and float it.[92a]

There has been no lack of attempts to provide for a direct utilization or a recovery of useful material from the great amounts of biomasses produced by waste water treatment. In the 1950s efforts taken to produce protein-containing vitamin concentrate led to the production of activated sludge from brown coal processing which contained up to 1000 μg of vitamin B_{12} per liter of sludge.

The vitamin content of baker's yeast as compared with that of a biomass obtained from the activated sludge of a brown coal processing plant is shown in Table 16.

Behrens and Augst[92b] attempted to use a carotene-forming yeast (*Rhodotorula glutinis*), which had been isolated from the trickling filter of a BCPWW purification plant, as a substrate in condensation waters for the production of vitamin A. After a retention time of 10 hr, however, the yeast contained only 3 μg = 5 IU of β-carotene per gram of dry matter. Moreover, foreign infections (white yeast) prevailed as the time of cultivation was increased.

First attempts at a utilization of waste waters from the brown coal industry for the production of feed protein have been described by Rieche et al.[45] A mixture of molasses as well as processes of phenosolvan extraction were used for the growth of *Torula utilis,* where the fermentation process was run continuously. The mixed-substrate effect caused by the sugar is of advantage for the utilization of the phenols and the fatty acids.

Investigations on the direct utilization of sludge from BCPWW purification plants for animal feeding have been described by Behrens.[93]

After a sludge conditioning, the use of decanters led to biomass concentrations which made it appear suitable for use as a feed supplement.[94] Exploratory feeding experiments with pigs and poultry did not show any detrimental effects on the health of the animals or on the taste of the meat, showing that activated-sludge biomass in combination with other proteins is well suited for fattening improvement. The amino acid composition of a biomass from a BCPWW plant is shown in Table 17. A continuation of these experiments was prevented for reasons of the content of heavy metals in the biomass as well as problems of an economical separation of the sludge.

The attempts made so far to utilize the biomass obtained from the process of waste water treatment showed that a systematic synthesis of the desired products by high-performance strains of bacteria will be preferred in any case.[93]

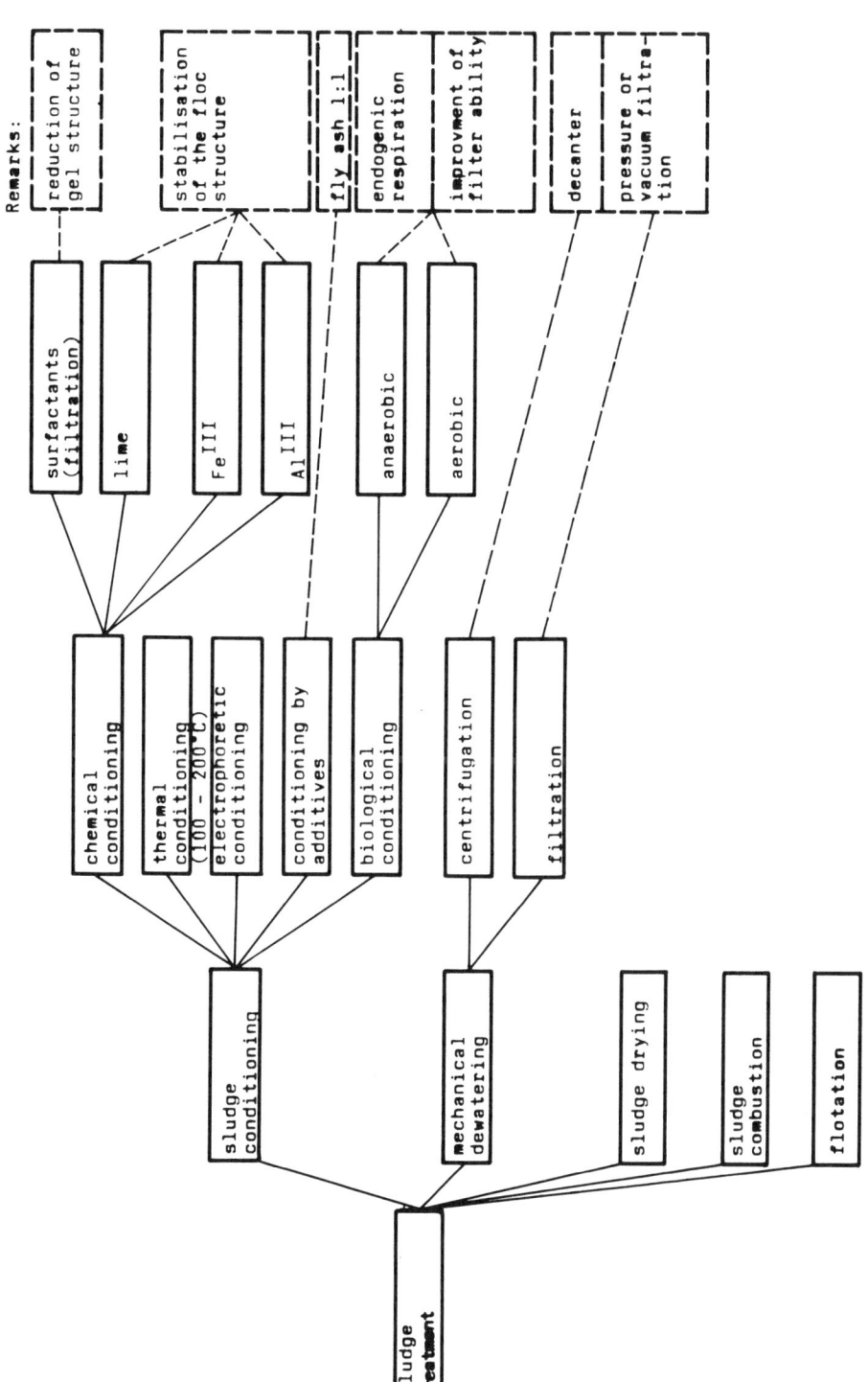

FIGURE 13. Sludge treatment methods.

Table 16
CONTENT OF VITAMINS IN FEED
YEAST AND IN SLUDGE FROM A
PLANT FOR LIGNITE CONDENSATION
WATER TREATMENT (γ PER GRAM
ASH-FREE DRY MATTER)

Vitamins	Feed yeast	Sludge
Cyanocobalamine	0	5.6
Thiamin	5—53	26
Riboflavin	26—70	70
Pyridoxine	35	18.7
Panthothenic acid	112.7	40—190
Nicotinic amide	3260	153—600
Biotin	3.13	0.5—3.6
Choline	552.5	2500—5000
m-Inositol	973.1	2700—3600
p-Aminobenzoic acid	6.35	17—62
Folic acid	31.3	4—31

Table 17
COMPARISON OF THE PROTEIN AND AMINO
ACID CONTENT IN FEED YEAST AND SLUDGE
FROM A PLANT FOR BCPWW

	Feed yeast (% in dry matter)	Sludge (% in dry matter)
Protein	**47.43**	**59.0**
Amino acid		
Arginine	3.6	3.6
Alanine	n.d.	4.1
Glycine	0.22	2.0
Isoleucine	3.75	2.6
Lysine	4.14	2.0
Phenylalanine	2.41	1.9
Tryptophane	0.66	2.0
Glutamic acid	6.9	2.0
Cystine	0.68	3.7
Histidine	1.31	1.2
Leucine	3.57	n.d.
Methionine	0.84	2.0
Threonine	2.85	2.5
Valine	2.98	1.7
Tyrosine	n.d.	1.0
Asparaginic acid	n.d.	2.0
Proline	n.d.	2.0
Serine	n.d.	2.0

Note: n.d. = not determined.

IV. TECHNOLOGY OF BIOLOGICAL WASTE WATER TREATMENT OF BROWN COAL PROCESSORY WASTE WATERS

A. Substrate-Specific Demands on the Technology

The specificity of the technology required for the microbiological degradation of waste waters of the brown coal processing industry results from the high organic loads, the specificity of the substances contained, and the involved strong foaming tendency. The technical realization of process is determined by the specific supply of nutrients offered by the waters and by the toxicity of the substances contained.

From considerations on the kinetics of degradation it has been found that a two-stage process is most favorable for the rapid degradation of the organic load (see Section III.B.5). In high-performance biological treatment, the foaming control plays a very important role because of the high oxygen feed required. A continuous process management makes it possible to match the degradation performance of the microorganisms to the content of toxic substances, and hence to control the actual substrate concentration. The reduction of high nitrogen leaching values is achieved by the combination of nitrification and denitrification stages with process steps of high performance biological treatment. The insufficient nutrient supply must be compensated by corresponding nutrient additions.

B. Processes of Biological Purification

1. The Magdeburg P-Process

In the first biological treatment plants, the insufficient nutrient supply of the BCPWW involved a mixing with municipal sewages. Nolte et al.[38] were the first to practice a biological purification of unblended BCPWW by charging mineral salts, especially phosphates and, if required, magnesium, potassium, and iron. He used a conventional activated sludge technology with bottom aeration, where the addition of 7 mg phosphorus per kilogram of BOD_5-load was required. The Magdeburg P-process required high dilutions, because only feed concentrations with a BOD_5 value of up to 500 mg O_2 per liter were tolerated.[95,96]

2. The Tower Trickling Filter Method

The erection of new plant complexes for brown coal processing by new technologies made it necessary to further develop the inadequate "classical" biological treatment plants in the G.D.R. The efforts taken to increase the feed concentration of BCPWW in the biological treatment plants initially led to the development of the tower-type trickling filters, which were intended to reduce a high initial load, while a final purification was to take place in a second step in a conventional activated sludge plant.

This process was realized in VEB Braunkohlenkombinat Lauchhammer.[97] Pretreated waste waters from the coking plant were directed to tower trickling filters which utilized the natural transfer of oxygen. A great number of investigations were carried out on these systems, and the results were published.[98,99] It turned out that in these systems the transfer of oxygen was relatively low. Therefore, it was necessary that the waste waters were highly diluted, because a stable operation was obtained only for a BOD_5-concentration of less than 2200 mg O_2 per liter. At a higher concentration the degree of degradation was drastically reduced. Already for BOD_5-concentrations above 1200 mg O_2 per liter it was no longer possible to detect dissolved oxygen in the tower outlet.[98]

For two tower trickling filters connected in series, with a height of 20 m, a degradation efficiency of 25 to 30% (as referred to the BOD_5 load) was achieved at the dilution required; the degradation rate per unit volume was found to be 1.1 to 2.0 kg BOD_5 per cubic meter per day.[100] Additional forced aeration did not appreciably increase the capacity but required an additional input of energy that amounted to 2.64 kWh per kilogram of additionally degraded BOD_5 load.[101] The total purification efficiency of a plant consisting of two tower trickling filters connected in series as a high-loaded stage, followed by a conventional activated sludge plant, is relatively low (Table 18).

In the concrete case investigated, tower trickling filters are suitable only for a relatively low load. The energy input required for delivering the liquids up to the tower level is considerable.

In the last few years a modified form of the trickling filters method, the rotating biological

Table 18

**DEGRADATION EFFICIENCY ACHIEVED IN THE
BIOLOGICAL WASTE WATER TREATMENT PLANT OF
THE LAUCHHAMMER COKING PLANT[91]**

Parameter	Inlet conc ($mg\ \ell^{-1}$)	Outlet conc ($mg\ \ell^{-1}$)	Degradation efficiency (%)
BOD_5	2500	1080	57
NH_4-N	345	304	12
Total phenols	167	109	35
Volatile phenols	8	5	37
Volatile fatty acids	1220	640	48

Table 19

**DEGRADATION EFFICIENCY OF A TECHNICAL PILOT
PLANT FOR HIGH PERFORMANCE BIOLOGICAL
TREATMENT WITH TURBINE AERATION**

Parameter	Inlet conc ($mg\ \ell^{-1}$)	Outlet conc ($mg\ \ell^{-1}$)	Degradation efficiency (%)
BOD_5	18,000	2,200	88
NH_4-N	2,700	860	68
Volatile phenols	100	0.75	99.3
Volatile fatty acids	4,500	160	96.5

From Zülke, H. J., Kontinuierliche technische Nachreinigung von Braunkohlen-Kokereiabwässern nach dem biologischen Intensivverfahren, *Fortschr. Wasserchem. Ihrer Grenzgeb.*, 5, 172, 1967.

contactor or the rotating disk contactor, gained an increasing importance for municipal sewage and nitrification stages. Today modern tower-type biological treatment systems are provided with a sophisticated and controlled forced aeration.[102,103]

3. Methods of High-Performance Biological Treatment

The further development of the purification processes led to the two-stage process with a high-loaded first stage, the so-called high-speed aerated, or high-performance biological treatment processes. The degradation rate per unit volume of these processes are eight to ten times higher than those of the classical ones.

It is necessary (see Section III.B.5) to operate with high inlet concentrations, where the amount of toxic substances contained does not exceed the degradation capacity of the microorganisms and the process is not run below the substrate limitation. The necessary final purification is accomplished by conventional methods, together with the sanitary sewage of the factory (see Section IV.B.4). The high oxygen supply to the high-performance stage was first realized using turbine aeration sets (of the Pfaudler, Dorr-Oliver, and Yeomah systems).[104,105] In a technical pilot plant with oxygen supply rates of up to 3.5 kg O_2 per cubic meter per hour in coking plant waste water, a BOD_5 degradation efficiency of approximately 85% and a degradation rate of 25 kg BOD_5 per cubic meter per day were achieved by this method[106] (Table 19).

There arose, however, cooling problems due to the highly positive heat tonality of the process. Moreover, the high abrasive wear of the main bearings proved unfavorable. An excessive foaming occurred, which was difficult to control even by chemical defoaming agents and surface nozzle spraying.[7]

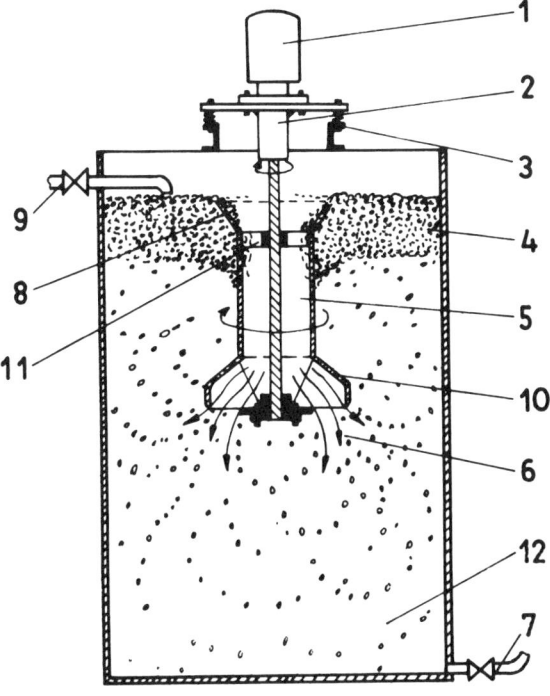

FIGURE 14. Principle of the centrifugal aerator with automatic foam-drawing device; according to Franz.[107] 1: drive, 2: bearing, 3: vibration absorber, 4: foam, 5: centrifugal aerator with automatic foam-drawing device, 6: oxygen feed, 7: outlet, 8: air and foam suction, 9: inlet, 10: radial impeller, 11: axial impeller, and 12: water/air emulsion.

On the basis of the good results of high-performance biological degradation, but avoiding the problems of turbine aeration, Franz[107,108] developed the centrifugal aerator with automatic foam-drawing device, where the oxygen supply rate was deliberately limited to the requirements of a standard high-performance biological degradation of approximately 0.5 kg O_2 per cubic meter per hour.[8]

By means of this aerator, oxygen yield values ranging between 2 kg O_2 per kilowatthour for low-emulsified liquids and 4 kg O_2 per kilowatthour for higher-emulsified liquids, and degradation rates of about 10 kg BOD_5 per cubic meter per day were achieved. The principle of the centrifugal aerator with automatic foam-drawing device (Figure 14) is realized by a radial rotor coupled rigidly to the motor, running at a peripheral speed of 20 to 30 ms^{-1}, which utilizes the cavitation effect to draw in air continuously and, after disrupting it into extremely small bubbles, to inject the latter downwards into the liquid. At the same time the developing foam is drawn in and destroyed by the centrifugal force in the rotating cylinder. The construction of the rotor, which is coupled rigidly to a suspended shaft without bearings in the liquid, ensures a long lifetime and enables repairs to be carried out without emptying the basin (Figure 14).

The positive balance of a small-scale application[108] led to the industrial application of this technique (see Section IV.C), which yielded the degradation values listed in Table 24.

Another modern aeration system that meets the requirements of high-performance biological treatment of BCPWW is the deep-jet aeration system designed by the principle of Bunsen's water-jet pump, which reaches extremely high degradation rates per unit volume,[109] where, however, an additional energy input is required for foam destruction. The construction and the mode of action of a deep-jet aeration device[110,111] are shown in Figure 15.

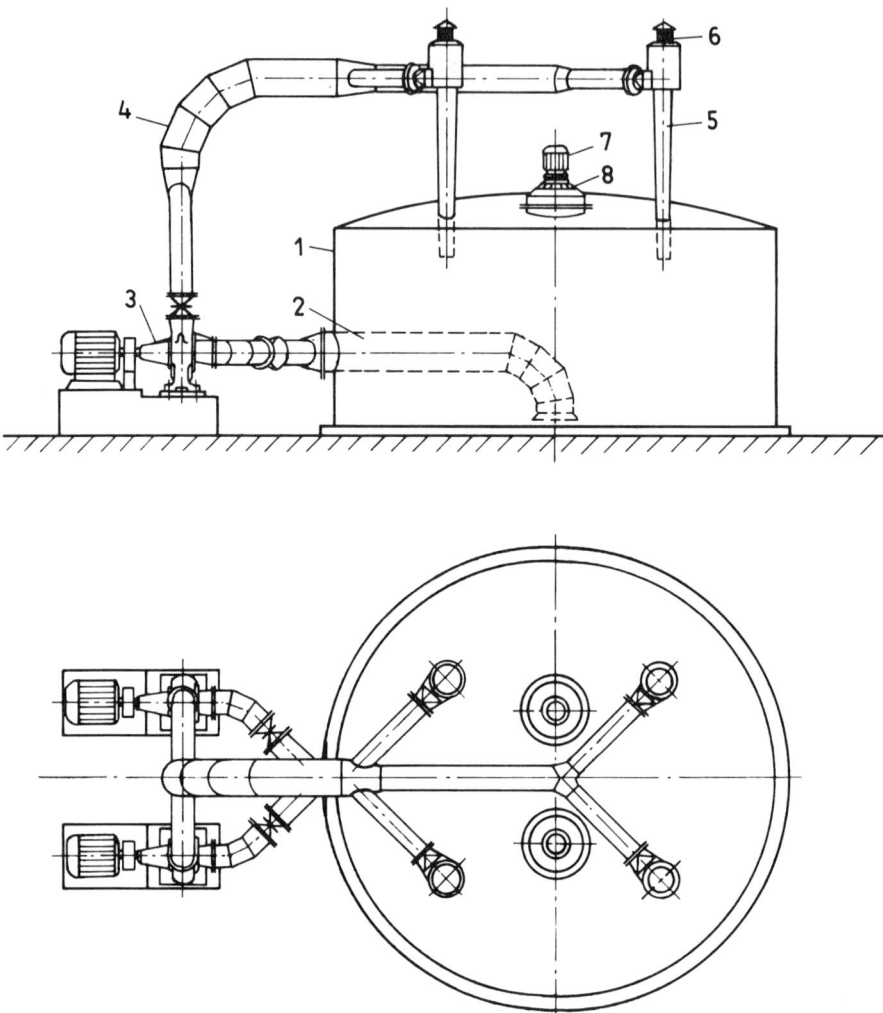

FIGURE 15.   Construction of an aeration tank using the deep jet aeration process for biological waste treatment (fixed-top tank). 1: fixed-top tank, 2: suction line, 3: circulation pump, 4: pressure line, 5: duct overflow, 6: fresh-air inlet, 7: foam-breaking unit, and 8: exhaust-air conduit.

The technique used is circulating aeration with external recirculation. From the reaction volume of the biotank reactor to be aerated,[112,113] the circulating medium is drained off through the tank bottom and delivered by a centrifugal pump and a pressure pipe to the gas-mixing device, where atmospheric air is drawn in and injected into the medium, which is then recirculated to the reaction volume. Thus, the jet dipping into the reactor liquid consists of a gas/liquid dispersion which can be produced in two kinds of devices, in so-called "duct overflows" (older form) and in so-called injectors (newer form).[114,115] In the duct overflows the circulated liquid rushes down in a free gravity flow through tubes (ducts) tapering in the direction of flow, whereby it entrains the gas. This presupposes that the tube section at the outlet is larger than that which the liquid would reach by the velocity of fall, so that the pressures at the inlet and outlet are approximately equal. Injectors, on the other hand, are deep-jet aerators with a jet suction effect. For that purpose they are funnel-shaped, with an air supply arranged concentrically in them, which ends at the most narrow cross-section of the funnel.

Table 20
BIOTANK REACTORS OF CLOSED CONSTRUCTION
(FIXED-TOP TANK)

Rated volume	m^3	500	1000	2000	3200
Reaction volume	m^3	400	750	1500	2750
Cylindrical height	m	5	8.5	10.5	12
Diameter	m	11.3	11.3	15.2	18.7
O_2 supply rate	kg hr^{-1}	>200	>400	>800	>1600

Thus, an air-drawing ring nozzle is formed at the point of the highest flow rate. This allows a substantial reduction of the constructional length of the injector as compared with the duct overflow. The development of the biotank reactors made it possible to change over from the conventional activated sludge plants, with high requirements of constructional work and space, to compact equipment units in all-steel construction. This ensures that the necessary conditions for an intensive matter transfer are fulfilled in an optimum way. Such conditions include, among other things, a high turbulence, finest gas dispersion, maximum retention time of air in the medium, and as high a degree of homogeneity as possible.

The closed construction also allows the treatment of waste waters with high organic loads and a strong tendency to foaming by means of high oxygen flow rates per unit volume for foaming control purposes; a mechanical foam breaker was developed.[116] This device operates on the basis of mechanical agitating elements having the form of rotating centrifugal disks. The foam accelerated radially and tangentially in this way is then broken by percussion, shearing, and impact forces. Further, it is possible to additionally adapt the biotank reactor to highly foaming waste waters by providing for a variable tank level and by withdrawing the reactor effluent close to the bottom on the delivery side of the recirculation pumps. For high oxygen supply rates per unit volume so-called two-phase recirculation pumps (special pumps for delivering liquids with a high gas content) were developed, which make it possible to degas the reactor liquid by means of back-blading, thus ensuring a stable operation with good efficiencies.[117,118] The closed construction of the biotank reactor not only prevents aerosol formation, but also ensures that the heat liberated from the biological conversion maintains a temperature level which exceeds the temperatures in open activated-sludge basins by 10 to 20°C. Thus, the rate of elimination of the organic substances contained in the waste waters is additionally increased by a factor of two to four.

To achieve a high turbulence, high oxygen transfer rates, and a high oxygen exploitation, jet aeration most recently has been further developed, where two principles have been taken into consideration:

1. Deflection of the jet aerators from the vertical in the same sense as the rotation of the earth (tangentially and, on the northern hemisphere, counterclockwise) in order to generate a rotation of the medium contained in the tank reactor, whereby the distance traveled by the gas bubbles in the liquid is increased
2. Increasing the length of the jet pipe and immersing it into the tank reactor liquid in order to increase the depth of penetration of the liquid jet with the entrained gas bubbles

The following batch parameters were used for the most important reactor units (Table 20).

Table 21 shows the essential performance parameters of the biotank reactors as compared with conventional activated-sludge plants.

Analogous reactor systems developed by other workers utilize "volume aerators" that require a supply air pressure to overcome the hydrostatic pressure. Due to this aeration system, however, the oxygen rate per unit volume is not sufficient to reach the high deg-

Table 21
PERFORMANCE PARAMETERS OF THE BIOTANK
REACTORS

Performance parameter	Unit $g\ell^{-1}$	Conventional plants	Biotank reactors
BOD$_5$ inlet conc		0.3—2.5 max	5, 10, and 20
BOD$_5$ degradation rate	$\dfrac{kg\ O_2}{m^3/day}$	2.0—3.5	5, 15, and 25
Oxygen supply rate	$\dfrac{kg\ O_2}{m^3/hr}$	0.15—0.25	0.5, 5, and 10
O$_2$ yield	$\dfrac{kg\ O_2}{kWh}$	1.5—2.5	2—4
Space requirements	%	100	50

radation rate per unit volume of the biotank reactors. The degradation values achieved by the use of the biotank reactor technique in the purification of BCPWW will be quoted in Section IV.C.2.

4. Final Purification Stage

According to the knowledge gained from the biological and kinetic investigations of the degradation of BCPWW (see Sections III.B.3 and III.B.5), the degree of purification (degradation) is limited to approximately 90% of the offered organic load in the high-performance stage, and, due to the excess of substrate and the floc-destroying turbulence of the intensive aeration, the biomass does not tend to sedimentation. Therefore, it is necessary that the high-performance biological treatment is followed by a standard-performance final purification stage operating on the basis of classical activated sludge basins, sedimentation basins, and oxidation ponds. Because of the substrate-dependent mode of reaction it is recommended here to use a multistage plant with a plug flow-like flow pattern.[119] This is realized by U-shaped, continuous-flow basins about 4 m deep, about 5 m wide, and up to 100 m long, with a bottom aeration through perforated latticed pipes or sintered plates and an aeration volume of 35 m^3 of air per kilogram of BOD$_5$ degraded as well as an energy yield of 1.0 kg O$_2$ per kilowatthour, or by a cascade of surface-aerated basins with aeration intensities ranging between 5 and 15 m^3 of air per square meter of basin bottom area per hour, and an energy yield of 1.5 to 2.0 kg O$_2$ per kilowatthour.[76]

The activated sludge plants reach degradation rate per unit volume of 0.3 to 2.5 kg BOD$_5$ load per cubic meter per day. The age of the sludge is adjusted to 2 to 3 days by means of the recycling rate in order to achieve a good settleability. For the final purification stage of the BCPWW treatment, a degradation efficiency (referred to BOD$_5$) of about 80% of the effluent of the high-performance stage is achieved.

5. Advanced Biological Purification of Sewage

The evaluation of the effluent quality of sewage not only by the organic load, but also by the nitrogen content of sewage requires the inclusion of biological nitrification and denitrification (see Section III.B.6), if outlet concentrations of less than 100 mg ℓ^{-1} shall be achieved. This is because in the normal biological purification nitrogen is eliminated only by incorporation into the biomass and by strip effects. Generally nitrification and denitrifiction are technologically realized by combining one or more nitrification and denitrification stages with one of the known high-performance biological techniques.

Here it is possible to realize a variety of different combinations, examples of which will be given in Section IV.C.3. The activated-sludge stage following the high-performance stage can be abandoned. Due to the relatively low matter-transfer rates with respect to oxygen

Table 22
SPECIFIC PARAMETERS FOR ADVANCED
BIOLOGICAL PURIFICATION OF BCPWW

Parameter	Specific value
Denitrification rate	2.5 NO_3-N $kgDW^{-1}$ day^{-1a}
Nitrification rate	0.1 kg NH_4-N $kgDW^{-1}$ day^{-1}
Oxygen yield	2.0 kg O_2 kWh^{-1}
Specific energy demand	0.75 kWh per kilogram BOD_5

[a] kgDW = kilogram dry weight

Table 23
SUITABLE WEAK LIQUOR QUALITY FOR
ADVANCED BIOLOGICAL PURIFICATION

Parameter	Concentration range ($g\ell^{-1}$)
BOD_5	3.5—15
COD_{Cr}	7.5—20
NH_3 (total)	$\leqslant 1.2$
Volatile phenols	$\leqslant 0.25$
Phenols (total)	1.5—3.0
Fatty acids	3.0—5.0
Extractable substances	$\leqslant 0.15$

supply and foam formation, the nitrification stage is well controllable (e.g., bottom aeration) but requires large basin volumes and, if possible, devices for biomass retention.

Therefore, besides the possibility of providing fixed devices for biomass retention in the aerated systems,[120-122] a great variety of other possibilities for biomass retention have also been proposed, e.g., the introduction of plastic chips,[123,124] activated carbon, and alumina, respectively.[125] Other proposals include the use of the fluidized bed reactor[126,127] and the application of the rotating biological contactor technique or the rotating disk contactor.[128-130] The denitrification stage requires anoxic conditions, and hence only a thorough mixing, where — as was mentioned in Section III.B.6.b — strictly anoxic conditions are not necessary. The utilization of the nitrate oxygen leads to a saving of energy — for a specific energy demand of 0.75 kWh per kilogram BOD_5 load — of about 25 to 30%, as referred to the total process. Depending on the quality of the weak liquor, and especially if the carbon to nitrogen ratio is unfavorable, the denitrification stage can be placed ahead, and a partial stream of nitrate-containing water from the effluent of the secondary sedimentation basin of the following nitrification stage can be returned to the denitrification basin (see Section IV.C.3). The multistage design of the process contributes to an improvement of the rate of nitrogen elimination, reduces the hydraulic load of the total process by reducing the cycles, and avoids toxic limiting concentrations. This presupposes the presence of an efficient common mixed population for all stages. The total process allows a nitrogen reduction of 70 to 90% with a simultaneous biological BOD_5 reduction of 95 to 99%. The specific parameters achieved here are as follows (Table 22).

The advanced biological purification is described in detail in Section IV.C.3. It is positively practicable with a weak liquor in which the concentrations of the substances contained vary within the ranges listed in Table 23.

The effluent data obtained in semitechnical plants are listed in Section IV.C.3, Tables 27 and 28.

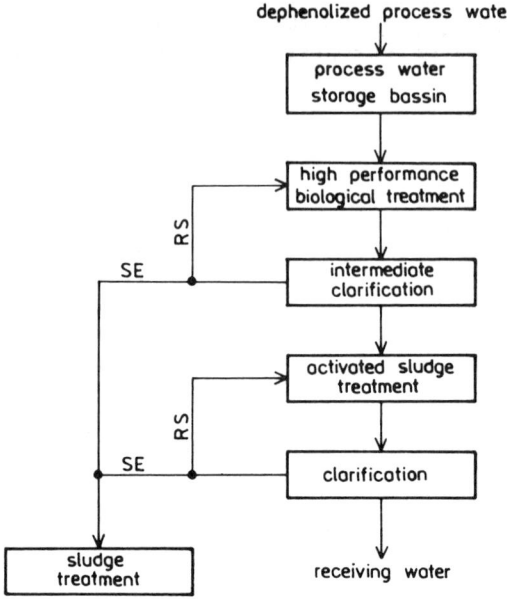

FIGURE 16.   General flow chart of the high-performance biological purification by means of the PKM centrifugal aerator with automatic foam-drawing device. RS: return sludge and SE: sludge excess.

Table 24
DEGRADATION EFFICIENCY OF HIGH-PERFORMANCE BIOLOGICAL PURIFICATION WITH AERATION BY A CENTRIFUGAL AERATOR WITH AUTOMATIC FOAM-DRAWING DEVICE, ACCORDING TO DATA FROM PKM

Parameter	Inlet conc (mg ℓ^{-1})	Outlet conc (mg ℓ^{-1})	Degradation efficiency (%)
BOD$_5$	5200	1050	80
NH$_4$-N	1000	800	20
Volatile phenols	25	3	88
Volatile fatty acids	2500	250	90

C. Technical Plants Realized in the German Democratic Republic for the Biological Purification of Brown Coal Processory Waste Waters

1. High-Performance Biological Purification by the PKM Process with Aeration by a Centrifugal Aerator with Automatic Foam-Drawing Device

The purification is carried out by the process flow sheet shown in Figure 16 for the total plant. The heart of the high-performance biological treatment system is the aeration by centrifugal aerators with automatic foam-drawing device. In an application of this process to the purification of waste waters from brown coal coking and brown coal elevated-pressure gasification, the degradation values listed in Table 24 were obtained in the high-performance stage.

The degradation results listed in Table 24 were obtained under the following specific process conditions (Table 25).

Table 25
SPECIFIC PARAMETERS OF HIGH-PERFORMANCE
BIOLOGICAL TREATMENT BY THE PKM PROCESS

Parameter	Value
Retention time	6 hr
Loading rate per unit volume	20 kg BOD_5/m^3/day
Sludge loading rate	5 kg BOD_5/kgDW/day
Sludge conc	4 kgDW/m^3
Oxygen yield (for aeration and foam)	0.8 kWh/kg BOD_5 degraded

2. High-Performance Biological Purification with CLG Biotank Reactors

The process flow chart for the total biological waste water purification by the CLG method is shown in Figure 17.

The mode of action of the biotank reactors shall be explained for an application to waste water treatment in a brown coal processing plant of the G.D.R.

Figure 18 shows the tank reactor stage with exhaust-air conduit, which was installed as the first biological purification stage (high-performance stage), consisting of four tank reactors each with a gross volume of 3200 m³. The closed construction not only offered an environmentally compatible solution, but also made it possible to maintain the temperature level between 35 and 40°C in the summer and between 20 and 30°C in the winter.

Accordingly, the rates of elimination are high; the values of these rates for a throughput of V = 1300 m³ hr⁻¹ in mixed water are shown in Table 26.

It is remarkable how large the concentration fluctuations in the influent waste water can be which are still tolerated by the biotank reactor stage. This can be attributed to the immediate reduction of the concentration of the influent substrate to the concentration level of the reactor content, which is accomplished by the rapid thorough mixing with an immediate supply of oxygen. This makes a preceding detection stage unnecessary. The arrangement of the individual active devices in the bioreactor is shown in Figure 19. It shows the two-phase circulation pumps with pressure and expansion lines, the injectors deflected counterclockwise, and the foam breaker arranged at the center of the reactor top. Depending on the height of the foam, it is switched on and off by means of conductivity detectors. The two-phase circulation pumps are switched on and off depending on the actual load or on the actual foam pressure.

3. The PKM Process for Advanced Biological Purification

As was already mentioned in Section IV.B.5, it is possible to choose different combinations of the individual degradation stages when biological nitrification and denitrification are included. Here we shall present two semitechnically tested variants.

Variant 1 — The principle of this variant can be observed from Figure 20, consisting of a high-performance biological purification followed by several nitrification and denitrification stages. This process was tested for waste waters from brown coal coking and brown coal elevated-pressure gasification. The degradation results obtained are shown in Table 27.

Variant 2 — As can be observed from Figure 21, this process works with a preceding denitrification stage and a corresponding recirculating process management; it is most suitable for waste waters with unfavorable carbon to nitrogen ratios. It was tested on waste water from brown coal low-temperature coking. The degradation results for the total process are shown in Table 28.

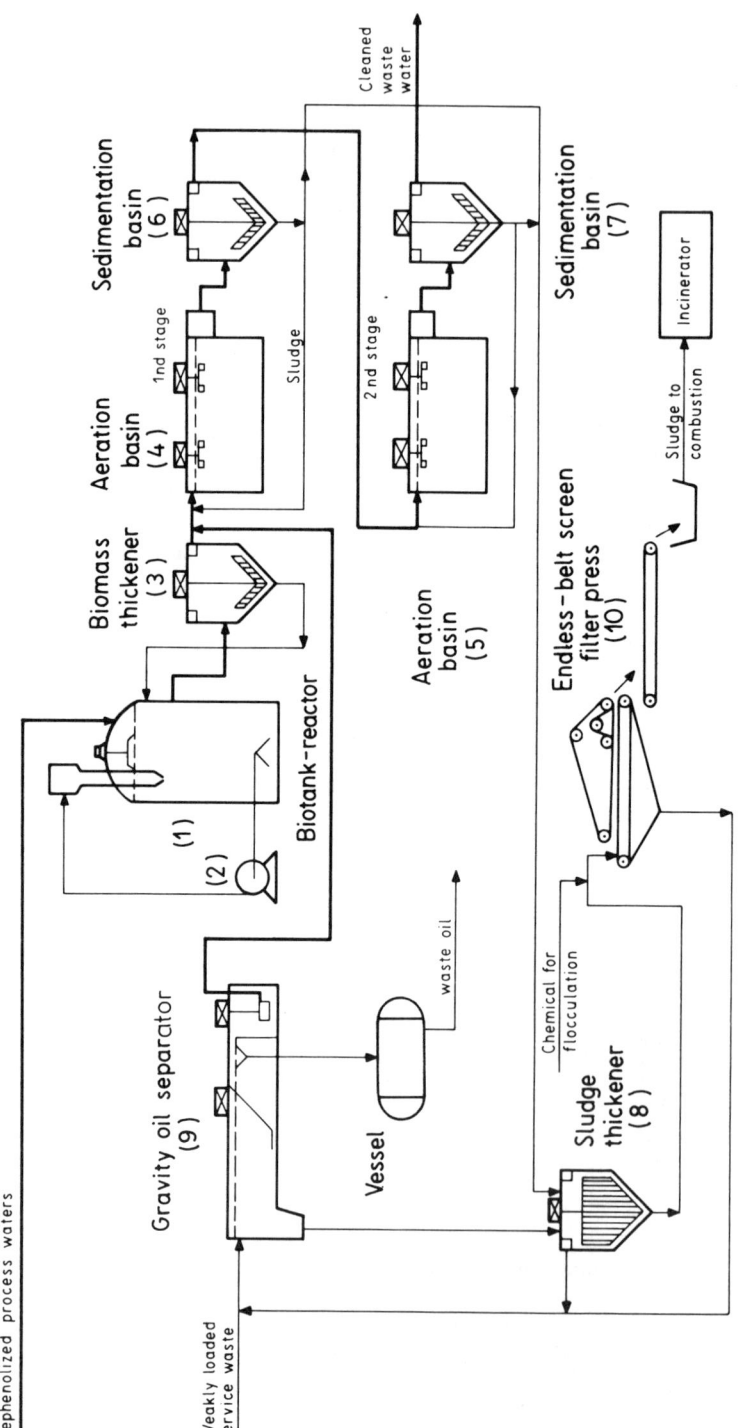

FIGURE 17. Process flow diagram: complete biological purification by means of the CLG biotank reactor.

FIGURE 18.   First biological purification stage (high-performance biological treatment) with tank reactors and exhaust-air conduit in a carbochemical factory in East Germany.

Table 26
**DEGRADATION EFFICIENCY OF HIGH-
PERFORMANCE BIOLOGICAL PURIFICATION
WITH BIOTANK REACTORS; ACCORDING TO
CLG DATA**

Parameter	Inlet conc (mg ℓ^{-1})	Degradation efficiency (%)
BOD_5	2,000—12,000	85
NH_4-N	500—3,000	35
Volatile phenols	250—600	99
Volatile fatty acids	1,000—2,000	90
Formaldehyde	1,000—3,000	45

FIGURE 19.   Biotank reactor with two-phase circulation pumps, pressure and expansion lines, and foam breaker.

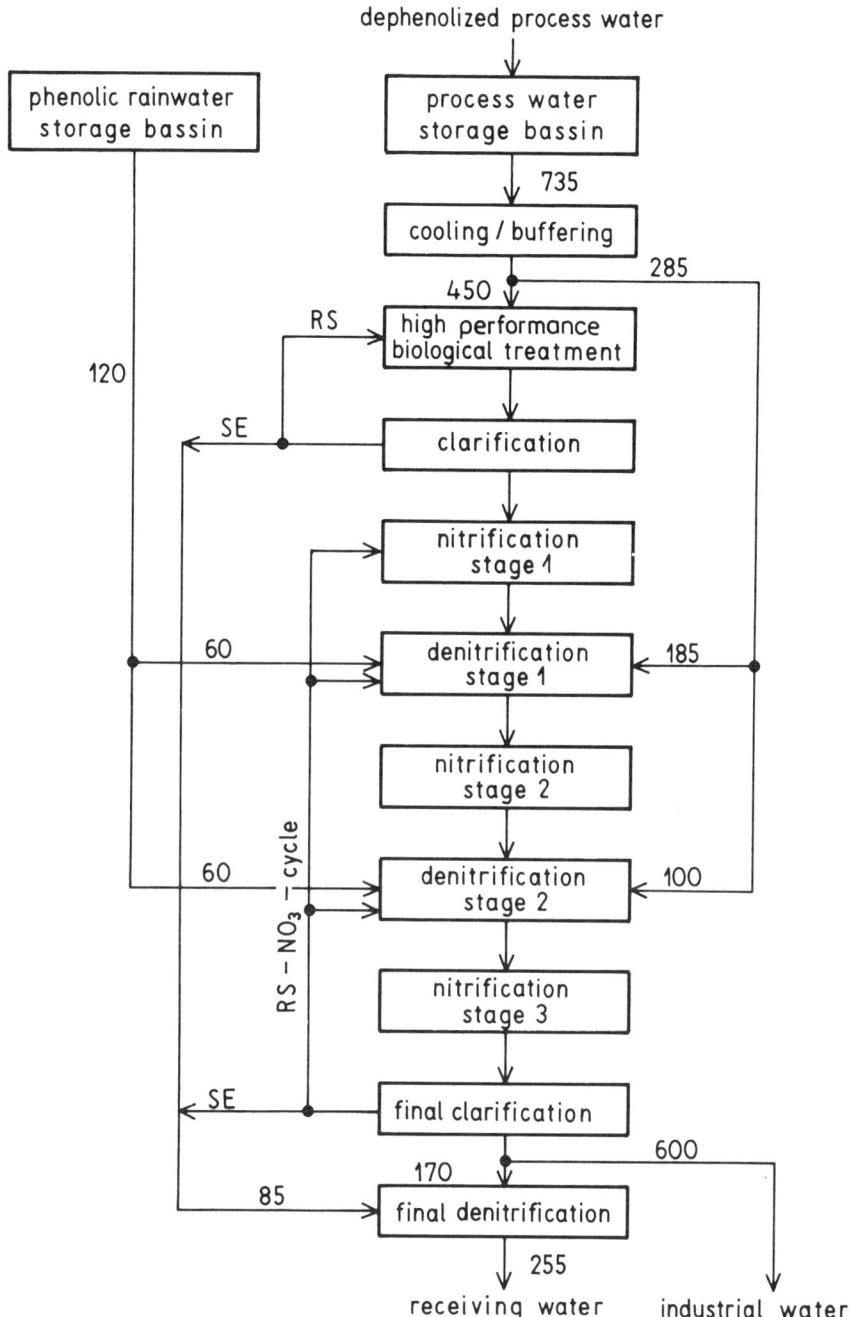

FIGURE 20. Mass flow chart of an advanced biological purification process followed by multistage nitrification/denitrification; according to PKM. RS: return sludge, SE: sludge excess, and 85: amounts in m³ hr⁻¹.

V. DISCUSSION AND PROSPECTS

The application of new technologies of brown coal processing, e.g., coal liquefaction, opens new possibilities of reducing the produced amounts of waste water. On the other hand, the production of coke from brown coal will, due to the special properties of the coke, always remain an important application. The development of new brown coal resources,

Table 27

DEGRADATION EFFICIENCY OF A PILOT PLANT FOR ADVANCED BIOLOGICAL PURIFICATION PRECEDED BY A HIGH-PERFORMANCE BIOLOGICAL TREATMENT; ACCORDING TO PKM DATA

Parameter	Inlet conc (mg ℓ^{-1})	Outlet conc (mg ℓ^{-1})	Degradation efficiency (%)
BOD$_5$	5000	≤50	99
Total N	700	≤150	80
Volatile phenols	50	≤1	98
Volatile fatty acids	2100	0	100

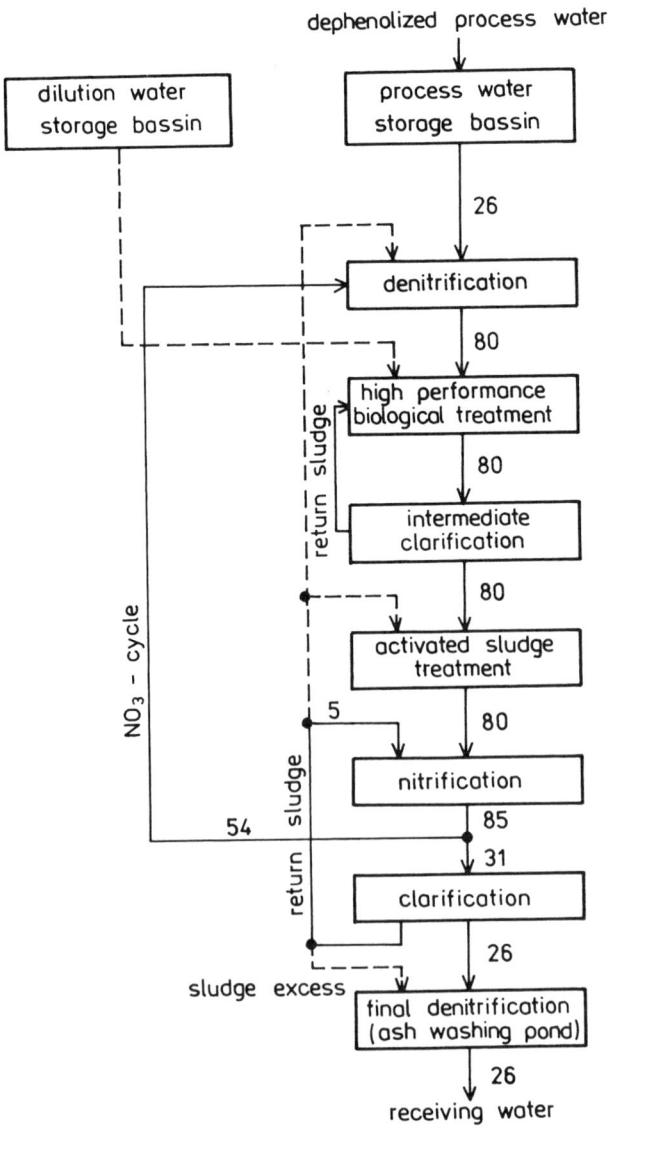

FIGURE 21.   Mass flow chart of an advanced biological purification process preceded by denitrification; according to PKM. ————: continuous flow, - - - - -: flow as required, and 80: amounts in m³ hr⁻¹.

Table 28
DEGRADATION EFFICIENCY OF A PILOT PLANT
FOR ADVANCED BIOLOGICAL PURIFICATION
PRECEDED BY DENITRIFICATION; ACCORDING TO
PKM DATA

Parameter	Inlet conc (mg ℓ^{-1})	Outlet conc (mg ℓ^{-1})	Degradation efficiency (%)
BOD$_5$	5200	≤ 30	99.5
Total N	1000	≤ 150	85
Volatile phenols	25	≤ 0.5	98

especially that of lower grades such as saliferous coals, raises new problems for waste water treatment. The full impact is not foreseeable today.

The matured technologies in physical waste water treatment processes offer only little reserves to be further developed.

The two-stage process management allows the purification of undiluted process waters, so that the amount of waste water can be reduced.

The use of closed tank reactors, leading to a further increase in capacity, raises problems due to the increased foaming and the high evolution of heat in the process. The development of thermophilic processes may be of advantage here. The results of investigations of the thermophilic degradation of BCPWW fall in line with the well-known tendencies of a thermophilic treatment:[74]

- An enhanced respiratory elimination of the substances contained, together with a reduced formation of biomass, and, correspondingly,
- A lower yield factor, the causes of which are likely to be found in an increased maintenance metabolism

On the other hand, the thermophilic regime involves a reduced diversity of species in the mixed population, which influences the flocculation. Furthermore, the investigations exhibit, that in the special cause of BCPWW the thermophilic degradation necessitates a significant higher concentration of dissolved oxygen (at least 1.5 mg ℓ^{-1}) than the mesophilic degradation (at least 0.35 mg ℓ^{-1}), the causes of which are certainly specific for the substrates contained.[74]

Complex BCPWW treatment plants will no longer be operable without nitrification and denitrification stages.

The supply of energy required by the aerobic purification for the elimination of the substances contained can be coupled with the generation of energy in anaerobic processes. The technology of the anaerobic processes is not yet developed. This may open new ways.

Special proposals, such as the elimination of phenols by enzymatic reactions, were discussed. At present it cannot yet be told whether they are technically relevant. Moreover, enzymatic reactions may be of interest in sludge conditioning.

According to the knowledge gained so far, the use of commercial adapted strains offers little advantages. For the degradation of synthetically produced, persistent substances contained in waste waters, the construction of new strains by plasmid transfer may offer decisive advantages.

ACKNOWLEDGMENTS

The authors wish to thank Mrs. Osbar (PKM), Dr. Franz (PKM), deceased, and Mr. Meyer (CLG) for many helpful discussions.

LIST OF ABBREVIATIONS

ADP:	adenosine diphosphate
ATP:	adenosine triphosphate
BCPWW:	brown coal processory waste water
BOD_5:	biological oxygen demand (after 5 days)
CoA:	coenzyme A
COD:	chemical oxygen demand
CV:	caloric value
D:	dilution rate (h^{-1})
DH:	dehydrogenase
DW:	dry weight
eE coal:	brown coal from the east of the Elbe River
FAD:	flavin adenine dinucleotide
$FADH_2$:	reduced form of flavin adenine dinucleotide
FC:	fixed carbon
F.R.G.:	Federal Republic of Germany
HTC:	high-temperature carbonization
LTC:	low-temperature carbonization
NAD:	nicotinamide dinucleotide
$NADHH^+$:	reduced form of nicotinamide dinucleotide
NAH:	naphthalene degradating plasmid
$N_{NO_3^-}$:	nitrogen calculated from nitrate
OCT:	octane degradating plasmid
PHG:	phenol degradating plasmid
Prot:	protein
SNG:	substitute natural gas
TCC:	tricarboxylic acid cycle
TOC:	total organic carbon
TOL:	toluene degradating plasmid
UASB:	upflow anaerobic sludge blanket
VM:	volatile matter
wE coal:	brown coal from the west of the Elbe River
X:	grams of biomass
XYL:	xylene degradating plasmid

REFERENCES

1. **Mackowsky, M. T., Hodek, W., Benthaus, F., et al.,** Kohle, in *Ullmanns Enzyklopädie der technischen Chemie,* Vol. 14, 4th ed., Bartholomé, E., Biekert, E., Hellmann, H., Ley, H., Weigert, W. M., and Weise, E., Eds., Verlag Chemie, Weinheim, West Germany, 1977, 288.

2. **Gundermann, E.,** *Chemie und Technologie des Braunkohlenteers,* Akademie-Verlag, Berlin, 1964.

3. **Rammler, E. and Von Alberti, H. J.,** *Technologie und Chemie der Braunkohlenverwertung,* VEB Deutscher Verlag für Grundstoffindustrie, Leipzig, East Germany, 1962.

4. **Bieger, F., Eisenlohr, K.-H., Herbst, G., et al.,** Kohle, Gaserzeugung aus Kohle und Kohlenwasserstoffen, in *Ullmanns Enzyklopädie der technischen Chemie,* Vol. 14, 4th ed., Bartholomé, E., Biekert, E., Hellmann, H., Ley, H., Weigert, W. M., and Weise, E., Eds., Verlag Chemie, Weinheim, West Germany, 1977, 357.

5. **Kurtz, R.,** Kohle, Schwelung und Verkokung von Braunkohle, in *Ullmanns Enzyklopädie der technischen Chemie,* Vol. 14, 4th ed., Bartholomé, E., Biekert, E., Hellmann, H., Ley, H., Weigert, W. M., and Weise, E., Eds., Verlag Chemie, Weinheim, West Germany, 1977, 491.

6. *Statistical Yearbook of United Nations,* International Publications Service, a division of Taylor & Francis, Philadelphia, 1981, 199.
7. **Mangold, K. H., Kaeding, J., Lorenz, K., and Schuster, H.,** *Abwasserreinigung in der chemischen und artverwandten Industrie,* VEB Verlag für Grundstoffindustrie, Leipzig, East Germany, 1979.
8. **Franz, B.,** Abwasserreinigung, in *Verfahren der Gasaufbereitung,* Schmidt, J., Ed., VEB Deutscher Verlag für Grundstoffindustrie, Leipzig, East Germany, 1970, 470.
9. **Wurm, H. J.,** Das Entphenolen von Kokereigaskondensaten nach dem Phenosolvan-Verfahren, *Glueckauf,* 104, 517, 1968.
10. **Neff, I.,** Möglichkeiten der Abwasserreinigung in der Kokereinebengewinnung, *Chem. Tech.,* 8, 938, 1982.
11. **Gravelle, D. V. and Panayiotou, C.,** Temperature effect on phenol extraction from water using mixed solvents, *J. Chem. Eng. Data,* 25, 23, 1980.
12. **Rieche, A. and Strankmüller, J.,** Über die Reinigung phenolhaltiger Abwässer nach dem Asche-Luft-Verfahren, *Wasserwirtsch. Wassertech.,* 8, 64, 1958.
13. **Schulmann, J., Fuchs, P., Kresta, J., Wimmer, P., and Zanova, V.,** Reinigung von Phenolabwässern in den Nordböhmischen Gaswerken in Uzin, *Wasserwirtsch. Wassertech.,* 18, 155, 1968.
14. **Anon.,** Phenols in refinery waste water can be oxidized with H_2O_2, *Oil Gas J.,* 73, 84, 1975.
15. **Klibanov, A. M.,** Enzymic removal of hazardous pollutants from industrial aqueous effluents, *Enzyme Eng.,* 6, 319, 1982.
16. **Atlow, S. C., Bonnadonna-Aparo, L., and Klibanov, A.,** Dephenolization of industrial waste waters catalyzed by polyphenol oxidase, *Biotechnol. Bioeng.,* 26, 599, 1984.
17. **Besselievre, E. B.,** *The Treatment of Industrial Wastes,* McGraw-Hill, New York, 1969.
18. **Meissner, B.,** Biological decomposition of phenol, *Wasserwirtsch. Wassertech.,* 3, 470, 1953.
19. **Meissner, B.,** Investigations of the disposal of phenol-containing wastes by biological procedures, *Wasserwirtsch. Wassertech.,* 5, 82, 1955.
20. **Meissner, B.,** Broadening the scope of total analysis procedures by using cascade distillation, *Wasserwirtsch. Wassertech.,* 8, 417, 1958.
21. **Meissner, B.,** The biological decomposition of organic substances. Processes of self-purification and biological sewage purification, *Wasserwirtsch. Wassertech.,* 8, 483, 1958.
22. **Thielemann, H., Behrens, U., and Leibnitz, E.,** Zur quantitativen Gas-Chromatographie durch Titration, *Chem. Technol.,* 13, 737, 1961.
23. **Stamoudis, V. C. and Luthy, R. G.,** Determination of biological removal of organic constituents in quench waters from high-BTU coal-gasification pilot plants, *Water Res.,* 14, 1143, 1980.
24. **Ringpfeil, M.,** Qualitative und quantitative papierchromatographische Untersuchungen der Polyphenole, Schwelwässern diploma work, Karl Marx University, Leipzig, East Germany, 1955.
25. **Leibnitz, E., Behrens, U., and Ringpfeil, M.,** Die qualitative papierchromatographische Bestimmung der im Schwelwässer vorkommenden Polyphenole, *Wasserwirtsch. Wassertech.,* 6, 299, 1956.
26. **Leibnitz, E., Behrens, U., and Ringpfeil, M.,** Die quantitative papierchromatographische Bestimmung der Polyphenole in Schwelwässern, *Wasserwirtsch. Wassertech.,* 7, 266, 1957.
27. **Blackburn, W. H., Barker, L., Catchpole, J. R., and Hollingworth, N. W.,** Composition of ammoniacal liquor. I. Continuous vertical retort liquor, *Gas Counc. (GB), Res. Commun.,* G1, 24, 1955.
28. **Behrens, U., Ringpfeil, M., and Gabert, A.,** Neuere Methoden der Schwelwässeranalytik, *Chem. Technol.,* 12, 670, 1960.
29. **Fisher, R. B., Parsons, D. S., and Morrison, G. A.,** Paper chromatography, *Nature,* 161, 764, 1948.
30. **Leibnitz, E., Behrens, U., and Winter, W.,** Bestimmung des biochemischen Sauerstoffbedarfes von toxischen Wässern mit der Warburg-Apparatur, *Wasserwirtsch. Wassertech.,* 8, 10, 1958.
31. **Meissner, B.,** East German Patent 19439, 1960.
32. **Thielemann, H.,** Beiträge zur Einführung gaschromatographischer Analysen-methoden in der technischen Mikrobiologie, Doctoral thesis, Karl Marx University, Leipzig, East Germany, 1962.
33. **Fachgruppe Wasserchemie in der Gesellschaft Deutscher Chemiker,** *Deutsche Einheitsverfahren zur Wasser-, Abwasser- und Schlamm-Untersuchung,* 3rd ed., Verlag Chemie, Weinheim, West Germany, 1960.
34. **Fowler, G. J., Ardern, E., and Lockett, W. T.,** The oxidation of phenol by certain bacteria in pure culture, *Proc. R. Soc. London, Ser. B,* 83, 561, 1910.
35. **Wagner, R.,** Über Benzolbakterien, *Z. Gaerungsphysiol.,* 4, 289, 1914.
36. **Kalabina, M. and Rogovskaya, C.,** Phenolzerfall unter Einwirkung von Mikroorganismen, *Z. Fisch.,* 20, 153, 1934.
37. **Sierp, F. and Fränsemeier, F.,** Biological decomposition of phenol, *Vom Wasser,* 8, 85, 1934.
38. **Nolte, E., Meyer, H. J., and Franke, E.,** Versuche zur Durchführung des Belebtschlammverfahrens bei gewerblichen Abwässern, *Vom Wasser,* 8, 126, 1934.
39. **Egorova, A. N.,** Einige physiologische Daten über thermophile phenoloxydierende Bakterien, *Z. Allg. Landwirtsch. Mikrobiol. Ind.,* 15, 467, 1946.

40. **Bringmann, G.**, Versuche zum biologischen Abbau von Phenolen im Belüftungsverfahren durch Mikroorganismen der Gattung Nocardia, *Gesund. Ing.*, 75, 252, 1954.

41. **Monod, J.**, *Recherches sur la croissance des cultures bacterienes*, Herman and Cie, Paris, 1942.

42. **Herbert, D.**, A theoretical analysis of continuous culture systems, in *Continuous Culture of Microorganisms*, Symp. Microbiol. Group, University of London, Monogr. No. 12, Society of Chemical Industry, London, 1961, 21.

43. **Kunst, S.**, Vergleiche der Phenolabbauleistungen zweier Belebtschlümme mit und ohne Zusatz von adaptierten Bakterien, *Gas-Wasser-Fach, Wasser Abwasser*, 125, 254, 1984.

44. **Landa, S.**, Biologischer Abbau von Phenolen, *Listy Chem.*, 47, 622, 1953.

45. **Rieche, A., Hilgetag, G., Martini, A., and Lorenz, M.**, Continuous production of yeast from phenol-containing substrates, in Proc. 2nd Symp. Cont. Ferm. Microorganisms, Prague, 1964, 293.

46. **Zimmermann, R.**, Über phenolspaltende Hefen, *Naturwissenschaften*, 45, 165, 1958.

47. **Zülke, H.-J.**, Erfassung mikrobieller Leistungskriterien als Bestandteil der Verfahrenswahl für die biologische Reinigung von Industrieabwässern, Doctoral work, Ernst Moritz Arndt University, Greifswald, German Democratic Republic, 1971.

47a. **Behrens, U.**, unpublished results.

48. **Behrens, U., Bürger, G., Stottmeister, U., Hoh, M., and Franz, B.**, Gewinnung biokinetischer Daten von Abwässern mit dem Respirationsfermentor. I. Methodische Grundlagen und die Bestimmung der spezifischen Wachstums- und Sauerstoffaufnahmerate, *Acta Hydrochim. Hydrobiol.*, 8, 443, 1980.

49. **Waller, M.**, Untersuchungen über die Reinigung von Abwässern der braunkohleveredelnden Industrie in einer Modelltropfkörperanlage und über das Vorkommen und die physiologische Leistung von Sproßpilzen, biologischen Rasen Doctoral work, Martin Luther University, Halle, East Germany, 1962.

50. **Neufeld, R. D., Mack, J. D., and Strakey, J. P.**, Anaerobic phenol biokinetics, *J. Water Pollut. Contr. Fed.*, 52, 2367, 1980.

51. **Klein, K., Steinberg, R., Fiethen, B., and Overath, P.**, Fatty acid degradation in *Escherichia coli*. Inducible system for the uptake of fatty acids and further characterization of old mutants, *Eur. J. Biochem.*, 19, 442, 1971.

52. **Tabuchi, T. and Serizawa, N.**, Production of 2-methylisocitric acid from odd-carbon normal alkans by a mutant of *Candida lipolytica*, *Agric. Biol. Chem.*, 39, 1049, 1975.

53. **Maesen, T. J. M. and Lako, E.**, The influence of acetate on the fermentation of baker's yeast, *Biochim. Biophys. Acta*, 9, 106, 1952.

54. **Samson, F. E., Katz, A. M., and Harris, L.**, Effects of acetate and other short-chain fatty acids on yeast metabolism, *Arch. Biochem. Biophys.*, 54, 406, 1955.

55. **Stanier, P. Y.**, Cleavage of aromatic rings with eventual formation of beta-ketoadipinic acid, in *Methods in Enzymology*, Vol. 2, Colowick, S. P. and Kaplan, N. O., Eds., Academic Press, New York, 1955, 281.

56. **Axell, B. C. and Geary, P. J.**, Metabolism of benzene by bacteria. Purification and properties of the enzyme cis-1,2-dihydroxy-3,5-cyclohexadiene (nicotinamide adenine dinucleotide) oxido-reductase (cis benzene glycol dehydrogenase), *Biochem. J.*, 136, 927, 1973.

57. **Claus, D. and Walker, N.**, The decomposition of toluene by soil bacteria, *J. Gen. Microbiol.*, 36, 107, 1964.

58. **Klibanov, A. M., Alberti, B. N., Morris, E. D., and Felshin, L. M.**, Enzymic removal of toxic phenols and anilins from waste waters, *J. Appl. Biochem.*, 2, 414, 1980.

59. **Raymond, R. L. and Jamison, V. W.**, Biochemical activities of *Nocardia*, *Adv. Appl. Microbiol.*, 14, 93, 1971.

60. **Perry, J. J.**, Microbial cooxidations involving hydrocarbons, *Microb. Rev.*, 43, 59, 1979.

61. **Jamison, V. W., Raymond, R. L., and Hudson, J. O.**, Microbial hydrocarbon cooxidation. III. Isolation and characterization of an α,α'-dimethyl-cis, cis-muconic acid-producing strain of *Nocardia corallina*, *Appl. Microbiol.*, 17, 853, 1969.

62. **Leibnitz, E., Behrens, U., Striegler, G., and Gabert, A.**, Zur mikrobiellen Oxydation der p-ständigen Methylgruppe in Xylenolen, *Z. Allg. Mikrobiol.*, 2, 81, 1962.

63. **Gaal, A. and Neujahr, H. Y.**, Metabolism of phenol and resorcinol in *Trichosporon cutaneum*, *J. Bacteriol.*, 137, 13, 1979.

64. **Chakrabarty, A. M., Gunsalus, C. F., and Gunsalus, I. C.**, Transduction and the clustering of genes in fluorescent pseudomonads, *Proc. Natl. Acad. Sci. U.S.A.*, 60, 168, 1968.

65. **Chakrabarty, A. M.**, Genetic basis of the biodegradation of salicylate in *Pseudomonas*, *J. Bacteriol.*, 112, 815, 1972.

66. **Heinaru, A., Kivisaar, M., Habicht, J., Mae, A., and Kasak, L.**, Plasmids encoding phenol degradation, in Conf. Metab. Plasmids, Tallinn, Estonia, U.S.S.R., 1982, 100.

67. **Bradley, D. E. and Williams, P. A.**, The TOL plasmid is naturally depressed for transfer, *J. Gen. Microbiol.*, 128, 3019, 1982.

68. **Chatterjee, D. K. and Chakrabarty, A. M.**, Genetic homology between independently isolated chlorobenzoate degradative plasmids, *J. Bacteriol.*, 153, 532, 1983.

69. **Negoro, S., Taniguchi, T., Kanaoka, M., Kimura, H., and Okada, H.,** Plasmid-determined enzymatic degradation of nylon oligomers, *J. Bacteriol.,* 155, 22, 1983.
70. **Chakrabarty, A. M.,** Great Britain Patent 1436573, 1976.
71. **Jeenes, D. J., Reinecke, W., Knackmuss, H. J., and Williams, P. A.,** TOL plasmid pWWO in constructed halobenzoate-degrading *Pseudomonas* strains: enzyme regulation and DNA structure, *J. Bacteriol.,* 150, 180, 1982.
72. **Jeenes, D. J. and Williams, P. A.,** Excision and integration of degradative pathway genes from TOL plasmid pWWO, *J. Bacteriol.,* 150, 188, 1982.
73. **Ringpfeil, M.,** Die Lösung verfahrenstechnischer Probleme bei der Projektierung biologischer Reinigungsanlagen, *Fortschr. Wasserchem. Ihrer Grenzgeb.,* 6, 21, 1967.
74. **Martius, G., Behrens, U., and Franz, B.,** Gewinnung biokinetischer Daten von Abwässern mit dem Respirationsfermentor. II, *Acta Hydrochim. Hydrobiol.,* 15, 357, 1987.
75. **Stottmeister, U.,** Beschreibung eines Fermentationssystems mit geschlossenem Gaskreislauf und die Anwendung als Respirometer und Bilanzapparatur, *Abh. Dtsh. Akad. Wiss. Abt. Math. Nat. Tech.,* 34, 113, 1978.
76. **Franz, B.,** Reinigung von Gaswässern aus der thermischen Kohleveredlung, in *Hydrotechnik im Bergbau und Bauwesen,* Strzodka, K., Ed., VEB Deutscher Verlag für Grundstoffindustrie, Leipzig, German Democratic Republic, 1975, 336.
77. **Koné, S. and Behrens, U.,** Zur Bedeutung der wahren maximalen spezifischen Wachstumsrate, *Acta Biotechnol.,* 3, 76, 1983.
78. **Koné, S. and Behrens, U.,** Die Bedeutung des Erhaltungsstoffwechsels und der ''maximalen'' Ausbeute, *Acta Biotechnol.,* 3, 73, 1983.
79. **Van Uden, N.,** Transport-limited growth in the chemostat and its competitive inhibition; a theoretical treatment, *Arch. Mikrobiol.,* 58, 145, 1967.
80. **Sharma, B. and Ahlert, R. C.,** Nitrification and nitrogen removal, *Water Res.,* 11, 897, 1977.
81. **Beccari, M., Passino, R., Ramadori, R., and Tandoi, V.,** Inhibitory effects on nitrification by typical compounds in coke plant waste waters, *Environ. Technol. Lett.,* 1, 245, 1980.
81a. **Behrens, U.,** private communication.
82. **Heinrich, D.,** Laboruntersuchungen zum Einfluß von Sauerstoff als limitierender Faktor bei der Nitrifikation, *Wasser Abwasser,* 122, 304, 1981.
83. **Luthy, R. G.,** Treatment of coal coking and coal gasification waste waters, *J. Water Pollut. Control Fed.,* 53, 325, 1981.
84. **Focht, D. D. and Chang, A. C.,** Nitrification and denitrification related to waste water treatment, *Adv. Appl. Microbiol.,* 19, 153, 1975.
85. **Beccari, M., Di Pinto, A. C., Passino, R., and Ramadori, R.,** La rimozione biologica dell'azots dalle acque di scarico, *Quad. Ist. Ris. Acque,* 53, 45, 1980.
86. **Blaszczyk, K. M., Mycielski, R., Jaworowska-Deptuch, H., and Brzostek, K.,** Effect of various sources of organic carbon and high nitrite and nitrate concentrations on the selection of denitrifying bacteria. I. Stationary cultures, *Acta Microbiol. Pol.,* 29, 397, 1980.
87. **Lewandowski, Z.,** Biological denitrification in the presence of cyanide, *Water Res.,* 18, 289, 1984.
88. **Koné, S. and Behrens, U.,** Zur Kinetik der Denitrifikation. II. Mischpopulation und Gaswasser als Kohlenstoffquelle, *Acta Hydrochim. Hydrobiol.,* 10, 243, 1982.
89. **Bürger, G. and Behrens, U.,** On the Kinetics of Denitrification, Part IV: Studies on denitrification with the pH auxostat, *Acta Hydrochim. Hydrobiol.,* 13, 691, 1985.
90. **Krüger, W.,** Erfahrungen mit der Bodenbehandlung der Abwässer der Großkokerei Lauchhammer in der Praxis, *Wasserwirtsch. Wassertech.,* 14, 106, 1964.
91. **Schulz, F.,** Bodenbehandlung von Abwässern der Braunkohlenveredlung unter Berücksichtigung der Erfordernisse industriemäßiger Pflanzenproduktion, Doctoral thesis, Humboldt University, Berlin, 1978.
92. **Leibnitz, E., Ringpfeil, M., and Klöppel, E.,** Uber die flotation von mikroorganismen aus biologisch behandeltem phenosolvandünnwasser, *Fortschr. Wasserchem. Ihrer Grenzgeb.,* 5, 189, 1967.
92a. **Behrens, U.,** unpublished results.
92b. **Behrens, U. and Augst, E.,** unpublished results.
93. **Behrens, U.,** Die submerse biologische Abwasserreinigung als mikrobiologische Biomassesynthese, Doctoral thesis B, Academy of Sciences, Berlin, 1970.
94. **Behrens, U. and Klappach, G.,** Abwasser als Rohstoff für mikrobielle Produktsynthesen, *Fortschr. Wasserchem. Ihrer Grenzgeb.,* 13, 1, 1974.
95. **Schmidt, H.,** Biologische Entphenolung von Schwelwässer nach dem Magdeburger P-Verfahren, *Braunkohle,* 40, 365, 1941.
96. **Nolte, E.,** Reinigungsanlagen für Phenolwässer nach dem Magdeburger-P-Verfahren, *Beitr. Wasser-, Abwasser- und Fischereichemie,* 2, 3, 1947.
97. **Meinck, F., Stooff, H., and Kohlschütter, H.,** *Industrieabwaesser,* Gustav Fischer Verlag, Stuttgart, 1968, 461.

98. **Götze, A. and Richter, H.,** Erfahrungen und Ergebnisse beim Betrieb einer hochbelasteten Turmtropf-körperanlage zur Reinigung vorentphenolter Abwässer der B T-Verkokung im BKK Lauchhammer, *Freiberger Forschungsh. A,* 170, 28, 1960.

99. **Richter, H.,** Neuere Erkenntnisse beim Betrieb der biologischen Nachreinigung von Phenosolverdünnwasser im VEB Braunkohlenkombinat Lauchhammer, *Freiberger Forschungsh. A,* 220, 87, 1962.

100. **Lorenz, K.,** Abwasserprobleme der Braunkohlenchemie, *Abh. Akad. Wiss. Kl. Chem. Geol. Biol.,* 1, 3, 1957.

101. **Richter, H.,** Versuche zur Ermittlung technologischer Kenngrößen am Turmtropfkörper, *Freiberger Forschungsh. A,* 295, 47, 1964.

102. **Mueller, J., Sell, E., and Leistner, G.,** Operating experience with the Biohochreactor, *Ber. Wassergütewirtsch. Gesundheitsingenieurwes. Techn. Univ. München,* 28, 75, 1980.

103. **Diesterweg, G. and Lingen, P.,** Zukunftsweisende Abwasserreinigungstechnik — die Bayer-Turmbiologie, *Chem. Tech.,* 10, 565, 1981.

104. **Kalinske, A. A.,** Oxygen absorption studies using mechanical air dispersers, *Sewage Ind. Wastes,* 27, 572, 1955.

105. **Ringpfeil, M., Mangold, K. H., and Zülke, H. J.,** Die intensivbiologische Aufbereitung von Braunkohlenreaktionswässern als komplexes verfahrenstechnisches Problem, presented at 1. Sitzung des A. A. "Flüssige Brennstoffe" der Brennstofftechnischen Gesellschaft der DDR, Markkleeberg, East Germany, June 1965.

106. **Zülke, H. J.,** Kontinuierliche technische Nachreinigung von Braunkohlen-Kokereiabwässern nach dem biologischen Intensivverfahren, *Fortschr. Wasserchem. Ihrer Grenzgeb.,* 5, 172, 1967.

107. **Franz, B.,** DD WP 82459, 1969.

108. **Franz, B.,** Einsatz von Saugkreiselbelüftern bei der biologischen Reinigung von Gaswässern, *Wasserwirtsch. Wassertech.,* 18, 164, 1968.

109. **Jagusch, L. and Püschel, S.,** Tauchstrahlbelüftung — ein neuartiges, im VEB Kombinat "Schwarze Pumpe" erprobtes Belüftungssystem zur biologischen Reinigung hochbelasteter Abwässer, *Wasserwirtsch. Wassertech.,* 18, 160, 1968.

110. **Jagusch, L., et al.,** G.D.R. Patent 56 763, 1967.

111. **Jagusch, L., et al.,** G.D.R. Patent 111 144, 1975.

112. **Wenige, L., et al.,** G.D.R. Patent 126 783, 1972.

113. **Wenige, L., et al.,** G.D.R. Patent 2 006 073, 1983.

114. **Liepe, F., et al.,** G.D.R. Patent 111 805, 1975.

115. **Liepe, F., et al.,** G.D.R. Patent 117 353, 1976.

116. **Liepe, F., et al.,** G.D.R. Patent 137 942, 1980.

117. **Jagusch, L., et al.,** G.D.R. Patent 101 947, 1973.

118. **Jagusch, L., et al.,** G.D.R. Patent 136 287, 1979.

119. **Ringpfeil, M.,** Die Lösung verfahrenstechnischer Probleme bei der Projektierung biologischer Reinigungsanlagen, *Fortschr. Wasserchem. Ihrer Grenzgeb.,* 6, 21, 1967.

120. **Harremoes, P.,** Criteria for nitrification in fixed film reactors, *Water Sci. Technol.,* 14, 167, 1982.

121. **Heinrich, D.,** Nitrification of waste waters in submerged fixed-bed reactors, *Stuttg. Ber. Siedlungswasserwirtsch.,* 81, 209, 1947.

122. **Sekoulov, J. and Heinrich, D.,** Pilot studies on nitrification in fixed bed reactors, *Wasserwirtschaft,* 71, 331, 1981.

123. **Stepmann, F.-W.,** DE 3 017 439, 1981.

124. **Huang, J. M., Wu, Y. C., and Molof, A.,** Nitrified secondary treatment effluent by plastic media trickling filter, in Proc. Int. Conf. Fixed-Film Biological Processes, Part 1, 1982, 2 (AD-A 126 377), Wu, Y. C., Ed., National Technical Information Service, Springfield, Va., 1982, 870.

125. **Fuchs, U. and Reimann, H.,** European Patent 52 855, 1982.

126. **Dunn, J. J., Tanaka, H., Uzman, S., and Denac, M.,** Biofilm fluidized-bed reactors and their application to waste water nitrification, *Ann. N.Y. Acad. Sci.,* 413, *Biochem. Eng.,* 3, 168, 1983.

127. **Jeris, J. S. and Owens, R. W.,** Biological fluidized beds for nitrogen control, in *Advances in Water and Wastewater Treatment: Biological Nutrient Removal,* Wanielista, M. P. and Eckenfelder, W. W., Jr., Eds., Ann Arbor Science, Ann Arbor, Mich., 1978, 199.

128. **Antonie, R. L.,** Nitrogen control with the rotating biological contactor, in Adv. Water Waste Water Treat.: Biol. Nutrient Removal, Pap. Conf., 1978, 263.

129. **Müller, J. A., Paquin, P., and Famularo, J.,** Nitrification in rotating biological contactors, *J. Water Pollut. Control Fed.,* 52, 688, 1980.

130. **Watanabe, Y., Ishiguro, M., and Nishidome, K.,** Nitrification kinetics in a rotating biological disk reactor, *Prog. Water Technol.,* 12, 233, 1980.

131. **Bürger, C. and Mauersberger, P.,** DD WP 274184.8, 1985.

132. **Cross, W. H., Chian, E. S. K., Pohland, F. G., Harper, S., Kharker, S., Cheng, S. S., and Lu, E.,** Anaerobic biological treatment of coal gasifier effluent, *Biotechnol. Bioeng. Symp.,* 12, 349, 1982.

133. **Suidan, M. T., Siekerka, G. L., Kao, S. W., and Pfeffer, J. T.,** Anaerobic filters for the treatment of coal gasification waste waters, *Biotechnol. Bioeng.,* 25, 1581, 1983.

134. **Suidan, M. T., Strubler, C. E., Kao, S. W., and Pfeffer, J. T.,** Treatment of coal gasification wastewaters with anaerobic filter technology, *J. Water Pollut. Control Fed.,* 55, 1263, 1983.

135. **Harper, S. R., Cross, W. H., Pohland, F. G., and Chian, E. S. K.,** Adsorption — enhanced biogasification of coal conversion wastewater, *Biotechnol. Bioeng. Symp.,* 12, 401, 1983.

136. **Fedorak, P. M. and Hrudey, S. E.,** Batch anaerobic methanogenesis of phenolic coal conversion wastewater, *Water Sci. Techn.,* 17, 143, 1985.

137. **Olthoff, M., Kelly, W. R., Wagner, G., and Oleszkiewicz, J.,** Anaerobic treatment of variety of industrial waste streams, Proc. 39th Ind. Waste Conf., 1985, 697.

138. **Fedorak, P. M. and Hrudey, S. E.,** Anaerobic treatment of phenolic coal conversion waste-water in semicontinuous culture, *Water Res.,* 201, 113, 1986.

139. **Stottmeister, U., Martius, G., Kuschk, P., Bürger, G., and Mauersberger, P.,** New results of brown coal processing wastewater treatment, *Acta Biotechnol.,* in preparation.

Chapter 2

LITERATURE STUDY ON THE FEASIBILITY OF MICROBIOLOGICAL DECONTAMINATION OF POLLUTED SOILS*

A. O. Hanstveit, W. J. Th. van Gemert, D. B. Janssen, W. H. Rulkens, and H. J. van Veen

TABLE OF CONTENTS

Editorial 67

Abstract 68

I.   Introduction 71

II. Inventory of Cases of Soil Pollution 71
     A.   Inventory According to Categories of Contaminants 72
          1.   Solvents (About 120 Locations) 72
          2.   Biocides (About 40 Locations) 72
          3.   Mineral Oil and Oil Products (About 150 Locations) 72
          4.   Contamination at Gasworks Sites (About 70 Locations) 72
          5.   Heavy Metals (About 200 Sites) 74
          6.   Other Miscellaneous Contaminants (About 150 Sites) 74
     B.   Some Examples of Serious Cases of Soil Contaminants 75

III. Factors Influencing the Choice of Soil Decontamination Procedures 75
     A.   Soil 75
          1.   Physicochemical Properties 75
               a.   Type of Stratification 75
               b.   Mineral Composition and Organic Compound
                    Fraction 76
               c.   Structure of the Soil 76
               d.   Composition (Physical and Chemical) 77
               e.   Ground Water Composition 77
          2.   Geohydrological Properties 78
               a.   Permeability 78
               b.   Ground Water Level 78
               c.   Ground Water Potentials 78
          3.   Mechanical Properties 78
               a.   Compressibility 78
               b.   Frictional Properties 80
     B.   Ground Water 80
     C.   Air 80
     D.   Contaminants 80
     E.   Volume and Geometry 80
     F.   Climatology 81
     G.   Technical Limitations 81
     H.   Topography of the Surroundings 81

---

* This chapter was edited by J. F. de Kreuk.

I.   Urgency with Aspect to Planned Use of the Site 82
J.   Legislation on Soil Contamination...................................... 82

IV.   Microbiology 82
        A.   Biodegradation of Pollutants.................................... 82
        B.   Recalcitrant Compounds... 83
        C.   Environmental Factors Affecting Biodegradation 85
        D.   Strategies for Stimulated Biodegradation 86
        E.   Application of Special Microorganisms for the Degradation of
                Xenobiotics.. 88
        F.   Biological Degradability of Important Environmental Pollutants......... 89
                1.   Crude Petroleum: Refinery Products and Aromatic
                        Hydrocarbons... 89
                2.   Miscellaneous Nonhalogenated Aromatics 89
                3.   Volatile Halogenated Hydrocarbons...................... 90
                4.   Halogenated Aromatics 90
                5.   Pesticides 90

V.   Microbial Soil Decontamination Techniques 90
        A.   *In Situ* Decontamination.. 91
                1.   Landfarming 91
                2.   Bioextraction.. 91
        B.   Decontamination After Excavation of the Soil.................... 91
                1.   Landfarming 91
                2.   Composting-Like Techniques 91
                        a.   Open Systems 91
                        b.   Closed Systems................................. 91
                        c.   Process Industrial Systems 92

VI.   Emission Control Measures Accompanying Microbiological Soil
        Decontamination 92
        A.   Installation of Barriers....................................... 92
        B.   Drilling of Wells 92
        C.   Collection of Gases.. 92
        D.   Treatment of the Gases Collected 92
        E.   Monitoring Systems 92

VII.   Nonmicrobiological Soil Decontamination Techniques...................... 93
        A.   *In Situ* Treatments... 93
        B.   Treatment of the Soil After Excavation......................... 93

VIII.   A Comparison of Microbiological Soil Decontamination Techniques 94

IX.   A Comparison of Microbiological and Nonbiological Soil Decontamination
        Techniques.. 96

Appendix A: Biodegradation and Metabolism of Important Environmental
Pollutants.. 97

I.   Introduction... 97

II.   Aromatic Compounds... 97

A.   Aromatic Hydrocarbons, Phenols, and Phthalates 97
    1.   Biodegradation of Aromatic Compounds......................100
    2.   Microbiology 100
    3.   Metabolism of Aromatic Compounds 101
    4.   Conclusions......................103
B.   Chlorobenzenes......................103
    1.   Biodegradation......................104
    2.   Metabolism 105
    3.   Enzymes 106
    4.   Dihalogenated Compounds......................107
    5.   Improvement of Strains 107
    6.   Genetics 109
    7.   Anaerobic Degradation......................110
    8.   Conclusions......................110
C.   Polychlorinated Biphenyls......................110
    1.   Biodegradation of PCBs 110
    2.   Biochemistry and Genetics......................111
    3.   Conclusions......................112

III.   Aliphatic Compounds......................112
A.   Halogenated C1 and C2 Hydrocarbons 112
    1.   Biodegradation......................112
    2.   Conclusions......................113
B.   Hexachlorocyclohexanes......................113
    1.   Biodegradation......................113
    2.   Biochemistry......................115
    3.   Conclusions......................115
C.   Aldrin and Dieldrin......................115
    1.   Degradation and Metabolism......................115
    2.   Conclusions......................116

Appendix B: Microbial Soil Decontamination Techniques......................116

I.   Introduction......................116

II.   *In Situ* Decontamination 117
A.   Landfarming......................117
    1.   Principle 117
        a.   Oxygen Content 118
        b.   Fertilizer......................118
        c.   Moisture Content 118
        d.   pH......................118
        e.   Temperature 118
        f.   Seeding with Microorganisms 118
    2.   Implementation 118
    3.   State of Development 118
    4.   Costs......................118
B.   Bioextraction 119
    1.   Principle 119
    2.   Implementation 120
    3.   Use in Soil Decontamination......................120
    4.   Costs......................120

III.   Decontamination After Excavation..121
    A.   The Excavation Process 121
    B.   Landfarming..122
        1.   Principle 122
        2.   Implementation 122
        3.   Use in Soil Decontamination...............................122
        4.   Costs..122
    C.   Composting-Like Techniques......................................122
        1.   Principle 122
        2.   Implementation 125
            a.   Open Systems 125
            b.   Closed Systems....................................125
            c.   Mixing 127
            d.   Current Uses 127
        3.   Use in Soil Decontamination...............................127
        4.   Costs..127
    D.   Industrial Processing Systems 129
        1.   Principle 129
        2.   Implementation 129
        3.   Use in Soil Decontamination...............................131
        4.   Costs..131

Appendix C: Emission Control Measures Accompanying Microbiological Soil
Decontamination 131

I.   Introduction..131

II.   Containment Techniques 131

III.   Removal of Gases 132
    A.   Ventilation Through Pipes...133
        1.   Principle 133
        2.   Applicability 133
        3.   Costs..133
    B.   Ventilation Through Trenches 134
        1.   Principle Description.....................................134
        2.   Applicability 134
        3.   Costs..135
    C.   Gas Barriers 135
        1.   Description...135
        2.   Applicability 136
        3.   Costs..136
    D.   Treatment of Gases..137

IV.   Monitoring Systems..137
    A.   Process Monitoring..137
    B.   Ground Water Monitoring..137
    C.   Monitoring of Gaseous Emissions (*In Situ*)......................138

Appendix D: Nonmicrobiological Soil Decontamination Techniques 139

I.   Introduction..139

II.   Treatment of Excavated Soil 142
    A.   Extraction 142
        1.   Principle and Background......................................142
        2.   Potential Applications...143
            a.   Heavy Metals...143
            b.   Cyanides 143
            c.   Hydrocarbons (Including Halogenated
               Hydrocarbons)...143
        3.   State of the Art 143
    B.   Thermal Treatment 143
        1.   Evaporation of the Contaminations by Means of Direct
           Heating with Hot Gases.......................................144
            a.   Principle and Background 144
            b.   Potential Applications 144
        2.   Evaporation of the Contaminants by Indirect Heating
           Via a Heat Exchanger...145
        3.   Thermal Destruction of the Contaminants.......................145
        4.   State of the Art of the Various Thermal Techniques 146
    C.   Steam Stripping 146
    D.   Chemical Treatment 146
        1.   Principle and Background......................................146
        2.   Potential Application...147
            a.   Heavy Metals...147
            b.   Cyanides 147
            c.   Hydrocarbons and Halogenated Hydrocarbons............147
        3.   State of the Art 147

III.   *In Situ* Treatment of Soil 148
    A.   Extraction (Leaching) 148
    B.   Chemical Treatment 149

IV.   Discussion 149

V.   Treatment of Excavated Soil..149
    A.   Extraction 149
    B.   Thermal Treatment 150
    C.   Steam Stripping 150
    D.   Chemical Treatment 150

VI.   *In Situ* Treatment of Soil 150

References...150

EDITORIAL

Microbiological soil decontamination will always require a multidisciplinary approach. Microbiologists, engineers, hydrologists, and soil scientists may be needed to tackle a specific case for decontamination.

This multidisciplinary character is reflected in the constitution of the group of authors:

- Dr.ir. W. J. Th. van Gemert (chemical engineering — TNO*)
- Drs. A. O. Hanstveit (microbiology — TNO)
- Dr. D. B. Janssen (microbiology — GBC-RUG**)
- Dr.ir. W. H. Rulkens (chemical engineering — TNO)
- Mr. H. J. van Veen (chemical engineering — TNO)

The report is "constructed" accordingly. Each of the sections was written mainly by one of the above-mentioned authors. This explains the differences in character between the various sections, and these differences are further increased by the natures of TNO and GBC-RUG; the former directs activities towards applied research and the latter towards more fundamental subjects.

Therefore, Dr. Janssen (GBC) is responsible for the greater part of Section IV and Appendix A (on microbiology) and the other authors for the other parts of the report.

For readability it was decided to edit the study in such a way that the technical information is given in Appendixes, while the report itself contains the summaries thereof.

ir. J. F. de Kreuk

ABSTRACT

The feasibility of microbial decontamination of soil polluted by organic chemicals was studied on the basis of literature references and the experience of the authors.

The following subjects were covered:

- The frequency of soil contamination and the nature of the contaminants
- The biodegradability of contaminants
- Soil decontamination procedures
- Methods of protecting the environment against pollution from a contaminated area or from the decontamination procedures

In spite of the existence of many cases of permanent soil pollution, most organic compounds appear to be inherently biodegradable under favorable conditions, which include availability of nutrients and an electron acceptor (oxygen), moderate temperature, pH 6 to 8, and the presence of degraders.

The rate of biodegradation may, however, be low because of the chemical nature or physical behavior of a compound. The latter may affect availability for microorganisms, for example, by a low solubility or a strong adsorption on soil particles. Biodegradation will probably also not occur, if special organisms are required or if different conditions are needed for the various steps of the degradation process.

By modifying the conditions and, if necessary by seeding with specific microorganisms, it is to be expected that biodegradation may be enhanced significantly.

In an inventory of cases of soil contamination over 700 contaminated sites were identified in the Netherlands. Although many contaminants were present at most of these sites, the following divisions, according to the prevailing contaminant, could be made:

1. Solvents, 120 sites, of which there were 45 with chlorinated solvents, 45 with non-chlorinated solvents, and 30 with nonspecified solvents
2. Biocides, 40 sites
3. Oil and oil-like products, 150 sites

---

* Organization for Applied Scientific Research.
** Groningen Biotechnology Center, State University of Groningen.

4. Gasworks (with polynuclear aromatics), 70 sites
5. Heavy metals, 200 sites
6. Unspecified, 150 sites

It is clear that microbiological decontamination techniques will not primarily be used to tackle cases of contamination with heavy metals, although microorganisms may be used for facilitating leaching of heavy metals. In particular, microbial decontamination may provide a valuable tool for remedial measures in "solvents and oil cases" and in some cases of biocide contamination and some gaswork sites. There are two approaches for tackling a specific case:

1. Decontamination of the soil *in situ:* the soil is not excavated, but the conditions in the soil are modified to facilitate decontamination. If the soil is sufficiently permeable, an aqueous solution containing nutrients, oxygen, microorganisms, etc. can be circulated through the contaminated soil (bioextraction). If contamination is only superficial (<0.5 m deep), landfarming techniques can be used whereby normal agricultural procedures such as plowing, harrowing, and fertilizing are used in order to obtain biodegradation of, for example, oil compounds.
2. Decontamination of the soil after excavation: a decontamination process, adapted to the nature of the contaminant and of the soil is carried out on site or at an off-site decontamination plant. Landfarming can also be used to decontaminate excavated soil after spreading in a layer of 0.2 to 0.3 m. Most other decontamination procedures resemble composting processes and are carried out either in the open in windrows (e.g., the Beltsville process) or in closed equipment (e.g., the Dano process in a rotating drum, or the Kneer system using a vertical tapered cylinder). A reactor can be used in special cases when a selected strain of microorganisms is required for degradation; in these cases the use of industrial mixers for solids may be feasible.

The following aspects must be considered before a choice of a microbial decontamination procedure can be made:

- The biodegradability of the contaminants
- The nutrient status of the soil and the possibility of improving this status (including oxygen)
- pH
- The presence and influence of other carbon sources (positive or negative) on the decontamination process
- The presence of microorganisms, which are able to degrade the relevant contaminants
- The physical behavior of the contaminants (adsorption to soil, leachability, permeability of the waste for water and microorganisms, melting points, etc.)
- Toxicity of the contaminants for microorganisms
- Soil characteristics
- Ground water and flow characteristics
- "Development status" of the technique
- Costs
- Environmental nuisance (dust, smell, noise, etc.)

The relative importance of these aspects is determined by the actual choice and the specific "properties" of the site. Insight into all of these aspects is, however, required.

Care should be taken that the decontamination process will not form a source of pollution. A number of containment techniques and venting and collecting systems for volatile compounds and/or products are, therefore, described in this report.

Horizontal transport of contaminants (e.g., via ground water) may be prevented either by screens of bentonite, plastics, grout, etc. or by hydrological measures. Vertical transport may have to be prevented when the contaminated soil is situated over a permeable layer. In such cases, the permeable layer may need to be blocked by injection of bentonite, resins, grout, etc. Plastics, foil, or bituminous products can be used for covering soil.

The effectiveness of such a "containment" system must be monitored; wells for this purpose are described in this report.

Gaseous products can be vented from soil through a collecting system, consisting of slotted pipes or gravel-filled trenches. Forced ventilation for treatment of the gases may be necessary.

Containment and controlled sanitary landfilling are alternatives to (biological) treatment. Further alternatives, when microbial decontamination is not feasible, are

* Extraction (either *in situ* or after excavation): the extraction process can be accompanied by chemical oxidation. It can be used for cases with heavy metal pollution or with cyanides.
* Thermal treatment: the soil then is heated by appropriate means in order to volatilize the contaminants. The waste gases should be treated.

The following conclusions can be drawn from this study:

1. Microbial decontamination of polluted soils is, in principle, feasible when favorable conditions for biodegradation are created.
2. The most promising procedures are expected to be

   * Landfarming of the excavated soil
   * Composting (either in the open or in closed equipment)

3. The techniques used for the above-mentioned techniques must be applied in such a way that contamination of the surroundings is prevented.
4. In special cases selected strains of microorganisms may be used in equipment such as industrial mixers for solids.
5. *In situ* treatment (without excavation) can only be used when the soil is sufficiently permeable.
6. The following groups of compounds may be treated biologically:

   * Nonhalogenated aromatic solvents and the lower polynuclear aromatics (naphthalene and anthracene) and compounds such as phenols, phthalic acids, and cresols
   * Mineral oil and products thereof (petrol, kerosene, etc.)
   * A number of halogenated compounds such as mono- and dichlorobenzenes, chlorophenols, and chlorinated aliphatic acids
   * Halogenated C_1 and C_2 compounds and pesticides (biodegradation of these compounds is likely to need particularly well-adapted strains)

7. Microbial decontamination is expected to be less expensive as compared to other treatments. Costs for microbiological techniques will range between Dutch florin (Dfl.) 10.00/ton for *in situ* treatment and Dfl. 50.00 to 60.00/ton for composting-like techniques in closed systems. The costs for thermal treatment range between Dfl. 100.00 and 250.00/ton and those for controlled tipping are approximately Dfl. 75.00/ton.

I. INTRODUCTION

In recent years many illegal dumping sites for chemical waste have been "discovered" in the Netherlands. Furthermore, (illegal) dumping of chemicals on sanitary landfills, and contaminated abandoned industrial sites such as old gasworks have also contributed to the list of contaminated sites. Most of these contaminated areas are located in or near the highly industrialized areas in the Netherlands. Space is a scarce commodity in these areas and housing estates are planned, or have already been built, on contaminated soil. In both cases this normally requires decontamination of the sites at great cost to the local or national authorities.

For this reason there is a great need for reliable and relatively cheap decontamination procedures. The biological treatment of sewage and wastes reveals that many organic compounds, otherwise considered to be rather difficult to degrade, can be degraded by microorganisms under the right conditions and that this biotreatment is cost-effective.

It is common knowledge that many of the chemical substances encountered in cases of serious soil pollution are, in principle, biodegradable. In fact it is surprising that compounds such as benzene have not already disappeared. These facts led both the Division of Technology for Society (MT-TNO) and the Groningen Biotechnology Centre of the "Rijksuniversiteit" at Groningen (State University at Groningen) (GBC-RUG) to pursue the development of microbiologically based soil decontamination processes.

This shared interest led to the joint implementation of the feasibility study described in this report. The GBC contributed knowledge of microbial metabolism and processes based on the tradition of the University of fundamental microbiological research. MT-TNO contributed experience on technical processes (i.e., waste handling) and on the biodegradation of chemical wastes and compounds by means of mixed cultures of microorganisms. The objectives of the study were to gain insight into:

- The size and nature of the problem
- The possibility of solving this soil contamination problem by microbiological techniques
- The mechanisms that prevent the degradation in soil of biodegradable compounds
- The possibility of developing microbiological decontamination processes and the limitations of these processes
- The identification areas where sufficient knowledge is lacking

The study was limited to the consideration of microbial soil decontamination. Photochemical reactions were, therefore, not taken into account, although it is known that atmospheric degradation of a number of compounds (chlorinated hydrocarbons) is the main degradation process. It should also be noted that the principles outlined in this study can also serve research areas outside that of soil decontamination; for example, the cleaning of hazardous waste by adsorbing it to a recyclable inert carrier.

II. INVENTORY OF CASES OF SOIL POLLUTION

This inventory has been limited to the number of soil pollution cases in the Netherlands and the nature of the pollutants involved. Although data on the actual concentrations of contaminants in soil and on the amount of contaminated soil are important, they were too scarce to include them in the inventory. Only those cases in which the contaminating chemical has been identified are included; the inventory is, therefore, not complete. Furthermore, new cases are still being discovered regularly. The inventoried cases are divided into the following categories (Section II.A):

1.   Solvents

    ● Chlorinated
    ● Nonchlorinated
    ● Not defined

2.   Biocides
3.   Mineral oils
4.   Contamination at gasworks sites
5.   Heavy metals
6.   Other miscellaneous contaminants

Contaminants from different categories were present on most sites. In such cases the category of the main contaminant was chosen. A number of special cases is dealt with in Section II.B.[5,47]

A. Inventory According to Categories of Contaminants
    Table 1 summarizes the results of this inventory. The cases (approximately 700) are discussed in the following sections according to the categories mentioned above.

1. Solvents (About 120 Locations)
    Chlorinated hydrocarbons were the main contaminants in 50% of the cases in which the solvents were specified. Specifications were not given in 30 of the 120 cases.
    The frequency of the occurrence of the ''important'' solvents at contaminated sites is shown in Table 2.
    The most frequently found chlorinated and nonchlorinated solvents were trichloroethane and toluene, respectively. In 18% of the cases the type of solvent was not specified.

2. Biocides (About 40 Locations)
    The biocides given in Table 3 were mentioned most. The majority of the locations contaminated by hexachlorocyclohexane (HCH) is found in Twente (eastern part of the Netherlands). In 24% of the cases the biocide involved was not specified. The great diversity of biocides dumped is evident from the high percentage for ''various'' biocides.

3. Mineral Oil and Oil Products (About 150 Locations)
    In more than 80% of the cases the nature of the contamination was only given as ''oil''. Petrol and diesel oil were quoted as examples of oil products found.

4. Contamination at Gasworks Sites (About 70 Locations)
    At sites of present or former gasworks, 66 cases of soil contamination have been identified. According to *De Kleine Gifatlas* (The Small Poison Atlas) there are, however, about 230 gasworks sites in the Netherlands.[47] If a great number of these sites are in fact contaminated, then this category forms one of the most frequently occurring causes of soil contamination.
    Gasworks waste may contain the following components:

●   Aromatics (benzene, toluene, trimethylbenzene, xylenes, styrene, naphthalene, anthracene, indene, fluoranthene, phenanthrene, chrysene, pyrene, phenols, and cresols), up to 10 to 40 g/kg of soil
●   Tar products
●   Cyanides and ferricyanids, up to 1 g/kg of soil

Table 1
RESULTS OF THE INVENTORY ON SOIL POLLUTION

	Solvents							
	Chlorinated	Nonchlorinated	Not further described	Total	Biocides	Oil	Gasworks waste	Heavy metals	Others
Number of sites	45	45	30	120	40	150	70	200	150

Note: Breakdown of the sites by category of principal contaminant.

Table 2
THE MOST IMPORTANT SOLVENT
CONTAMINANTS

Chlorinated solvents (%)		Nonchlorinated solvents (%)		Various (%)
Trichloroethane	(19)	Toluene	(23)	(18)
Perchloroethylene	(8)	Xylene	(13)	
Tetrachloromethane	(8)	Benzene	(5)	
Methylenechloride	(3)			
Unspecified	(3)			

Note: 100% = 120 of such cases.

Table 3
THE MOST IMPORTANT
BIOCIDE CONTAMINANTS
(%)

Hexachlorocyclohexane	21
Pentachlorobenzene	7
2,4,5-T	7
Organomercury	7
Arsenic compounds	7
Unspecified	24
Various	27

Note: 100% = 40 of such cases.

Table 4
CASES OF CONTAMINATION BY "OTHER
MISCELLANEOUS" CONTAMINANTS (%)

Tar products and aromatics	30
Chromium-containing waste	10
Fly and bottom ash from combustion of coal	10
Cyanides	8
Accumulator waste	6
Developer	5
Unspecified	30

Note: 100% = 150 of such cases.

5. Heavy Metals (About 200 Sites)

Although contamination of soil with heavy metals occurs frequently, it will not be dealt with here because the manipulation of heavy metals by microbiological means is not considered to be part of this study.

6. Other Miscellaneous Contaminants (About 150 Sites)

A breakdown according to classes of compounds (if known) is given in Table 4. Table 4 shows that tar products and aromatics made the greatest contribution to this group. Since the contaminants were not specified in 30% of the cases, a great variety of contaminants may be expected to be found in this group.

B. Some Examples of Serious Cases of Soil Contamination

Three particular cases, which are exceptional by nature, extent, or urgency are given below.[44] These are by no means the only serious cases; they are merely given to illustrate the nature of the problem being found.

Volgermeerpolder — In this case[51] 2 to 4 × 10^6 m³ of waste is distributed over 600 ha, at depths up to 5 m. The contamination consists of halogenated aromatics (di-, tri-, tetra-, penta-, and hexachlorobenzenes, chlorophenols, 2,4,5,-T, and TCDD) and HCHs (α-HCH, lindane). Geohydrological containment is applied to check further spreading of the contaminants.

Krimpen a/d IJssel — A company (EMK) "worked up" chemical waste on this site. This resulted in serious soil contamination, because one of the "procedures" of the firm consisted of pouring the waste onto the soil under the buildings.

Gouderak — In this case[5] about 150,000 m³ of contaminated soil is situated under a housing estate. The contamination consists of:

- Polycyclic aromatics (about 135,000 kg)
- Benzene (110 kg)
- Aldrin, dieldrin, and endrin (about 14,000 kg)
- PCBs (about 13,000 kg)

III. FACTORS INFLUENCING THE CHOICE OF SOIL DECONTAMINATION PROCEDURES

When a case of soil contamination has been discovered, a number of decisions have to be taken, resulting in either a commitment to environmental protection and/or decontamination measures, or to a reasoned decision not to undertake action. The decision-making process may follow the model presented in Figure 1. Various aspects relating to a specific case will influence the final decision. These aspects are listed in Table 5 and are further discussed below (in Section IV).

The steps under the dotted horizontal line in the algorithm are the subject of the study described here (Figure 1).

A. Soil

The composition and properties of the soil concerned are important for the selection of the decontamination procedure and the correct process conditions. The following properties are important in this context: composition (including ground water), type of stratification, permeability, and the compressibility. The soil-related aspects (Table 5) are discussed below.

1. Physicochemical Properties

a. Type of Stratification

Stratification of the soil is an important factor in relation to

- Reaction to mechanical activities
- Direction and velocity of the ground water flow
- Distribution profile of the contaminants
- Possibilities for *in situ* decontamination

Information on (type of) stratification can be obtained from soil maps and from measurements at trial borings.

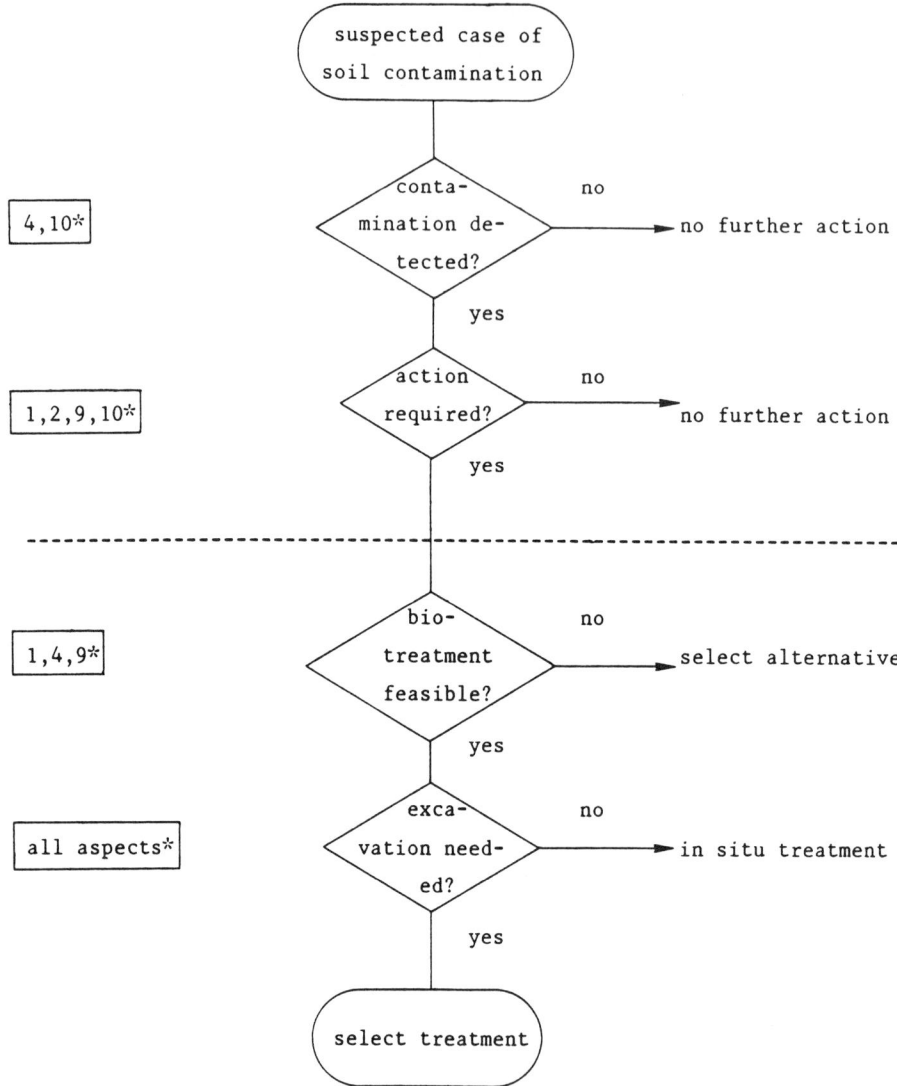

FIGURE 1.   Algorithm for the decision process after the discovery of a suspected case of soil pollution. The aspects which will guide the decision-making process, which are given in Table 5, and further discussed in the text are shown in the boxes on the left.

b. Mineral Composition and Organic Compound Fraction

These properties are of great importance both for the distribution of the contamination and for possibilities of decontamination of the soil. Fine mineral fraction and organic substance (humus) restrict the spreading of contamination by ion exchange and adsorption phenomena. Furthermore, it must be taken into account that in decontaminating soil (in particular by extraction) small mineral particles can often only be dewatered with much difficulty, and this can make decontamination by extraction difficult.

c. Structure of the Soil

The structure of the soil and in particular the porosity plays an important role in determining permeability.

Table 5
FACTORS WHICH MAY
STRONGLY INFLUENCE THE
CHOICE OF SOIL
DECONTAMINATION
PROCEDURES

Soil
    Physicochemical properties
    Geohydrological properties
    Mechanical properties
    Microbiology
Ground water
    Mobility
Air
    Emissions to the air
Contaminants
    Concentration and nature
    Physical state
    Distribution
    Biodegradability
Volume and geometry
Climatology
Technical limitations
Topography of the surroundings
Urgency with respect to planned use of the site
Legislation on soil contamination

d. Composition (Physical and Chemical)

The behavior of contaminants in the soil is determined, to a considerable extent, by the composition of the soil. Soil is a multiple phase system, consisting of about 50% solid matter, 25% liquid, and 25% gaseous phase (depending on the type and moisture content of soil). Interactions between the various components in soil are shown schematically in Figure 2.[40]

Gaseous products may mix with the air present in the soil (Figure 2, [11]) but may also dissolve in the aqueous phase (Figure 2, [12]). Plants and microorganisms in the soil consume O_2 and produce CO_2. The transport (by diffusion) of O_2 and CO_2 is often limited by various barriers; this results in low O_2 concentrations in soil and, consequently, in reducing conditions, sometimes even in nonwaterlogged top soils.

Knowledge of the chemical composition is required to determine natural contents of the various elements. This knowledge is important for the evaluation of the effectiveness of a decontamination operation and for establishing target levels. Although the composition of soil is highly dependent on location, it is possible to indicate the average elemental composition of the lithosphere (Table 6).

e. Ground Water Composition

The ground water composition is an important factor in relation to

- Environmental load of xenobiotic compounds
- Evaluation of the possibility of well drilling and of the necessity of discharge
- Evaluation of the necessity for containment

Insight into the general composition of ground water can be obtained from ground water charts. Analysis is normally required for detailed information.

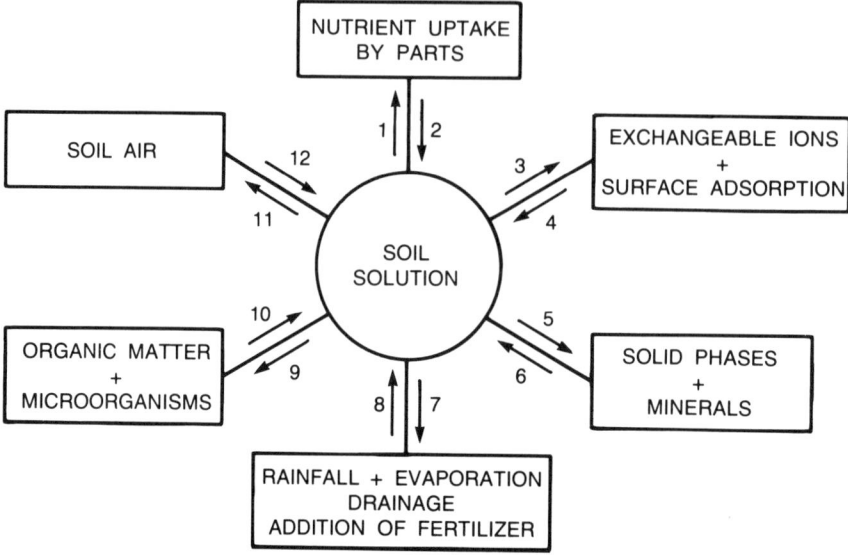

FIGURE 2.   The dynamic equilibria in soils.

2. Geohydrological Properties

a. Permeability

The permeability of the soil is of importance for a number of processes taking place in it. These processes include ground water flow and transport of contaminants.

The permeability is determined by factors such as composition, stratification, and shape and size of particles and pores in the soil. Furthermore, it is important to know if layers of differing permeability are present. General information on permeability and soil structure can be obtained from records and from geological and soil charts. More detailed information must be obtained from field or laboratory measurements.

b. Ground Water Level

The level of the ground water is an important determinant for possible leaching of contaminants and the implementation of earth-moving operations.

Information on ground water levels can be obtained from soil charts or from measurements for more detail.

c. Ground Water Potentials

Ground water potentials and distribution over the various soil strata determine the direction and velocity of ground water flows. Ground water potentials are important determinants of the extent to which contaminated ground water effects the surroundings. Ground water potentials can be measured in the field.

3. Mechanical Properties

a. Compressibility

If ground is loaded mechanically, the various ground layers are compressed to a greater or lesser extent. This compression is proportional to the load. Soft grounds, such as clay and peat are highly compressed when loaded and consequently may show significant settling. In such a case the ground water is said to be "overstrained". This may result in expulsion of the pore water. Soil loading can be adsorbed by the pore water when a low permeability restricts pore water flow. A ground layer with "overstrained" ground water forms a distinct blockade against natural ground water flows and thus may hamper the transport of contaminated ground water.

Table 6
ELEMENTAL COMPOSITION OF THE LITHOSPHERE AND SOILS

Element	Atomic weight (g)	Content in lithosphere (ppm)	Common range for soils (ppm)
Ag	107.87	0.07	0.01—5
Al	26.98	81,000	10,000—300,000
As	74.92	5	1—50
B	10.81	10	2—100
Ba	137.34	430	100—3,000
Be	9.01	2.8	0.1—40
Br	79.91	2.5	1—10
C	12.01	950	
Ca	40.08	36,000	7,000—500,000
Cd	112.40	0.2	0.01—0.70
Cl	35.45	500	20—900
Co	58.93	40	1—40
Cr	52.00	200	1—1,000
Cs	132.91	3.2	0.3—25
Cu	63.54	70	2—100
F	19.00	625	10—4,000
Fe	55.85	51,000	7,000—550,000
Ga	69.72	15	5—70
Ge	72.59	7	1—50
Hg	200.59	0.1	0.01—0.3
I	126.90	0.3	0.1—40
K	39.10	26,000	400—30,000
La	138.91	18	1—5,000
Li	6.94	65	5—200
Mg	24.31	21,000	600—6,000
Mn	54.94	900	20—3,000
Mo	95.94	2.3	0.2—5
N	14.01	—	200—4,000
Na	22.99	28,000	750—7,500
Ni	58.71	100	5—500
O	16.00	465,000	
P	30.97	1,200	200—5,000
Pb	207.19	16	2—200
Rb	85.47	280	50—500
S	32.06	600	30—10,000
Sc	44.96	5	5—50
Se	78.96	0.09	0.1—2
Si	28.09	276,000	230,000—350,000
Sn	118.69	40	2—200
Sr	87.62	150	50—1,000
Ti	47.90	6,000	1,000—10,000
V	50.94	150	20—500
Y	88.91		25—250
Zn	65.37	80	10—300
Zr	91.22	220	60—2,000

From Lindsay, W. L., *Chemical Equilibria in Soils,* John Wiley & Sons, New York, 1979. With permission.

b. Frictional Properties

If loading of ground layers disturbs the force equilibrium within these ground layers, displacement may result in loss of cohesion of the soil particles. The resistance of the ground to shifting, the so-called frictional resistance, is of importance for the consequences of earth-moving activities (e.g., excavation). The frictional properties of soils depend on properties, such as texture, extent of permeability, and type of stratification.

B. Ground Water

The influence of ground water flows on soil properties has been described above (Section A).

Microbiological decontamination of soil may result in movement of contaminated ground water to surrounding areas. The acceptability of this contamination will depend on the extent and environmental quality requirements imposed on the ground water.

The "Leidraad bodemsanering" (soil rehabilitation directive) gives standards for maximum tolerable pollution levels for soil and ground water in relation to (future) soil use.[48]

C. Air

Air pollution can result from soil decontamination processes due to emission of compounds with "sufficient" vapor pressure under the process conditions.

Environmental limits on the emission of gases determine whether decontamination of gaseous emissions will be required. Biodegradation normally results in harmless compounds such as carbon dioxide, water, and inorganic halogens.

D. Contaminants

As much information as possible on the contamination present should be available. Information concerning nature, concentrations, concentration gradients, rates of biodegradation, and concentrations inhibiting biodegradation is required. The information on these aspects will have to be obtained by chemical analysis and by performing tests with contaminated soil. In some cases the test methods required will still have to be developed.

E. Volume and Geometry

The geometry of the contaminated site forms an important criterion for selection of the appropriate rehabilitation method. These will depend on the quantity and nature of the contaminant present, and on the soil type, water content, and ground water level.[26] A liquid will penetrate soil (in particular porous sandy soils) until it reaches an impermeable layer or the ground water table. It then will spread, mostly in the direction of the ground water flow. Soluble compounds may dissolve in ground water and be transported in solution. Liquids which are heavier than water will only be halted by an impermeable layer (such as clay). During penetration, a portion of soil first becomes saturated before penetration continues. Soil may hold between 15 and 40 ℓ of oil per cubic meter (Table 7). In a homogeneous ground without any stratification or perceivable variations in the size distribution of the pores, the infiltration is pear-shaped (see Figure 3). The vertical component is the result of gravity and the horizontal component is the result of capillary forces.

If the soil is poorly permeable, the capillary forces will play a relatively large role and the contaminant will penetrate faster in a horizontal then a vertical direction. Furthermore, heterogeneities in the soil exert a great influence on the shape of the contaminated area. In practice soil is not homogeneous, so that the actual distribution pattern is irregular. Penetration of a contaminant is generally less deep than would be predicted on the basis of theoretical considerations.

Table 7
TYPICAL OIL RETENTION CAPACITIES FOR
VARIOUS SOILS[24]

Soil	Oil retention capacity (ℓ/m^3)
Stones and coarse gravel	5
Gravel and coarse sand	8
Coarse sand and medium sand	15
Medium sand and fine sand	25
Fine sand and silt	40

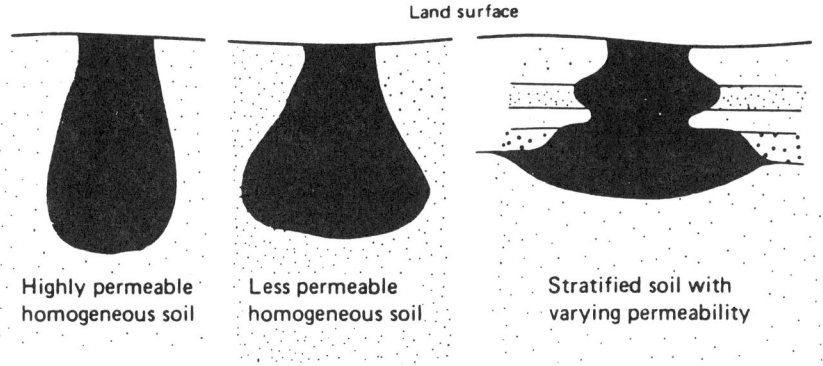

FIGURE 3.   Generalized shapes of spreading bodies.

F. Climatology

If the method chosen implies a treatment under "ambient conditions", the local climatological conditions will form a natural critical value. Temperature fluctuations, periods of rain, and prolonged drought must be taken into account.

G. Technical Limitations

The dimensions of the process equipment required, in particular, to allow the necessary residence time when contaminated soil is treated in a plant, must not exceed practical limits. In addition, the technical facilities for the realization of the required process conditions (e.g., addition of oxygen, nutrients, and heat) must be available.

H. Topography of the Surroundings

In particular when *in situ* treatment is considered, a number of topographical aspects must be considered. These include:

- The seasonal accessibility of the site
- The regularity of the surface, e.g., the presence of course irregularities such as small dunes and streams
- The hilliness of the terrain (slopes)
- The distance of the terrain from drinking water extraction areas or residential districts

I. Urgency with Aspect to Planned Use of the Site

If technical considerations allow *in situ* treatment, the degree of urgency may be the deciding factor for or against actual implementation of the treatment. This urgency will depend on the degree of hazard of the contamination, on plans for the contaminated area, and the time scale for execution of the plans.

J. Legislation on Soil Contamination

The choice of the treatment will always fall within the relevant legislation. Among others, the following acts are relevant in the Netherlands: the Public and Private Nuisance Act, the Interim Soil Rehabilitation Act, the Chemical Wastes Act, the Air Pollution Abatement Act, and the Surface Water Pollution Act.

IV. MICROBIOLOGY

The turnover of xenobiotics in soils (and waters) is mainly dependent on the degradative activity of microorganisms. Other processes, such as evaporation and photodestruction are of minor importance and contribute only to the disappearance of certain classes of chemicals from the upper layer of soils (and surface waters). Microorganisms, especially bacteria, yeasts, and fungi, are capable of degrading many kinds of man-made chemical compounds, but despite enormous versatility there are numerous cases in which long-term pollution of soils and waters has been observed. Apparently, the biological turnover of the respective chemicals is not high enough under the prevailing environmental conditions. On the basis of an investigation into the factors that affect the biodegradation rates, it should be possible to understand the causes of the observed accumulations and develop methods that allow an increased rate of biological turnover.

A. Biodegradation of Pollutants

The biodegradation of man-made chemical compounds has attracted the attention of many microbiologists and environmental scientists, because the degree of sensitivity to microbial attack greatly determines the fate of xenobiotics in the environment. The severe environmental persistence that is shown by several pesticides and other industrial chemicals is caused by the inability of microorganisms to metabolize these compounds under the prevailing environmental conditions.

Since many chemical compounds can have severe toxic effects, it has become apparent that the persistence of synthetic chemical compounds in the environment is very undesirable. For this reason, microbiological studies were carried out leading to the identification of several organisms which were capable of complete degradation, partial conversion, or co-metabolism of polluting chemicals.[12,39] A number of degradative pathways for individual compounds as carried out by pure cultures of organisms have been established. In recent years, much effort has been devoted to understanding the genetic basis for the degradation of xenobiotic compounds. The identification and genetic analysis of several catabolic plasmids have given increased insight into evolutionary development, and modern genetic techniques have opened the way to increasing the degradative capabilities of selected microorganisms.

These developments point to the attractive possibility that microorganisms could be used for degradation processes with greater efficiency and in a broader area than has been the case so far.

Regarding microbial interactions with chemical compounds it is possible to distinguish between several groups of polluting compounds (Table 8). Most organic compounds can be degraded under aerobic conditions into carbon dioxide and water. In some cases, organic compounds are not utilized as carbon and energy sources, but only partially oxidized, often

Table 8
INTERACTION OF MICROORGANISMS WITH CHEMICAL
COMPOUNDS[12,37,39,45]

Compounds	Interactions
Organic compounds	Oxidation to CO_2 and H_2O; assimilation; fermentation, e.g., to CH_4; transformation by cometabolism; and conjugation, polymerization, and accumulation
Heavy-metal ions	Solubilization by acid producers, precipitation, reduction, accumulation, and conversion into organometallic compounds
Nitrogen-containing compounds	Degradation and assimilation, reduction/oxidation, and release as N_2
Halogenated organics	Degradation and dehalogenation with release of Cl^- or Br^- accumulation and conjugation
Organic phosphates	Assimilation, accumulation, and conjugation

by nonspecific enzyme systems. This process is called cooxidation or cometabolism. Under appropriate conditions, nitrate, ammonia, and organic nitrogen may be converted into molecular nitrogen. Degradation of halogenated compounds may be initiated by halide release.[3]

Heavy-metal ions may undergo reactions such as precipitation, adsorption, reduction, and incorporation into organic compounds.

Many chemical compounds that are frequently found as pollutants are in principle biodegradable (as shown by the isolation of mixed or pure cultures of microorganisms that utilize them as the sole source of carbon and energy for growth). If these compounds remain in the environment for long periods of time, this indicates that the prevailing conditions inhibit turnover. On the other hand, a number of xenobiotics are recalcitrant to biodegradation because the chemical structure of xenobiotics makes them resistant to biodegradation. Thus, the absence of biodegradation of chemicals may be attributed to:[1-3]

● The chemical structure and/or the physical behavior of a compound
● The unfavorable environmental conditions

B. Recalcitrant Compounds
Since it has been observed that a number of synthetic chemicals may persist in the environment for years after application, it has become apparent that xenobiotics may have molecular structures that are not or hardly recognized by microbial enzymes. Many plastics and other synthetic polymers, chlorinated aliphatic and aromatic hydrocarbons, and several pesticides have been found to resist microbial degradation. The turnover of such compounds is negligible under a variety of environmental conditions, whether aerobic or anaerobic, in soil, fresh water, and marine systems. Then, recalcitrance is an inherent characteristic of the xenobiotics themselves (Table 9).

One explanation for recalcitrance comes from an examination of the structural characteristics of the pollutant molecules. The presence of one or more of a number of chemical structures or groups may cause remarkable resistance, as is the case for extensive halogenation of organic solvents and the organochlorine pesticides. Methyl branching of alkanes, for instance, in petroleum hydrocarbons and alkylbenzene sulfonate detergents, also causes recalcitrance to microbial degradation, as does the presence of substituent nitro groups. Such groups, which cannot be attacked by available microbial enzymes, when present, may lead to complete inertness, or the formation of dead-end products during microbial transformation. Such compounds, which are not further degraded, accumulate. Therefore, the xenobiotic

Table 9
SOME PERSISTENT XENOBIOTIC
CHEMICALS

	Ref.
Volatile halogenated hydrocarbons	
Chloroform, 1,2-dichloroethane, 1,1,1-trich-	39, 42
loroethane, 1,1,2,2-tetrachloroethane,	
trichloroethylene, and tetrachloroethylene	
Halogenated aromatics	
Trichlorobenzenes, tetrachlorobenzenes,	49, 50
pentachlorobenzene, hexachlorobenzene,	
3,4,5-trichlorophenol, 2,3,4,5-tetrachloro-	
phenol, pentachlorophenol, and 2,3,6-	
trichlorobenzoate	
Pesticides	
2,4,5-T, 2-(2,4-DB), 2(2,4,5-TP), aldrin,	29, 38
and dieldrin	
DDT, heptachlor, chlordane, toxaphene,	1, 38
atrizine, diuron, and monuron	
Miscellaneous	
PCBs, particularly those with more than	39
three Cl	
TCDD and other chlorinated dioxins	
Alkylbenzene sulfonates (branched)	
Synthetic dyes	
Phthalate esters	

compound may have chemical structures that are not recognized by the transport, enzyme, or induction systems of microorganisms.

Among the most persistent chemicals are a number of compounds that are used for pest control in agriculture. These include chlorinated hydrocarbon insecticides, phenoxy herbicides, phenylureas, and triazines. Other classes of recalcitrant chemicals include the polychlorinated biphenyls (PCBs) and dioxines, among which TCDD is one of the most toxic and persistent chemicals present on earth. For many compounds the degradation rates in the environment and the susceptibility to microbial attack are unclear. Polycyclic aromatic hydrocarbons and several components of crude oil, particularly the branched and condensed cyclic aliphatics, are suspected to be only slowly degradable. Current knowledge, however, is insufficient to exclude extensive microbial transformation under favorable conditions.

The physical properties of a compound may minimize the susceptibility to microorganisms. Crude oil and oil products tend to form insoluble clumps and aggregates, and it has been demonstrated that treatment of oil pollution with dispersants increased biodegradation rates.[7,19] Microorganisms that produce emulsifying factors, as is the case with a number of hydrocarbon utilizers, may possibly be used to stimulate solubilization and biodegradation. The poor turnover of polycyclic aromatic hydrocarbons and ferricyanides may also be caused by extreme low solubility in water.

Degradation rates in soils are dependent on the presence of organisms that are able to convert xenobiotic products under the actual environmental conditions. In contaminated soils, there is often an increase in the number of organisms that degrade particular pollutants, which results in increased conversion. Therefore, in laboratory experiments inoculation of soils with bacteria that can degrade the xenobiotic has been shown to stimulate biodegradation.[13] These results were obtained with naturally occurring organisms isolated from soil samples, with organisms added as activated sludge, and with genetically altered bacteria. The latter were selected under conditions that not only allowed mutation and recombination

Table 10
CAUSES OF POOR BIODEGRADATION OF POLLUTANTS UNDER ENVIRONMENTAL CONDITIONS AND POSSIBLE REMEDIES

Compound	Cause of recalcitrance	Remedy
Petroleum (crude) (marine environment)	Clumping and insolubility[7,8,19,33]	Dispersing and emulsifying agents or microorganisms
	Lack of microbial activity[7,8,19,33]	Seeding with microorganisms or adapted sludge
	Nutrient limitation[7,8,19,33]	Addition of oleophilic fertilizers
Petroleum (soil environment and ground water environment)	Insolubility[1-3]	Emulsifying agents or microorganisms
	Nutrient limitation[7,8,19,33,148,149]	Addition of fertilizers
	Oxygen limitation[7,8,19,33,148,149]	Injection of air and frequent mixing or tilling
Aromatic hydrocarbons phenols (soil environment)	Nutrient limitation[7,8,19,33]	Addition of fertilizers
	Oxygen limitation[7,8,19,33]	Injection of air and frequent mixing or tilling
Halogenated hydrocarbons (soil environment)	Recalcitrant structures[1-3,42,150,152]	Seeding with selected organisms
	Low concentration[1-3,10]	Addition of a cosubstrate or seeding with selected microorganisms
Pesticides (soil environment)	Recalcitrant structures[1-3,29,38,151]	Seeding with selected microorganisms
	Low concentration[1-3,10]	Inducer addition or seeding with selected microorganisms
	Adsorption to soil or humus[1-3,29,38,151]	Wetting and mixing
Polycyclic aromatic hydrocarbons (soil environment)	Insolubility[1-3,148,149]	?
	Oxygen limitation	Injection of air and frequent mixing or tilling
Ferricyanides (soil environment)	Stable structures	Acid-producing microorganisms
	Insolubility	?
General	Unfavorable pH[1-3,7,8,19,33,148,149]	Addition of limestone

of genetic material, but also favored the survival of organisms with increased degradative capabilities.

Genetic engineering of genes involved in the catabolism of xenobiotics may, therefore, result in organisms capable of completely degrading compounds previously thought to be inert.[25]

C. Environmental Factors Affecting Biodegradation

A number of environmental factors affect the rate of biodegradation (see also Table 10):

- Availability of an electron acceptor such as (molecular) oxygen
- Levels of degrading microorganisms
- Water content of the soil
- Availability of nutrients such as phosphorus and nitrogen
- Temperature
- Concentrations of natural organic compounds
- Presence of toxic compounds

The absence of oxygen (anaerobiosis) or of another suitable electron accepter is the main cause of poor biodegradation of organic compounds. In fact, major accumulations of organic remains, as found in several natural deposits, are generally present at sites where they developed under anaerobic or oxygen-limited conditions. The poor mineralization in the absence of oxygen may be the result of a lack of a suitable electron acceptor to sustain microbial growth; having an acceptor is a strict requirement for oxygen at some of the degradative steps. Although a number of oxidative reactions are known to be carried out with nitrate or sulfate as electron acceptors, the concentration of these sediments and soils is too low to enable the oxidation of vast amounts of organic matter. Furthermore, some conversions require the action of microbial dioxygenases, which are enzymes that obligately require the presence of molecular oxygen for activity.

Biodegradation may also be hindered by a lack of other factors that are essential for growth, such as water, phosphorus, and nitrogen.

Biodegradation of oil in marine waters has, on a small scale, been found to increase when the oil was seeded with nitrogen- and phosphorus-containing nutrients,[7] and the same has been found for oil degradation after landfarming. (Raymond et al., 1976; Dibble and Bartha, 1979). Attempts to stimulate the biodegradation of oil in ground water by injection of nutrients and oxygen have been described.[7,33]

In soils, the availability of organic compounds may be severely decreased by the adsorption to or embedding within particulate or colloidal matter. This results in a microenvironment which precludes microbial attack due to inaccessibility of the substrate, particularly in clay and humus-rich soils.[1]

Adsorption, however, may also promote biodegradation when this causes the concentration of a compound to fall below the level at which it may inhibit microbial growth. In general, humidification, mixing of the soil, and dispersion of the chemical may stimulate biodegradation rates when poor turnover is caused by the inaccessibility of substrates.[1]

Inhibition of microorganisms and extracellular enzymes by environmental factors may also cause poor decomposition rates. These factors include acidity, high salt concentrations, extreme temperatures, and the presence of toxic compounds, such as heavy-metal ions.

Apart from the contaminant, contaminated soils may contain ''natural'' organic matter. The latter is known to have both stimulating and inhibiting effects on biodegradation. Stimulation is caused by:

- Supplying an extra carbon source for the development of a population of degraders[14,17,32,41]
- Cometabolism (or ''analogue metabolism'') causing the degradation of related slowly degradable compounds[31]
- Aiding the induction of relevant enzyme systems[14,30]

and inhibition by:

- Competitive inhibition, e.g., by competition for nutrients and growth factors[28]
- Diauxy leading to the preferential metabolism of easily degradable compounds[8,14]

Diauxy, in particular, may play an important role in the unexpected ''persistence'' of some xenobiotics.

In microbial decontamination of a particular soil it is of primary importance to establish which environmental factors limit biodegradation, because this enables the choice of an adequate treatment.

D. Strategies for Stimulated Biodegradation

In order to discuss the application of stimulated biodegradation for the cleanup of hazardous wastes and soils that are contaminated with chemicals, it is necessary to distinguish between current and potential (future) strategies. Processes that are currently used for the biological treatment of chemical and other wastes include:

1. Aerobic or anaerobic digestion in reactors or waste water purification plants
2. Semi-aerobic degradation in lagoons and ponds
3. Aerobic or anaerobic "composting", which is mainly applied for municipal wastes
4. Landfarming and application of wastes

Aerobic or anaerobic digestion in reactors or waste water purification plants — Aerobic biological treatment is used for the purification of both domestic and industrial waste waters. A well-adapted activated sludge is able to degrade many compounds such as nonchlorinated solvents, detergents, and quaternary ammonium compounds. Substances such as chlorinated solvents normally do not degrade.[45] Nondegradable compounds, e.g., heavy metals, which are toxic to microorganisms, may in higher concentrations inhibit the biodegradation process.[6] Treatment of chemical waste is, therefore, limited to wastes which are soluble at the mg/ℓ level or can be efficiently dispersed and which will not inhibit biodegradation. Anaerobic treatment of waste water is limited to concentrated fermentable effluents. Anaerobic dehalogenation is, however, known to occur and may, therefore, be used as part of a treatment procedure.

Semi-aerobic degradation in lagoons and ponds — Treatment of domestic waste in lagoons is applied in arid regions. It is not likely that this technique will be feasible in the Netherlands.

Aerobic or anaerobic "composting" — Composting is a well-known technique for the treatment of solid municipal waste. Decontamination of soil may be carried out in the same way (e.g., during temporary storage of excavated soil).

Landfarming and application of wastes — Landfarming operations are promising for the disposal of easily degradable wastes. It is often necessary to establish a favorable pH and to add mineral fertilizers, while frequent tilling may have to be used to increase the availability of oxygen.[7] Landfarming has already been used for the disposal of oily wastes[11] and degradable compounds, such as those present in gasworks effluents, and simple aromatic hydrocarbons may also be susceptible to this technique. However, soil that is used for landfarming is unsuitable for the production of crops, particularly when persistent chemicals or heavy metals are present. Migration of chemicals may occur, and in general soil application of stable hazardous chemicals is at odds with the cleanup of soils polluted with xenobiotic compounds.

All these techniques rely upon the manipulation of environmental conditions in order to stimulate biodegradation, without the addition of defined or selected cultures. Future techniques applicable to solid wastes and soils may include the following:

1. *In situ* stimulation of biodegradation in soils and underground waters by alterations of environmental conditions or by application of microorganisms
2. Utilization of reactors for the biological detoxification of contaminated solid wastes

***In situ* stimulation of biodegradation in soils and underground waters** — Stimulation of the biodegradation capabilities of soils and underground waters could possibly involve the injection of oxygen to air to obtain aerobic conditions, the injection of microorganisms that are selected for this purpose, and the administration of water, substrates, growth factors, and/or an electron acceptor. Jamison et al.[33] reported the use of forced aeration and nutrient addition to stimulate the *in situ* biodegradation of gasoline in ground water, a process which has been patented.[7]

Utilization of reactors for the biological detoxification of contaminated solid wastes — The use of bioreactors for the decontamination of solid wastes and contaminated soils could be characterized as a biological incineration process. Mixing and application of air or oxygen can easily be achieved, and if necessary, microorganisms, water, and substrates can be added. A process in which total combustion of organics into water and carbon dioxide can be achieved, with the elimination of substituent halogen, nitrogen, and sulfur-containing groups would be most advantageous. The energy costs for the cleanup of solid wastes with a process involving a bioreactor may be considerably higher than for composting, land disposal, or *in situ* procedures. However, a reactor process would be easier to control and a high rate of degradation might be obtainable. Although no experiments with such a reactor have been reported so far, it is conceivable that the operation could be cheaper than the current incineration techniques, which are carried out at high temperatures and require large amounts of energy. In addition to promoting biodegradation, microorganisms may also liberate chemicals from the soil matrix, which could be of advantage for the cleanup of hazardous wastes. The leaching of heavy metals, e.g., from fly ash and harbor sludge, is currently under study. Acid-producing microorganisms such as *Thiobacillus* can produce low pH conditions which result in the dissolving and transfer of heavy-metal ions to a liquid phase, from which they may be recovered. The technique may be integrated with temporary storage.

E. Application of Special Microorganisms for the Degradation of Xenobiotics

In situations where microbial activity towards polluting chemicals occurring in soils is low, it may be beneficial to add microorganisms with the required capabilities in the form of activated sludge, mixed cultures, or selected laboratory strains. In general, organisms capable of degrading simple hydrocarbons such as alkanes, cycloalkanes, substituted benzenes, and phenols (i.e., compounds that are frequently observed as environmental pollutants in oily wastes and gasworks wastes) are present in soils in sufficient amounts and thus do not have to be added. If they are not present, it is possible to increase microbial activity by the addition of adapted sludge or freeze-dried inocula. Organisms that can degrade poorly degradable pollutants are either present in soils in very low amounts or totally absent. In such cases it will be necessary to seed with specially selected organisms that have greater capabilities regarding the degradation of xenobiotic chemicals than naturally occurring strains.

It has been demonstrated in laboratory experiments that the addition of microorganisms to soils may stimulate the conversion of xenobiotic compounds[13,20,22] (see Table 10). Only a few experiments concerning the use of microbial inocula for biodegradation of oil under field conditions have been reported.[7]

However, from the results obtained so far it can be expected that the use of laboratory strains for the cleanup of soils contaminated with recalcitrant pollutants is possible, provided microbial cultures with sufficient activities can be obtained. Microbial enzymes obtained from selected cultures have also been considered for use in procedures such as pesticide container cleanup and soil-spoil cleanup.[43]

The evolutionary development of degradative pathways is dependent on the selection of strains with favorable capabilities that are formed during genetic recombination and mutation processes.

On the basis of these principles, it is possible to isolate strains with increased activities in the laboratory. Classic and modern genetic techniques enable a more direct alteration of heritable characteristics, and microbiological techniques such as continuous cultivation in a chemostat allow the isolation of desired strains under strongly selective conditions.[18] Improved strains may have developed new enzymatic activities by an alteration of substrate specificities or induction pattern. Furthermore, alterations in regulatory mechanisms may enable the formation of higher levels of degrading enzymes or allow catabolic systems to be formed in the absence of high levels of inducers.[18,36]

Since such laboratory strains with increased degradative capabilities may be of great value for cleanup purposes, it is essential to isolate microbial cultures that are capable of growing on recalcitrant xenobiotic compounds or of metabolizing them to harmless products.

Attention should particularly be paid to the isolation (or construction) of strains that can degrade industrial chemicals such as volatile halogenated hydrocarbons, chlorobenzenes, chlorophenols, nitroaromatics, nitriles, and related compounds. The biodegradation of pesticides has been the subject of intense investigation since it became apparent that many of these compounds may persist in the environment and accumulate in food chains.

Unfortunately, much less attention has been paid to the above chemicals, which are precursors or intermediates in the industrial synthesis of various petrochemical products. The latter compounds in particular are often found to be environmental pollutants or are encountered in waste streams.[49]

Halogenated C1 and C2 compounds, hexachlorocyclohexanes, chlorobenzenes, and chlorophenols are often found as pollutants in the Netherlands.

F. Biological Degradability of Important Environmental Pollutants

The realization of biological treatment for the degradation or detoxification of chemical compounds requires that during the purification process microorganisms are present that can metabolize the compounds concerned. Therefore, a description of the susceptibility of some important environmental pollutants towards microbial transformation is given below. Detailed information is presented in Appendix A.

1. Crude Petroleum: Refinery Products and Aromatic Hydrocarbons

Crude petroleum is composed of a saturated hydrocarbon fraction (e.g., alkanes and cycloalkanes), an aromatic hydrocarbon fraction (e.g., benzene, naphthalene, and numerous substituted derivatives), and an asphaltic fraction, which includes several nitrogen- and sulfur-containing compounds. Refined oils contain little or no asphaltic components.

Alkanes and (substituted) cycloalkanes are known to be biodegradable compounds under aerobic conditions.[8,21,35] Molecular oxygen is required for hydroxylation (the first metabolic step), while further oxidation may be coupled with the reduction of nitrate. Branched alkanes are less susceptible to degradation than *n*-alkanes.

The degradation of aromatic hydrocarbons (benzene, toluene, xylenes, trimethylbenzenes, ethylbenzene, and naphthalene) is also only known to proceed under aerobic conditions. Microbial enzymes that require molecular oxygen (oxygenases) are involved in the conversion of side chains and the fission of the aromatic ring. Detailed information about the biochemistry of the catabolism of both aromatic and aliphatic hydrocarbons by microorganisms is available. Although some reports about the anaerobic degradation of hydrocarbons have been published, this process is of little environmental importance and "at best proceeds at negligible rates in nature".[8]

Little information is available about the biodegradation of the more polar components of crude petroleum. Low boiling fractions containing aliphatic sulfides and mercaptans may be degradable, but the metabolic routes are unknown.

The ability to degrade hydrocarbons is widely distributed among microbial populations,[8] and hydrocarbon utilizers are generally present in soils, waters, and sediments. The presence of oil generally causes an increase in the population of hydrocarbon-utilizing microorganisms.

2. Miscellaneous Nonhalogenated Aromatics

The microbial degradation of aromatic compounds such as phenol, cresols (present at gasworks sites), and phthalates (plasticizers) has been demonstrated. Degradation occurs under aerobic conditions, but may also proceed under anaerobic conditions by nitrate respiration (e.g., phenol) or during methanogenesis (benzoates).[23]

Several microorganisms that rapidly utilize phenol, cresols, or benzoates for aerobic growth have been obtained in pure culture and the metabolic pathways have been studied.

Phthalate esters are less susceptible to microbial attack and are poorly degraded in various environments. However, some microbial cultures that are capable of degrading these chemicals have recently been obtained.

3. Volatile Halogenated Hydrocarbons

Most halogenated C1 and C2 hydrocarbons are poorly degradable in the environment and persist for long periods of time. Degradation is mainly dependent on evaporation followed by photolysis.

Microorganisms that can degrade these compounds are not generally present in the environment, although some biological conversion has been observed under laboratory conditions. (Leisinger, 1981; Motosugi and Soda, 1983). Chloroform and some chlorinated ethylenes may undergo slow reductive dechlorination in methanogenic cultures in the absence of oxygen and at a low redox potential. Furthermore, microorganisms that can aerobically degrade dichloromethane and 1,2-dichloroethane have recently been obtained in pure culture.

4. Halogenated Aromatics

The degradation of chlorinated benzenes and phenols, which are important industrial chemicals, is not well understood. (Motosugi and Soda, 1983). In general, recalcitrance increase with increasing chlorine substitution. The chlorinated phenols and benzoates are less recalcitrant than chlorinated benzenes. Degradation of chlorobenzenes, chlorophenols, and chlorobenzoates may occur under aerobic conditions.

Only a few microorganisms that can degrade mono- and dichlorophenols and benzoates have been obtained in pure culture and the metabolism of the compounds has been studied. Chlorobenzoates with one or two chlorine atoms can also be degraded under anaerobic conditions in methanogenic cultures. The PCBs are extremely resistant to microbial transformation; only cultures that can degrade the less chlorinated components have been described. This is in agreement with the relatively low levels of these components in environmental samples as compared to the higher chlorinated PCBs.

5. Pesticides

HCHs and drins are the main pesticides that are found as soil pollutants in the Netherlands. The drins (aldrin, dieldrin, and telodrin) are extremely recalcitrant and are hardly degraded by microorganisms in the environment. Laboratory experiments have yielded some cultures that can partially transform these compounds under aerobic conditions. HCHs are somewhat less resistant to microbial transformation, particularly under anaerobic conditions; anaerobic bacteria that can degrade these compounds are known.[38]

V. MICROBIAL SOIL DECONTAMINATION TECHNIQUES

There are essentially two ways to decontaminate polluted soils using microbiological methods. These are

1. *In situ* decontamination, without excavation of the soil
2. On- or off-site decontamination, after excavation of the soil which may be redeposited after decontamination

The choice of the technique to be used in any particular case will be based on a number of considerations. The most important of which have already been discussed in Section III. The techniques which may be considered are summarized below.

A. *In Situ* Decontamination

1. Landfarming

This technique is applicable when the contamination is concentrated in a top soil layer having a maximum depth of 50 cm. Sufficient oxygen is provided by regularly plowing or harrowing the soil layer. Nutrients are added by conventional methods. The costs of treatment depends on (among other factors) the time scale required for sufficient decontamination, and will be approximately Dfl. 5.00/ton of soil per year. Additional costs for containment, etc. are not included.

2. Bioextraction

Bioextraction is a process by which contaminants present in soil are biodegraded by adding biomass, nutrients, and air. Drainage pipes are installed into the soil body and water containing the necessary compounds is recirculated. The applicability is restricted to soils which are impervious to water. Costs of treatment are very site-specific and can hardly be estimated in general. Treatment costs of Dfl. 2.00/ton of soil per year have been mentioned for a soil body contaminated to a depth of 5 m. Additional costs for containment and chemicals (nitrate, etc.) are not included.

B. Decontamination after Excavation of the Soil

1. Landfarming

The main difference from *in situ* use of this technique is that in the present context landfarming is carried out at a suitable location elsewhere. In this way external conditions can be influenced more widely. Costs of treatment are strongly determined by the costs of transportation of the contaminated soil from the site to the treatment area. For a transportation distance of 50 km the costs of treatment are estimated to be Dfl. 70.-/ton. This is exclusive of additional costs for buying the necessary site, installing impermeable layers, etc.

2. Composting-Like Techniques

Composting has been applied for treatment of wastes, such as domestic waste, sewage sludge, or piggery waste, that contain compostable organic compounds.

Contaminated soils may be advantageous mixed with compostable materials prior to treatment. It is expected that the usual composting techniques can be carried out. The most relevant processes are as follows.

a. Open Systems

Windrow system — Soil is piled up in mounds, which are periodically turned to facilitate aeration. Costs of operation are dependent on the time scale required for degradation and can be estimated as Dfl. 30.-/ton, exclusive of transport costs.

Beltsville system — As in the Windrow system, soil is piled up in mounds. Aeration takes place, however, by introduction of air via perforated pipes passing through the mound. Costs of operation are again estimated as Dfl. 30.-/ton, exclusive of transport costs.

b. Closed Systems

The operation costs for closed systems depend strongly on the process conditions needed but can be estimated to Dfl. 50.-/ton, exclusive of transport costs.

Dano system — In this system soil is treated in horizontally rotating drums; part of the treated material is recycled.

Kneer system — In this system soil is passed downward through a column while air is blown in counterflow through the column.

Schnorr system — This system is similar to the Kneer system, although the column is divided into stages by horizontal plates. The soil is periodically transferred to a lower stage.

Triga system — This system is similar to the Kneer system, although the column is divided into compartments by vertical separation walls.

c. Process Industrial Systems

These include reactors for solids that are used in the process industry for carrying out physical and chemical operations. If the residence times needed are small (maximum of several days), then these systems may be suitable for decontamination of soils. Costs of operation, exclusive of transport costs, are comparable to those for closed systems.

All the techniques given above are described in more detail in Appendix B. This appendix also contains information on applicability, operating costs, and specific restrictions for the various techniques.

VI. EMISSION CONTROL MEASURES ACCOMPANYING MICROBIOLOGICAL SOIL DECONTAMINATION

A number of emission control measures may be needed when microbiological decontamination of soil is carried out. These techniques include the use of installations to avoid specific emissions and the monitoring of specific parameters. The following activities may be necessary.

A. Installation of Barriers

Barriers are installed in the soil to isolate the contaminated body from the surroundings. Vertical or horizontal barriers of either natural or synthetic materials are possible. Well-developed techniques are available now.

B. Drilling of Wells

Drilling of injection and/or drainage wells prevents contaminated ground water from reaching the surrounding soil.

C. Collection of Gases

Gases are collected when gases are emitted in concentrations which are considered to be hazardous. Moreover, they must be removed from the soil when the concentrations of gases in the soil body are so high that they inhibit biodegradation. Controlled transport of gases can be achieved by:

- Ventilation through horizontal and vertical perforated pipes placed in the soil body.
- Ventilation by trenches; these trenches are dug in the soil in order to create preferent transport directions of gases within the soil. The use of trenches may be combined with that of pipes.
- Gas barriers; horizontal barriers to prevent uncontrolled emissions of gases to the air may be applied. Clay, concrete, or synthetic materials can be used for constructing these barriers.

The well-developed techniques described are available now.

D. Treatment of the Gases Collected

Several techniques are available for further treatment of the gases in such a way that environmentally acceptable products are formed.

E. Monitoring Systems

Monitoring systems may have to be installed for:

- Analytical monitoring of the decontamination process
- Warning of sudden modifications in the process
- Protecting the ground water against pollution
- Warning of emissions to the air

Appendix C describes the various emission control techniques which are available, and gives an estimation of the costs.

VII. NONMICROBIOLOGICAL SOIL DECONTAMINATION TECHNIQUES

Nonmicrobiological soil decontamination techniques are as follows.

A. *In Situ* Treatments
Extraction — Extractant is distributed over the surface of the soil and subsequently recovered via wells. The extractant may be reused after treatment again.
Chemical treatment — A chemical species which reacts with the contaminants is added to the soil.

B. Treatment of the Soil after Excavation
Extraction — The soil is mixed with a suitable extractant which can be regenerated and reused. The extractant can be either water-based or organic. This technique is being developed at the moment.
Thermal treatment — Three processes are possible:

- Volatilization of the contaminants by direct heating by combustion gases; the contaminants are carried out of the soil by the exhaust gases.
- Volatilization of the contaminants by an indirect heat transfer via heat exchangers; in this case stripping gas must also be carried through the soil.
- Thermal destruction or incineration of the contaminants; in general, temperatures of 700 to 1200°C are necessary.
- Steam stripping.
- Chemical treatment.

A comparison of these various techniques regarding stages of development, field of application, energy consumption, and by-products is given in Table 11.
The treatment costs can be very roughly estimated as follows:

1. *In situ* treatment: not yet estimatable
2. Treatment after excavation:

   - Extraction: Dfl. 100 to 200/ton of soil
   - Thermal treatment: Dfl. 100 to 300/ton of soil
   - Steam stripping: Dfl. 100 to 200/ton of soil
   - Chemical treatment: Dfl. 30 to 100/ton of soil

A detailed description of the nonmicrobiological techniques and applicability and stage of development is given in Appendix D.

Table 11
COMPARISON OF NONMICROBIOLOGICAL DECONTAMINATION TECHNIQUES

Technique	Stage of development	Field of application	Consumption of chemicals	Consumption of energy	By-products
In situ					
Extraction	·	·	··	···	···
Chemical treatment	·	·	·	···	····
After excavation					
Extraction	··	····	··	····	·
Thermal treatment	···	····	····	·	···
Steam stripping	·	·	····	·	···
Chemical treatment	·	·	·	····	····

Note: The number of dots is a measure of the favorability of the judgment.

VIII. A COMPARISON OF MICROBIOLOGICAL SOIL DECONTAMINATION TECHNIQUES

In this section the various techniques for biotechnological decontamination discussed in Section V are compared according to the following criteria:

- Possibilities for the control of process conditions such as temperature, degree of humidity, oxygen, and nutrient concentrations
- The stage of development, the need for research prior to practical use, and the chance that practical use will be possible
- Biodegradation products
- Expected biodegradation rates which determine the size of equipment and facilities needed (investments)
- Important limitations in use
- Running costs, excluding investments and costs for, for example, analytical and toxicological monitoring

It is assumed that the contaminant under consideration is in part biodegradable; this is, of course, a prerequisite for biological decontamination. The following techniques are compared in Table 12:

1. *In situ* decontamination:

    a. Landfarming
    b. Bioextraction

2. Decontamination after excavation:

    a. Landfarming
    b. Composting-like techniques — open systems (Windrow and Beltsville) and closed systems (Dano, Kneer, Schnorr, and Triga)

Table 12
COMPARISON OF BIOLOGICAL DECONTAMINATION TECHNIQUES CRITERION

Technique	Possibilities for control of process conditions	Stage of development	Maximum periods of residence (realistic estimates)	Residues	Limitations in use	Most important limitations	Cost estimate (Fls./ton[a])
In situ treatment							
Landfarming[b]	·	·	Several years[c]	None	·[d]	Depth of contamination <50 cm	<Dfl. 10.-
Bioextraction	··	·	Up to 8 months	Waste water	··[e]	Type of soil and climate	<Dfl. 10.-
Treatment after excavation							
Landfarming	··	···	One to several years	None	···;	Large surface area required	Dfl. 50—75.-
Composting-like techniques[f]							
open systems							
Windrow	···	···	Months		···	Restricted turning up	Dfl. 75—125.-
Beltsville	···	···	Months	Compost-like[g]	···		
closed systems							
Dano	···	··	Days		··[h]	Restricted capacity	Dfl. 100—150.-
Kneer	···	···	1 day—1 week		··[h]		
Schnorr	···	··			···[h]		
Triga	···	··			··[h]		

a Rough estimates based on a good functioning of the various systems.

b This technique has been widely used for decontamination of soil with oil and oil products. No experience is yet available for processing other types of contamination.

c In some cases the time available may even be unlimited.

d Only applicable if sufficient time is available.

e Cannot be applied if the soil is inhomogeneous or impermeable. The possibilities are, therefore, restricted if much clay or peat is present in the soil.

f Much experience in the use of composting-like techniques for processing of domestic refuse is available. These techniques have not yet been tested for processing suitability for contaminated soil.

g If compostable material is added to the soil before processing, a compost-like end product will be produced.

h Because the approach is highly dependent on the type of process, the size of the plant available restricts the capacity. The process can, therefore, only be applied if the residence time of the contaminated soil in the system is not too great (longer than a few days). Once degradation has been started it may be completed in an open system.

Note: The number of dots is a measure of the favorability of the judgment.

The results can only be qualitative or semiquantitative because only little information on the nature and quantity of the contaminants is available; process conditions may, therefore, vary considerably from case to case.

Landfarming after excavation offers good prospects on both economic and technical grounds; it can be carried out in greenhouses when optimum process conditions are needed or when dangerous gases may be emitted. *In situ* landfarming is limited to those cases where only the first 0.5 m of the soil is contaminated. With special equipment soil may be treated to a greater depth.

A composting technique (using either the Schnorr system or composting in the open) appears to be a good alternative; with the Schnorr system, in particular, conditions (temperature, humidity, air flow, etc.) can be controlled and optimized.

IX. A COMPARISON OF MICROBIOLOGICAL AND NONBIOLOGICAL SOIL DECONTAMINATION TECHNIQUES

At the moment over 700 locations with chemical contamination of the soil have been identified. These locations have been divided into the following groups:

1. Solvents	→ 120 locations
2. Biocides	→ 40 locations
3. Mineral oils	→ 150 locations
4. Gaswork sites	→ 70 locations
5. Heavy metals	→ 200 locations
6. Others (tar products, fly ash, cyanides, etc.)	→ 150 locations

If a decision to rehabilitate the site is taken then a choice from the following alternatives can be made:

● Containment of the polluted soil, combined if necessary, with ground water drainage and treatment
● Removal of the polluted soil for temporary or permanent storage
● Microbial, physical, or chemical decontamination of soil (*in situ* or after excavation)

In those cases where heavy metals are the dominating contaminants (5) microbial decontamination is not appropriate; this probably also applies for the group 6 (the "other" compounds). Microbiological decontamination methods may be basically suitable for the groups 1 through 4. A number of criteria will determine whether microbiological techniques are to be preferred to physical or chemical techniques, in these cases, and which of the alternatives to biological treatment offers the best prospects. These criteria which will have to be considered separately for each case concluded the following:

● The stage of development, the need for research prior to practical use, and the chance that practical use will be possible
● Potential area of application
● Amount of additives required
● Amount and nature of the residues produced

- Results expected regarding the soil decontaminated
- Limitations in use, including aspects that are specific to each case of decontamination
- Costs

The following potential techniques are compared in Table 13:

1. *In situ* treatment

    a. Biological (landfarming and bioextraction)
    b. Physicochemical (extraction and chemical treatment)

2. Treatment after excavation

    a. Biological (landfarming and composting-like techniques — open and closed systems)
    b. Physicochemical (extraction, thermal treatment, steam stripping, and chemical treatment)

APPENDIX A: BIODEGRADATION AND METABOLISM OF IMPORTANT ENVIRONMENTAL POLLUTANTS*

I. INTRODUCTION

The inventory of soil contamination cases (Section II) shows a prevalence of certain classes of compounds in these cases. For that reason information on the microbial transformation and degradation of these compounds is given in this appendix as background information for the feasibility of microbiological decontamination of soils.

II. AROMATIC COMPOUNDS

A. Aromatic Hydrocarbons, Phenols, and Phthalates

Aromatic hydrocarbons and related compounds are a frequent cause of environmental pollution. They are widely used in industry and are often present in effluents and solid wastes.

1. Compounds such as benzene, toluene, ethylbenzene, and xylenes are used as organic solvents in the petrochemical, print, and paint industries.
2. Oil refinery products contain aromatic hydrocarbons: benzene, toluene, ethylbenzene, *iso*-propylbenzene, more complex alkylbenzenes, xylenes, and trimethylbenzenes. These are present in combination with aliphatic hydrocarbons and in some cases with polycyclic aromatic hydrocarbons, tar-like products, and sulfur- or nitrogen-containing asphaltic compounds.
3. Coal-tar products and wastes from coke ovens and gas plants may contain phenols, cresols, ethylphenols, xylenols, and also polycyclic aromatic hydrocarbons and several substituted benzenes.
4. Aromatic compounds are used in chemical synthesis. Phthalate esters are used as plasticizers in the production of plastics. Phenols are precursors of a variety of dyes, pharmaceuticals, odors, etc.

---

* This appendix was prepared by D. B. Janssen.

Table 13

COMPARISON OF MICROBIOLOGICAL AND NONBIOLOGICAL SOIL DECONTAMINATION TECHNIQUES

Technique	Stage of development	Potential area of application	Additives required	Energy required	Residues	Results	Restrictions	Costs
In situ treatment								
Biological								
Landfarming	.	.[a]	...	...	...	—	.[b]	...
Bioextraction	.	..[a]	...	...	...	—	.[c]	...
Physical/chemical								
Extraction	.	.	.	...	..	Dead soil	.[c]	...
Chemical treatment	.	.	.	...	...	—	.[c]	...
Treatment after excavation								
Biological								
Landfarming	..[d]	..[a]	...	...	...	—	..[c]	...
Composting-like techniques								
Open system	..[f]	..[a]	..[g]	...	..[h]	Soil + compost	...	...
Closed system	..[f]	..[a]	..[g]	...	..[h]	Soil + compost	...[j]	..
Physical/chemical								
Extraction	..	..[j]	.	..	.	Dead soil	...[k]	..
Thermal treatment	..[l]	..	...	.	...	Dead soil	..[m]	.
Steam stripping	.	.	...	.	.	—	.[m]	.
Chemical treatments	.	.	.	..	...	—	.	..

^a The contaminants must be biodegradable.

^b The contamination must be limited to a depth of about 0.5 m.

^c Cannot be applied if the soil is inhomogeneous or impermeable. The possibilities are, therefore, restricted if much clay or peat is present in the soil.

^d This technique has been widely used for decontamination of soil with oil and oil products. No experience is yet available for processing other types of contamination.

^e A large surface area is probably required.

^f Much experience in the use of composting-like techniques for processing domestic refuse is available. These techniques have not yet been tested for suitability for contaminated soil.

^g Bark or a compostable material may have to be added to facilitate aeration.

h If compostable material is added to the soil before processing, a compost-like end product will be produced.

i The reaction periods needed must not be too long.

j Basically there is a wide field of application, although drainage problems may arise, if soil has a high content of clay or peat.

k Clay or peat must not be present in large quantities.

l This technique has already been applied in the Netherlands.

m Thermal volatilization, incineration, or degradation of the contaminants must be possible.

Note: The number of dots is a measure of the favorability of the judgment.

Table 14
DEGRADATION OF AROMATICS BY MICROORGANISMS

Compound	Organisms	Ref.
Benzene and toluene	Several *Pseudomonas*	75
Toluene	*Nocardia corallina*	67
Xylene	*Pseudomonas*	75
Xylenols	*Pseudomonas*	67
Phenol	*Pseudomonas* and *Brevibacterium*	75
Cresols	*Nocardia, Candida tropicalis,* and *Bacillus stearothermophilus*	67
o-Diethylbenzene	*Pseudomonas*	82, 84
Propylbenzene	*Pseudomonas*	
Ethylbenzene	*Pseudomonas*	
Phthalates	*Micrococcus* and *Nocardia*	94
Phthalate esters	*Micrococcus* and *Nocardia*	71
Phenol and cresols	*Trichosporon cutaneum*	67
Toluene	*Trichosporon cutaneum*	
Phthalates	Denitrifying cultures	53
Phenol	Denitrifying cultures	56

1. Biodegradation of Aromatic Compounds

Since compounds that contain the benzene ring as a structural unit are widely distributed in nature, microorganisms capable of growth on aromatic compounds can easily be selected from soil samples. The majority of the monocyclic aromatic compounds is biodegradable by soil microorganisms, and these organisms seem to be generally present in soils. Therefore, compounds such as benzene, toluene, xylene, phenols, and cresols can be regarded as degradable. The character of the substituent group on the benzene ring may strongly increase or reduce biodegradation rates. Phthalates and phthalate esters are more recalcitrant, and the latter may persist in soil for years. Biodegradation of aromatic hydrocarbons is an aerobic process, which requires molecular oxygen. Hydroxylated aromatic compounds and phthalates may also be degraded by anaerobic oxydation coupled to nitrate respiration.

2. Microbiology

A number of reports describing organisms that are capable of aerobic growth on aromatic compounds have been described (Table 14).

From the literature, it appears that some strains may be particularly suitable for the degradation of certain classes of aromatic compounds. *Pseudomonas putida* mt-2 degrades toluene and xylenes, and the extensive biochemical and genetic characterization of *P. putida* could be of value for further improvement of the properties.[118] *P. putida* NCIB10015 is able to convert phenols and the three cresol isomers. The organism was isolated by enrichment in medium containing *m*-cresol as the sole carbon source. A *P. acidovorans* strain, 256-1, is able to oxidize several phthalate esters, e.g., diethylphthalate.

Enrichment with phthalates has also yielded *Micrococcus* strains. These could convert a variety of phthalate esters, e.g., diethylphthalate and dibutylphthalate.

Although most *Nocardia* strains show incomplete metabolism of aromatic compounds, some strains have been described that can grow with these compounds as the sole carbon and energy source. *Nocardia* DSM4351, which could be a *Rhodococcus*, may be very suitable for the degradation of phthalates; phenol is also converted by this strain. Both the *meta*- and *ortho*-pathways were found to be present, at well as the gentisate and homogentisate cleavage pathways (see below). *N. corallina* V-49 shows good growth with benzene, toluene, and hexadecane, and thus may also be able to convert *n*-alkyl-substituted aromatics.

Two yeast species show a good ability to degrade aromatic compounds. *Candida tropicalis*

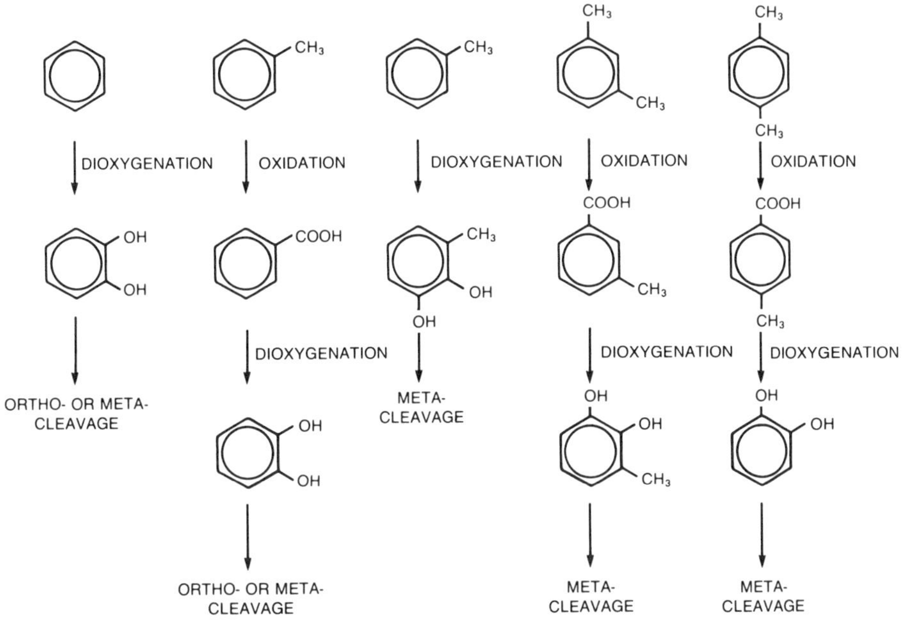

FIGURE 4. Formation of catechols from benzene, toluene, and xylene in *Pseudomonas*.

strains that grow with phenol and cresols have been described, but the properties of *Trichosporon cutaneum* are even more interesting; toluene, xylenes, *m*- and *p*-cresol, methylcyclohexane, and some *n*-alkanes can be used as carbon sources, and phenol appears to be preferred over glucose, as was shown by uptake experiments and sequential utilization in batch cultures.

3. Metabolism of Aromatic Compounds

The aerobic degradation of aromatic compounds proceeds via the formation of dihydroxylated compounds, which are the intermediates that undergo ring fission. Thus, the formation and ring cleavage of catechols are key steps in the metabolism of aromatic compounds.

Toluene and xylene may be degraded to catechol by two routes. Some *Pseudomonas* strains first convert toluene and xylene by oxidative ring fission without oxidation of the methyl group. Other strains first oxidize the methyl to carboxyl, and decarboxylate the benzoic acid formed prior to ring fission, as is the case with strains carrying the well-studied TOL plasmid pWW20, such as *P. putida* mt-2[118] (Figure 4).

Benzene, benzoates, phenol, and cresols can be converted into catechols by dioxygenase reactions, which are often accompanied by the removal of substituent carboxyl groups.

Ring fission in *Pseudomonas* may proceed in two ways: rapidly metabolized compounds are cleaved by intradiol- or *ortho*-cleavage between two carbon atoms each carrying a hydroxyl group. For the degradation of poorly utilized xenobiotics, the alternative *meta*- or extradiol-cleavage pathway may be more important. Finally, after *ortho*-cleavage further oxidation may proceed by a hydrolytic pathway and also by an oxidative route. The oxidative pathway probably functions only in the conversion of 3-carbon-substituted catechols.

The pathway for *ortho*-fission is given in Figure 5. The enzymes of this pathway are highly substrate-specific. From the induction pattern and the substrate specificity of the enzymes, it can be concluded that this pathway is of main importance during the degradation of easily metabolized compounds, such as mandelate, tryptophan, benzoate, and anthranilate. Catechol and derivatives of catechol produced during the degradation of xenobiotics are usually cleaved by the *meta*-pathway enzymes, with the exception of some halogenated

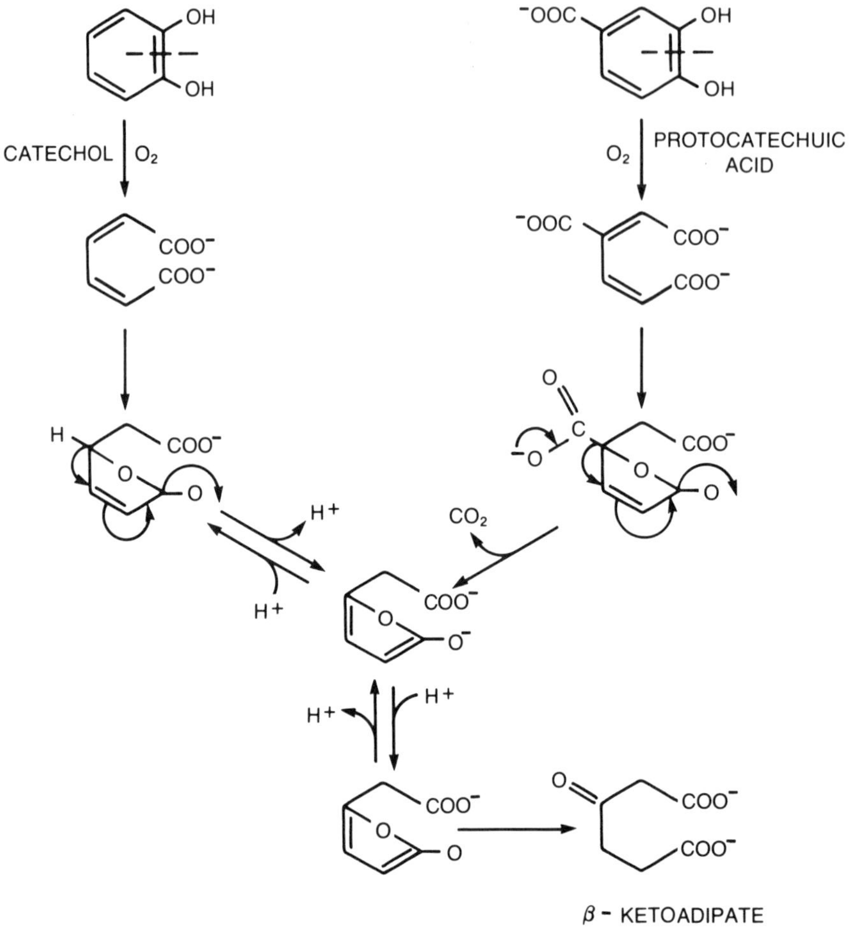

FIGURE 5. *ortho*-Fission of catechols. (From Chapman, P. J., Proceedings of the Workshop: Microbial Degradation of Pollutants in Marine Environments, Bourquin, A. W. and Pritchard, P. H., Eds., U.S. Environmental Protection Agency, Gulf Breeze, Fla., 1979, 28. With permission.)

compounds, and in the case of degradation of aromatic compounds in fungi.[66,69] The reason for this is that catechol-1,2-dioxygenase, the first enzyme of the *ortho*-pathway, has to be induced by high concentrations of *cis,cis*-muconate, the product of the first enzyme. Due to the high specificity of the enzyme, these high concentrations cannot be generated during growth on poorly utilized substrates.

meta-Cleavage is catalyzed by a catechol-2,3-dioxygenase, which is, together with the enzymes, responsible for the production of catechols from naphthalene, salicylate, toluene, and xylenes, a plasmid-encoded enzyme in *Pseudomonas*. Catechole-2,3-dioxygenase is characterized by broad substrate specificity, and it enables cleavage of several substituted derivatives of catechol. For the catabolism of halogenated aromatic (haloaromatic) compounds, however, this is hardly advantageous, since *meta*-fission often leads to dead-end products that still contain the halogen substituent. The product of *meta*-cleavage is 2-hydroxymuconic semialdehyde, or a derivative. As stated above, this compound may be further degraded by two routes:

1.   A hydrolytic pathway, which splits off formate and is followed by aldol-cleavage of 4-hydroxy-2-oxovalerate

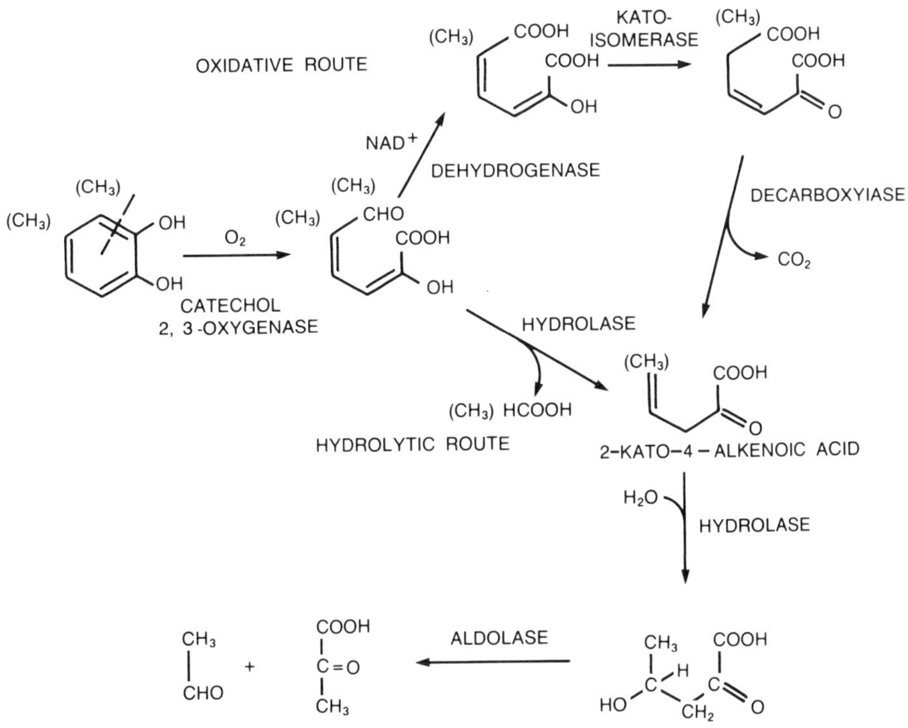

FIGURE 6. *meta*-Fission of catechols. (From Chapman, P. J., Proceedings of the Workshop: Microbial Degradation of Pollutants in Marine Environments, Bourquin, A. W. and Pritchard, P. H., Eds., U.S. Environmental Protection Agency, Gulf Breeze, Fla., 1979, 28. With permission.)

2. An oxidative pathway, which produces 2-hydroxymuconic acid, which is further metabolized by decarboxylation and aldol-cleavage (Figure 6)

The hydrolytic pathway of *meta*-cleavage is preferred during the oxidation of 3-methylcatechol, an intermediate in toluene and *o*-xylene degradation. Most other aromatics are predominantly catabolized by the oxidative pathway, e.g., catechol, 4-methylcatechol, and precursors of 4-methylcatechol: phenol, *p*-cresol, benzene, toluene, and *p*-xylene.

In addition to the catechols, gentisate, 3-methylgentisate, and homogentisate are important ring-fission substrates. The gentisate pathway may be the principal route for the degradation of aromatics in *Nocardia*. The final products are malate and pyruvate.

4. Conclusions

Many aromatic compounds are readily degradable. Phenols and cresols, which represent the mobile fraction and the main pollutants at the sites of old gasworks, are biodegradable, particularly under aerobic conditions. Aromatic solvents such as benzene, toluene, and xylenes are also degradable in the presence of oxygen and sufficient microbial activity. The same holds for the aromatic components of petroleum refinery products. Thus, improvement of biodegradation rates of these classes of pollutants must be achieved by optimization of environmental conditions. These include sufficient aeration, moderate temperature ($\sim 25°C$), and the presence of nutrients, water, and microorganisms.

B. Chlorobenzenes

Chlorinated benzenes, chlorinated phenols, and derivatives of chlorinated benzenes and phenols are important environmental pollutants. These compounds are formed as by-products

Table 15

MICROBIAL DEGRADATION OF CHLOROBENZENE DERIVATIVES

Compounds	Organisms	Conditions	Comments	Ref.
Chlorobenzene	Unidentified bacterium	Ae		104
3- and 4-Chlorobenzoate, 3,5-dichlorobenzoate, and 4-chlorophenol	*Pseudomonas*	Ae		36
2- and 4-Chlorophenol, 2,4- and 2,6-dichlorophenol, and 2,4,6-trichlorophenol	Soil organisms *Pseudomonas* *Alcaligenes*	Ae		55 36 73
Pentachlorophenol	Mixed culture *Arthrobacter*	Ae		15 22
1,2,3- and 1,2,4-Trichlorobenzene	Soil organisms and mixed culture	Ae		98
Monochlorobenzoate and 3,4-dichlorobenzoate	Sewage	Ae		68
2-, 3-, and 4-Chlorobenzoate	Methanogenic consortia	An	Methane produced	111
2,4-, 2,5-, 3,4-, and 3,5-Dichlorobenzoate	Methanogenic consortia	An	Monochlorobenzene produced	111

Note: Ae = aerobic and An = anaerobic.

during the chemical synthesis of pesticides, PCBs, and other haloaromatic compounds. The presence of these compounds in industrial waste waters has caused contamination of Rhine water, and improper disposal has led to soil pollution. Furthermore, chlorobenzoates are intermediates in the degradation of chlorophenoxy herbicides and chlorinated biphenyls, and pentachlorophenol has been widely applied as a pesticide. Chlorobenzenes are used as solvents (e.g., as dye carriers) in great quantities (thousands of tons annually).

1. Biodegradation

Few microbial cultures that are capable of degrading haloaromatic compounds have been described. Complete degradation, as judged by the stimulation of microbial growth, has been demonstrated for some *Pseudomonas*[72] and *Alcaligenes* strains[36] that could aerobically utilize chlorobenzoates and chlorophenols.

An organism that degrades chlorobenzene was recently described.[104] Organisms growing on chlorinated phenoxy herbicides and chlorinated biphenyls, which are degraded via chlorobenzoates, have also been described.[65,69] Pentachlorophenol was found to support growth of a mixed culture[15,117] and a strain of *Arthrobacter*.

A number of halogenated hydrocarbons which are biodegradable, as shown by the disappearance in soil and enrichment cultures, are given in Table 15.

The metabolism of haloaromatic compounds is dependent on the activities of catabolic enzymes that can cope with the steric effects of the substituents, and that can be produced under the appropriate conditions. Furthermore, in order to obtain detoxification a reaction sequence must be followed that allows the removal of the halogen substituent at some stage. In particular the arene-halogen bond in unsubstituted haloaromatics such as chlorobenzene is very stable. Halide elimination under aerobic conditions may occur either before or after ring cleavage of halocatechols, the majority of haloaromatic compounds being catabolized by the latter route. This explains the importance of the *ortho*-cleavage pathway for the catabolism of haloaromatic compounds, because displacement of halogen groups is carried out during lactone formation. Dehalogenation is, from an environmental point of view, the most important step during the conversion of haloaromatic compounds. Growth of microorganisms on phenoxyacetate herbicides does not necessarily require transformation of the halogenated ring structure, and could merely be the result of esterase activity and subsequent utilization of the acetate or glyoxylate produced as a carbon and energy source. Actually, accumulation of chlorobenzoates is frequently observed during the metabolism of organochlorine pesticides by microorganisms. On the other hand, some pathways in which there is an elimination of halides have also been described.

The evaluation of the biodegradation potentials of the organisms described is severely hindered by a lack of quantitative data. For most organisms, it has neither been established which specific growth rates can be obtained in batch or continuous flow cultures, nor have the rates of substrate conversion or dehalogenation been described. Also, data on the tolerance of degrading microorganisms to high concentrations of xenobiotic substrates are lacking.

Several organisms capable of degrading chlorinated phenoxy herbicides have been isolated in enrichment experiments (Table 15). CPA, MCPA, and 2,4-D can be degraded with the release of up to 90% of the original chlorine content. Strains of *Pseudomonas, Flavobacterium, Corynebacterium, Arthrobacter*, and others are found in such experiments. Fungi do not seem to participate in the degradation of 2,4-D. The fact that organisms that utilize 2,4-D are easily isolated from soils is consistent with the low persistence of this herbicide.

Shelat and Patel[107] isolated a *Bacillus polymisea* from sewage sludge which could grow on monochlorobenzene as the sole source of carbon. A *P. putida* initially grown on toluene (25 hr) could oxidize monochlorobenzene to 3-chlorocatechol.[82] *Pseudomonas* strains are also known to degrade dichlorobenzenes.[86,115]

Strains which degrade the substituted chlorobenzoates have also been isolated: *Pseudomonas* B13 and some derivatives of *Pseudomonas* B13 grow on 3-chlorobenzate, as will some other *Pseudomonas* strains. Some of the other isomers could be utilized by strains produced by genetic recombination or by mutation. 3-Fluorobenzoate hydrolysis and oxidation is also mediated by some of the *Pseudomonas* strains.

Chlorobenzoates are important intermediates in the degradation of PCBs.

Alcaligenes euthropus ATCC17697 and some *Pseudomonas* strains are capable of growing on 4-chlorophenol, and a *Pseudomonas* that utilizes 2,4-dichlorophenol has also been identified. Little information on the degradation of other halogenated phenols is available.

So far, only one organism that is capable of complete metabolism of 2,4,5-T has been isolated. This is a *P. cepacia* developed in the laboratory that can bring about 2,4,5-T degradation in soils in laboratory experiments (Table 14).

Our knowledge of the capacity of microorganisms to degrade synthetic chlorobenzene derivatives is at present far from complete. Many more organisms that are capable of degrading haloaromatics such as chlorobenzenes, a number of halogenated phenols, benzoates, and naphthalenes may be obtainable from contaminated environmental samples with the help of continuous culture enrichment techniques. Furthermore, genetic techniques have demonstrated value for the isolation of organisms that can take part in haloaromatic degradation in complex continuous cultures containing a mixture of these xenobiotics. This suggests that such organisms can be of value in sewage systems receiving haloaromatic-containing effluents.[60]

2. Metabolism

Some principles regarding the biochemistry of the degradation of haloaromatic compounds will be outlined below, and illustrated with examples.

Haloaromatic compounds are very toxic to microorganisms. This is generally due to the capability of these compounds to metabolize into halocitrate, which is a potent inhibitor of the tricarboxylic acid cycle. Detoxification of, and growth on, haloaromatics is thus dependent on the removal of the halide ion at some step preceding the formation of haloacetate and halocitrate.

The oxidative catabolism of aromatic compounds is initiated by the insertion of two *ortho-* (or *para-*) hydroxyl groups and the subsequent ring fission by *ortho-* or *meta-*cleavage. The *meta-*cleavage pathway, however, does not result in the degradation of halogenated compounds; although 4-chlorocatechol, for example, can be cleaved by catechol-2,3-oxygenase, this results not in chloride release, but in the accumulation of 5-chloro-2-hydroxymuconic semialdehyde as the dead-end product in the growth medium. The other isomers, 3-halocatechols, are suicide substrates for catechol-2,3-oxygenase, since extradiol-cleavage yields acylhalides, which are potent inhibitors of the enzyme.[36]

This failure of the *meta-*cleavage pathway to detoxify haloaromatics, together with the fact that the majority of aromatic compounds are cleaved by *meta-*fission in most organisms, is responsible for the toxic effects of haloaromatics when they are present in sewage influents. Such an explanation is consistent with the observation that the elimination of *meta-*cleavage enzymes by genetic techniques can increase the capability of organisms to degrade haloaromatic compounds. Almost no *meta-*cleavage activity is detectable in stable communities that grow on haloaromatic compounds.[36]

Elimination of halogen from haloaromatic compounds can occur either before or after ring cleavage. Dehalogenation preceding ring fission has only been rarely observed, e.g., the dioxygenase reaction that converts 2-fluorobenzoate into catechol in *Pseudomonas* strains (Figure 7).

The benzoate dioxygenase preferentially results in 1,2-dioxygenation, although some 5,6-dioxygenation also occurs.

Mutation of the enzyme[36] may result in a further increase in the selectivity for 1,2-dioxygenation, so that about 97% of 2-fluorbenzoate undergoes 1,2-dioxygenation.

Release of halide ions after ring cleavage by *meta-*fission is the most important route for the detoxification of haloaromatic compounds. Initial dioxygenation in monooxygenase reactions result in the formation of catechols, chiefly 3-haloderivatives. Ring cleavage by intradiolfission catalyzed by a suitable catechol-1,2-dioxygenase then follows. This reaction produces a halogen-containing muconic acid. Halide elimination is coupled to the formation of muconolactones and occurs during or immediately after lactonization (Figure 8).

Examples of such a sequence are the conversion of 3-chlorobenzoate in *Pseudomonas* B13,[36] of 4-chlorocatechol in a *Pseudomonas*,[74] and the degradation of 4-fluorobenzoate in a *Pseudomonas* strain.[85]

The herbicide MCPA is also degraded by a similar route in *Pseudomonas*.[81]

3. Enzymes

The degradation of chlorocatechols by *ortho-*fission and further degradation is not catalyzed by the normal *ortho-*cleavage pathway enzymes. The normal catechol-1,2-oxygenase is inefficient for ring cleavage of halocatechols, and the subsequent cycloisomerization (lactonization) and hydrolysis of the dienelactones that are formed after halide release are not carried out by the normal *ortho-*pathway enzymes. It appears that the limited substrate specificity of ordinary *ortho-*pathways enzymes does not allow the acceptance of the halocatechols as a substrate. *Pseudomonas* B13 contains three specific enzymes that convert chlorocatechol into maleylacetate: pyrocatechase II (a catechol-1,2-oxygenase), cycloisomerase II (a lactonizing enzyme), and hydrolase II are induced during growth on 3-chlorobenzoate. The final metabolite of this pathway is maleylacetate, in contrast to the normal *ortho-*pathway, which yields 2-ketoadipate as the key metabolite.

The oxidation of maleylacetate proceeds by one of the plasmid-encoded enzymes that plays a role in chlorobenzoate metabolism:[65] the central metabolism intermediates fumarate and acetate being produced.[36]

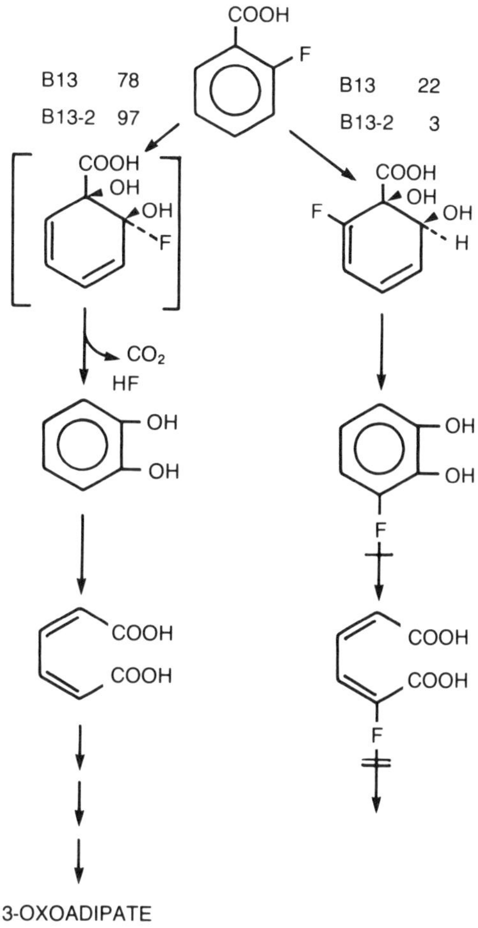

FIGURE 7. Release of fluoride during dioxygenation in *Pseudomonas*. (From Knackmuss, H. J., *Microbial Degradation of Xenobiotics and Recalcitrant Compounds* Leisinger, Th., Cook, A. M., Hütter, R., and Nüesch, J., Eds., Academic Press, London, 1981, 189. With permission.)

4. Dihalogenated Compounds

Several organisms are known to grow on dihaloaromatic compounds, these include *Pseudomonas* and *Arthrobacter* strains for which routes for complete degradation have been proposed[64,70] (Figure 9).

A first dehalogenation as described above occurs during or immediately after lactonization. Removal of the second halogen has been suggested to be partly dependent on the formation of intermediates in which the carbon-halogen bond is labilized, followed by a spontaneous release of halide. Such intermediates are formed when, for example, chlorinated compounds undergo the same reaction sequence as the analogues of these compounds derived from monohalogenated compounds. However, all the intermediate steps have not yet been demonstrated, and every intermediate has not yet been isolated.[36]

5. Improvement of Strains

Knackmuss[36] has pointed out that the successful application of microbial and genetic techniques increasing the capacity of microorganisms to degrade haloaromatic compounds will depend on the possibility of modifying the rate-limiting steps.

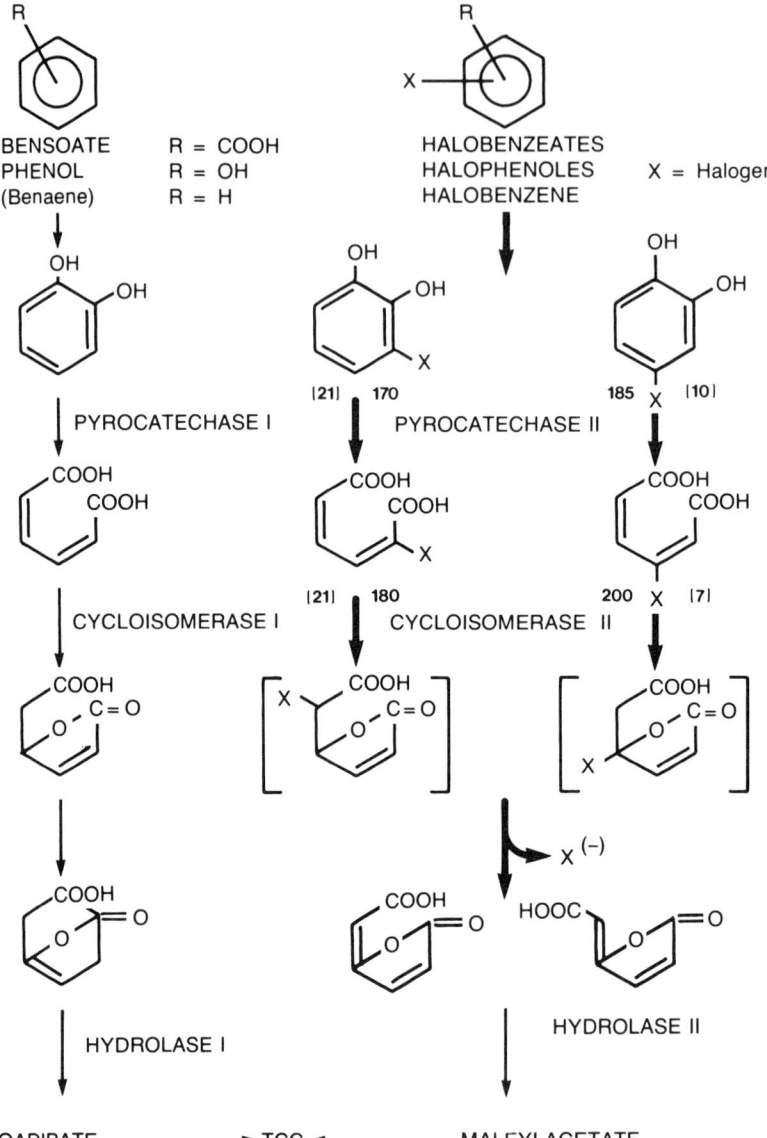

FIGURE 8.   Degradation of halo-substituted aromatic compounds in *Pseudomonas*. (From Knack-muss, H. J., *Microbial Degradation of Xenobiotics and Recalcitrant Compounds*, Leisinger, Th., Cook, A. M., Hütter, R., and Nüesch, J., Eds., Academic Press, London, 1981, 189. With permission.)

meta-Cleavage, which is a dead-end pathway, yields toxic products, and does not include dehalogenation, must be eliminated. All chlorobenzoate-degrading strains isolated do indeed lack *meta*-cleavage activity.

Second, the natural *ortho*-cleavage enzymes must be modified, or new activities must be introduced, to increase the activity of the *ortho*-pathway for halogenated substrates. The normal *ortho*-enzymes have very low activities with halocatechols. *Pseudomonas* B13, which utilizes 3-chlorobenzoate, has different catechol-1,2-oxygenase, cycloisomerase, and hydrolase enzymes from those found in strains which utilize benzoate.[36]

Third, enzymes that mediate the conversion of several haloaromatics to halocatechols must have a broad substrate specificity. For example, the TOL-plasmid-encoded enzyme

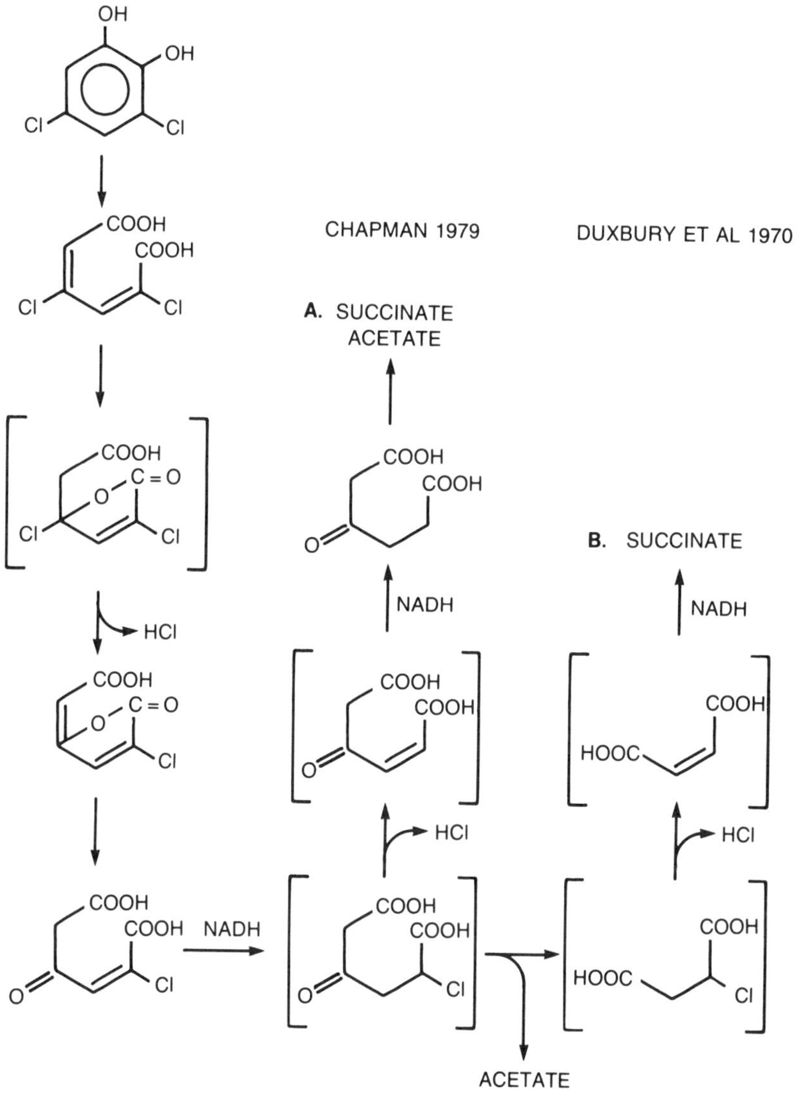

FIGURE 9. Degradation and dehalogenation of 3,5-dichlorocatechol by 2,4-D degrading organisms. (From Knackmuss, H. J., *Microbial Degradation of Xenobiotics and Recalcitrant Compounds,* Leisinger, Th., Cook, A. M., Hütter, R., and Nüesch, J., Eds., Academic Press, London, 1981, 189. With permission.)

that converts toluate into catechol[118] can be acquired by *Pseudomonas* B13 through conjugations, and this enables the organism to utilize not only 3-chlorobenzoate, but also 4-chloro- and 3,5-dichlorobenzoate. The toluate oxidase has a broad substrate specificity and can also convert the latter two compounds into halocatechols.[36]

Alteration of a number of other functions, such as transport, induction pattern, and dehalogenase activities might lead to a further improvement in the capacity of microorganisms to degrade haloaromatic compounds.

6. Genetics

It has recently been shown that the genes specifying bioconversion of 3-chlorobenzoate in *Pseudomonas* are located on a degradative plasmid. The ability of another strain to utilize 4-chlorobenzoate was also found to be mediated by a plasmid.[65] The genes for degradation of 2,4-D are also located on a plasmid.[76]

7. Anaerobic Degradation

It has recently been reported that haloaromatic compounds may be degraded anaerobically by mixed cultures of methanogenic bacteria. Suflita et al.[111] have described complete mineralization of 3-chlorobenzoate, bromobenzoates, and iodobenzoates to methane and carbon dioxide, although 2- and 4-chlorobenzoate were not converted. Complete degradation of 3,5-dichlorobenzoate was also found, whereas 2,5- and 3,4-dichlorobenzoate were partially dehalogenated, and 2,4- and 2,6-dichlorobenzoate not at all. It may well be that degradation of more compounds could be obtained under anaerobic conditions after long-term adaptation. The dehalogenation is apparently a reductive reaction, but the precise mechanism of dehalogenation remains unclear.

8. Conclusions

A number of haloaromatic compounds can be biodegraded and pure cultures have been isolated in the laboratory, and strains with improved capacity for biodegradation have been isolated. These improved strains continuously degrade influent haloaromatic compounds in reactor systems. Furthermore, microorganisms which degrade haloaromatic compounds may be maintained on activated carbon columns; such a system continuously degrades chlorobenzenes. The use of selected microorganisms with extended activities for the degradation of chlorinated benzene derivatives in polluted soils, waters, and industrial effluents is, therefore, a promising prospect.

C. Polychlorinated Biphenyls

Commercial preparations of polychlorinated biphenyls are a mixture of up to 100 different isomers of biphenyls containing one to ten chlorine atoms on the aromatic rings. Up to 2 million tons of PCBs have been produced worldwide since 1929, mainly under the commercial trade name Aroclor®. PCBs have been used in transformators, capacitors, cooling systems, plastics, paper, printing inks, and resins because they are chemically inert and have good temperature stability. Environmental persistence of PCBs[46] has led to restrictions on use and application; improper disposal in the past has, however, led to severe cases of soil and water pollution with PCBs. Furthermore, removal of PCBs still in use in capacitors and transformers is troublesome, and removal from small capacitors is not possible.

1. Biodegradation of PCBs

The PCBs generally present in the environment are the higher chlorinated biphenyls, that is, those containing five of more chlorine atoms per biphenyl molecule. The products manufactured at present are, however, mainly compounds with two chlorine atoms per biphenyl molecule. It is believed that this variation explains why they have higher susceptibility to biodegradation.[80]

This explanation is in agreement with the observation that mono- and dichlorinated biphenyls readily undergo biodegradation in activated sludge systems, whereas tri-, tetra-, and pentachlorinated biphenyls are much more resistant. Furthermore, isolated bacterial cultures which degrade PCBs mainly attack the mono- and dichlorinated biphenyls, whereas the tetra- and pentachlorinated biphenyls are not attacked (Table 16).[77,78]

The effect of chlorine substitution in PCBs on biodegradability can be summarized as follows:

1. Degradation decreases with increasing number of chlorine atoms. Biodegradation is very slow if four or more chlorinate atoms are present.
2. PCBs containing all chlorine atoms on one ring are less resistant than those having the same numbers of chlorine atoms but distributed over two rings.
3. PCB isomers having two chlorine atoms on either ring are less recalcitrant when these chlorine atoms are in the 2,3-position on at least one ring.

Table 16
DEGRADATION OF CHLORINATED BIPHENYLS BY MICROORGANISMS

Compound	Organisms	Comment	Ref.
Di- and trichloro-bi-phenyls	*Alcaligenes* and *Arthro-bacter*		78
p-Chlorobiphenyl	*Achromobacter*		52
Chlorinated biphenyls	*Nocardia, Pseudomonas*	Cooxidation	57
Dichlorobiphenyl	Mixed bacterial culture		17
Mono- and dichlorobi-phenyls	Activated sludge		114
p-Chlorobiphenyl	*Klebsiella*		90

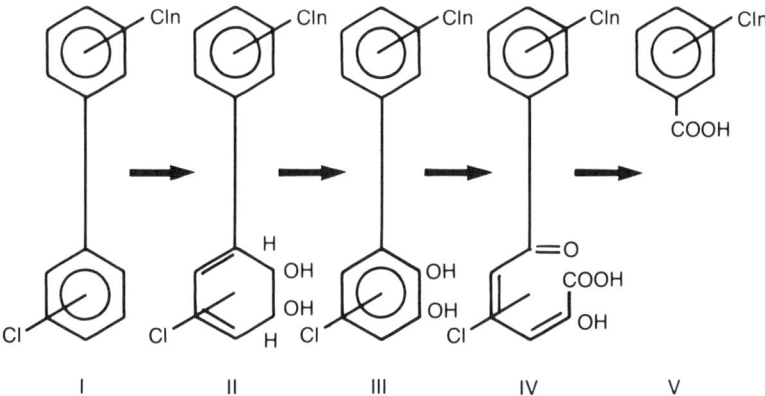

FIGURE 10.   Microbial conversion of PCBs by *Arthrobacter* and *Alcaligenes*. (From Furukawa, K., *Biodegradation and Detoxification of Environmental Pollutants*, Chakrabarty, A. M., Ed., CRC Press, Boca Raton, Fla., 1982, 33.)

4.   The first ring fission preferently occurs on the lesser chlorinated ring.

Degradation of chlorinated biphenyls occurs through the action of aerobic microorganisms and is an oxidative process which requires molecular oxygen. Pure cultures often only degrade PCBs partially, resulting in the accumulation of chlorobenzoates or related compounds. Complete conversion may more easily be obtained with mixed cultures.[80] Cooxidation may play a role in PCB degradation, particularly for the more highly chlorinated components.

2. Biochemistry and Genetics

The first step in the metabolism of PCB isomers by microorganisms is believed to be a dioxygenase-catalyzed reaction leading to the formation of 2′,3′-dihydroxy compounds. This is probably followed by *meta*-fission between C1 and C2, and conversion of the cleavage products to chlorobenzoic acids (Figure 10). Related routes may be followed during the conversion of other PCB isomers. An accumulation of either chlorinated benzoates or hydroxylated derivates of chlorobiphenyls is often seen during the degradation of commercial PCB preparations. Detailed information about the enzymology or genetics of PCB degradation is still lacking. However, it has recently been shown that PCB degradation may be plasmid-determined in at least some bacterial strains. A *Klebsiella* strain capable of degrading 4-chlorobiphenyl,[90] *Acinetobacter* sp. strain A6, and *Arthrobacter* sp. strain M5 were shown to contain catabolic plasmids that harbor genes involved in the degradation of chlorinated biphenyls.[79] The combined use of strains containing plasmids for chlorobenzoate degradation and strains containing plasmids for mono- and dichlorobiphenyl degradation has enabled the complete degradation of some chlorinated biphenyls.[79]

3. Conclusions

Mono- and dichlorinated biphenyls are biodegradable. The higher chlorinated components of PCB mixtures, particularly those with five or more chlorine atoms per biphenyl molecule are recalcitrant and microorganisms that can degrade them are not commonly present in nature. Biodegradation of PCBs is an aerobic process. Cooxidation may play a role, which means that a general increase in microbial activity may stimulate turnover. Incomplete degradation may occur. Recent developments suggest that more active bacterial strains may be isolated or constructed in the laboratory and that these may be useful for biological removal of PCBs.

III. ALIPHATIC COMPOUNDS

A. Halogenated C1 and C2 Hydrocarbons

1. Biodegradation

Only limited information on the biodegradation of chlorinated and brominated C1 and C2 compounds is available, although these chemicals are important environmental pollutants.[103]

Most of these compounds are very resistant to biodegradation, as indicated by extreme persistence in the soil environment.

The halogenated methanes chloroform, bromodichloromethane, and dibromochloromethane may be degradable under anaerobic conditions.[59] In the absence of oxygen and at low redox potential, dehalogenation is a reductive reaction that results in the replacement of a chlorine by a hydrogen substituent. From an environmental point of view, dehalogenation is an important step in the degradation of halogenated organic compounds because in general it results in a loss of, or decrease in, toxicity. No information is available about the mechanism of anaerobic dehalogenation, but it may be a partly nonenzymatic reaction.[37]

Organisms that can grow anaerobically on halogenated compounds have not yet been isolated. Mixed cultures that degrade such compounds under denitrifying and methanogenic conditions have been described.[61,62]

Cultures that are able to degrade halogenated C1 and C2 hydrocarbons under aerobic conditions have been described for dichloromethane and 1,2-dichloroethane.[108] Pure cultures of a *Pseudomonas* strain and a *Hyohomicrobium* strain capable of growth on dichloromethane have been obtained. Dichloromethane degradation proceeds via formaldehyde.[109]

Recently, cultures that utilize 1,2-dichloroethane have been isolated.[89,110] One organism was identified as *Xanthobacter autotrophicus* and could degrade a number of halogenated aliphatic hydrocarbons.[89a]

No mixed or pure cultures that can grow with 1,1-dichloroethane, 1,1,1-trichloroethane, trichloroethylene, or tetrachloroethylene have yet been described.

Immobilized cells of cultures that aerobically degrade dichloromethane have been used for the treatment of liquid streams containing this compound.[105,108] Biodegradation of chloroform, 1,1,1-trichloroethane, and perchloroethylene could not, however, be obtained in a biological column.[60]

Trichloroethylene is also believed to be very stable. The isolation of organisms that can attack all these compounds would be of great value, not only for the cleanup of contaminated soils, but also for the removal of halogenated compounds from drinking water.[112] This may be achieved by incorporating the organisms in the activated carbon columns that are currently used for water treatment.[101] Furthermore, the treatment of effluent wastes with columns or reactors in which suitable organisms are active would enable the biological detoxification of industrial effluents containing volatile halogenated hydrocarbons (Table 17).

Table 17
MICROBIAL DEGRADATION OF HALOGENATED C1 AND C2 HYDROCARBONS

Compound	Microorganisms	Comment	Ref.
Trichloromethane	Activated sludge,	Anaerobic	59
Bromodichloromethane	ground water, and	conditions	
Dibromochloromethane	mixed cultures		
Tribromomethane			
Dichloromethane	Mixed cultures		91
	Pseudomonas		108,109
	Hyphomicrobium		
	Activated sludge		105
Chloroform	Mixed methanogenic		61
Carbon tetrachloride	cultures		
1,2-Dichloroethane			
Tetrachloroethylene			
1,1,2,2-Tetrachloroethane			
1,2-Dichloroethane	Unidentified organism		89
	Xanthobacter		

2. Conclusions

Dichloromethane and 1,2-dichloroethane can be degraded aerobically by microorganisms, but the organisms involved are not generally present in nature. Tetrachloroethylene, chloroform, and carbon tetrachloride may be degradable under anaerobic conditions by methanogenic or dinitrifying consortia. However, organisms capable of rapid turnover of these compounds are at best very scarce in the environment.

B. Hexachlorocyclohexanes

The γ isomer of hexachlorocyclohexanes (also called lindane, γ-HCH) shows high insecticidal activity due to interference with the ionic permeability of nerve cell membranes. Lindane, consequently, has been widely used in agriculture for the control of insect-mediated pests. Commercial preparations of lindane and HCH wastes have been found to contain not only the γ isomer, but also the α, β, and δ isomers. Soil contamination with HCH-containing wastes has occurred at several locations in the Netherlands, e.g., in Twente and in the Volgermeerpolder.

1. Biodegradation

Lindane is generally much less persistent in the soil environment than DDT, aldrin, or dieldrin are. This is attributed to microbial degradation, which is greatest under anaerobic conditions.[38,99] Pure cultures of a number of anaerobic and facultatively anaerobic bacteria have been shown to metabolize lindane and isomers of lindane under anaerobic conditions (Table 18). Anaerobic organisms belonging to the genus *Clostridium* seem to play a major role in lindane degradation.[106]

The γ isomer of HCH is more sensitive to microbial degradation than the other components. Furthermore, α-HCH can be produced from lindane by microbial conversion. This fact, together with the large amounts of α-HCH present in commercial lindane preparations, explains the observation that particularly α-HCH is found as an environmental contaminant. The biodegradation of HCH by mixed or pure microbial cultures often does not lead to complete mineralization. Several intermediates (Figure 11), particularly chlorobenzenes and γ-TCCH (tetrachlorocyclohexane), may accumulate. Complete degradation by anaerobic soil microorganisms has, however, been observed.[38]

Table 18
MICROBIAL DEGRADATION OF HCH

Compound	Organisms	Condition	Comments	Ref.
γ-HCH	*Clostridium rectum* *Bacillus* Enteric bacteria	An	γ-TCCH and chlorine-free metabolites produced	88
α-, β-, γ-, and δ- HCH	Soil microorganisms	An	CO_2 produced	96
γ-HCH	*Clostridium* and *Bacillus*	An	Chlorobenzenes produced	54
γ-HCH	Fungi	An	Chlorinated benzenes and phenols produced	93
γ-HCH	*Clostridium*	An		102

Note: An = anaerobic.

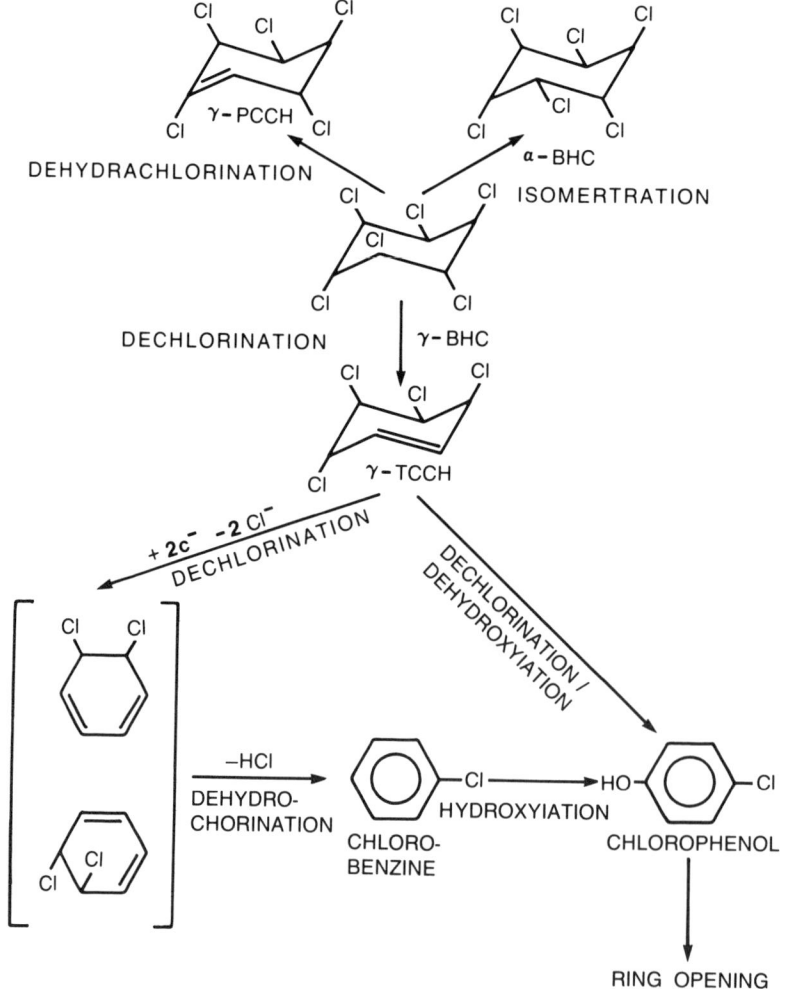

FIGURE 11.   Metabolism of lindane by microorganisms.

Table 19
MICROBIAL CONVERSIONS OF ALDRIN AND
DIELDRIN

Compound	Organism	Products	Ref.
Aldrin	*Aspergillus*	Dieldrin	92
Telodrin	*Penicillium*		
Aldrin	Several soil microorganisms	Dieldrin	113
Dieldrin	*Pseudomonas*	Photodieldrin	95
	Bacillus	*tr*-Aldrindiol	
	Trichoderma		
Dieldrin	*Pseudomonas*	Photodieldrin and other products	100
Dieldrin	*Pseudomonas*	Ketoaldrin	116
	Neurospora		
Dieldrin	*Trichoderma*	CO_2	58
Dieldrin	Soil microorganisms	CO_2	87

2. Biochemistry

The known metabolic conversions of lindane are outlined in Figure 11. The main pathway starts with reductive dechlorination of lindane to γ-TCCH. A second dechlorination yields dichlorocyclohexadiene, which presumably decomposes spontaneously to monochlorobenzene by dehydrochlorination. Ring opening occurs after hydroxylation.[38,99] Sound information on the enzymology of these conversions is still lacking. In some cases internal metabolites such as NADH or FADH may donate electrons for the dechlorination reaction, but other metabolites also may play a role.[38]

3. Conclusions

Microbial degradation of lindane and isomers of lindane is most efficient by anaerobic microorganisms under anaerobic conditions. The formation of α-HCH and the degradation of the intermediates γ-TCCH and chlorobenzenes are of critical importance. Under favorable conditions, complete microbial mineralization and release of all organic chlorine as chloride may occur. This will probably require the use of selected cultures of adapted microorganisms.

C. Aldrin and Dieldrin

The organochlorine insecticide aldrin has been produced in large quantities for the control of pests in agriculture. The improper disposal of industrial wastes has led to severe cases of soil pollution in the Netherlands, e.g., in Schiedam, Gouderak, and Rotterdam (all in the province of South Holland).

1. Degradation and Metabolism

From an ecological point of view, the most important transformation of the insecticide aldrin is the conversion to dieldrin by epoxidation (Table 19). Aldrin is biologically transformed into dieldrin and disappears by volatilization from soil. Dieldrin, however, is persistent and nonvolatile; it has been detected in soil for several years after the application of aldrin. Many different microorganisms, particularly fungi, have been found to be capable of this conversion.[38,99]

The degradation of dieldrin is a slow process. *trans*-Aldrindiol, photodieldrin, and ketoaldrin have been identified as metabolic products (Figure 12). Formation of CO_2 from dieldrin was found after attack by *Trichoderma*[58] and other bacterial and fungal cultures.[87] Information on the pathway of metabolism of *trans*-aldrindiol into water soluble metabolites and CO_2 is not yet available.

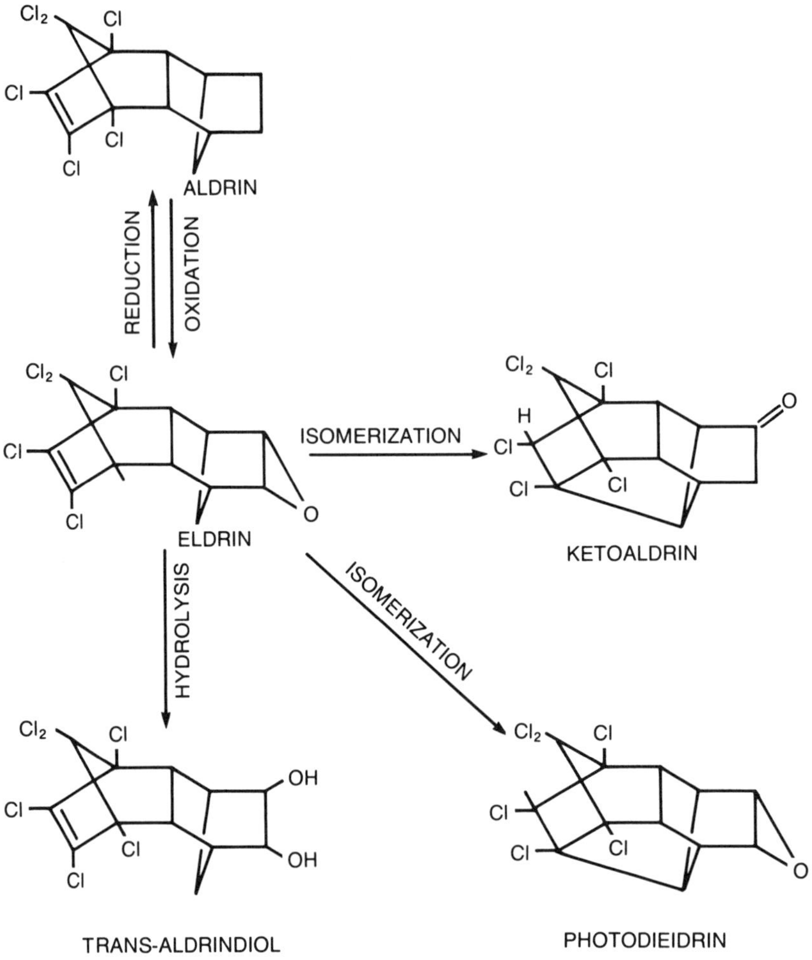

FIGURE 12.   Microbial conversion of aldrin.

2. Conclusions

Although aldrin is not persistent in soil, the microbial conversion product of aldrin, dieldrin, is degraded only very slowly. Dieldrin metabolism has been poorly studied, but it seems that microbial cultures capable of complete degradation can be isolated.

APPENDIX B: MICROBIAL SOIL DECONTAMINATION TECHNIQUES*

I. INTRODUCTION

This appendix describes and discusses various techniques for obtaining biodegradation of organic chemicals in order to decontaminate soil. Whenever possible, the costs of these techniques are (roughly) estimated. These costs are frequently obtained from U.S. literature; conversion of prices into Dutch guilders from U.S. dollars using current exchange rates may, however, lead to overestimation of the real costs under Dutch conditions. It should also be noted that the actual costs of a decontamination operation very much depend on:

---

* This appendix was prepared by W. J. Th. van Gemert and H. J. van Veen.

- The quantity of the soil to be treated
- The nature and the concentration of the pollutants
- The rate of biodegradation which can be obtained
- The degree of purification required
- Investments
- Transport distances

In addition, the costs given do not include those of analyses for process control, preliminary studies and investigations for technique selection, or checking the results.

The final choice of a technique will be determined by the costs, by the aspects mentioned in Section II, and by the limiting conditions given below. These limiting conditions can be divided into conditions imposed by the actual situation and by nature, and conditions imposed by technical limits.

Limiting conditions imposed by the actual situation and by nature are

- Presence of compounds which may inhibit biodegradation at the actual concentration level
- Biodegradability of the contaminants and the rate of biodegradation
- Climate
- Topography

Limiting conditions imposed by technical limits are

- Size of process equipment (in relation to biodegradation rate)
- Availability of technical means
- Criteria for the "end product" of the decontamination
- Protection of the surroundings against pollution, smell, etc.
- Personal safety

The soil decontamination techniques to be considered can be divided into two groups:

1. Techniques which do not require the excavation of the contaminated soil (*in situ* decontamination technique)
2. Techniques which do require the excavation of the contaminated soil; excavation is followed by treatment on or off site

II. *IN SITU* DECONTAMINATION

In agricultural practice soil conditions (pH, oxygen content, nutrient content, permeability, etc.) are often influenced by means of techniques such as liming, plowing, and fertilizing. The effect of these techniques is generally limited to the top 50 cm of the soil.

Application of these agricultural techniques to waste processing is described as "landfarming".

A second method of influencing the important parameters for biodegradation is a technique that in this report is called "bioextraction". Compounds are dissolved or suspended in water and circulated through the soil packet to be decontaminated.

A. Landfarming
1. Principle

In landfarming chemical waste is spread thinly over agricultural soil. Biodegradation is then promoted by plowing, fertilizing, and other standard farming procedures. Landfarming can be used when soil contamination is superficial (not deeper than 0.5 m). The technique is based on the aspects summarized below and more fully dealt with in Section III.

a. Oxygen Content
Aerobic degradation of waste is generally faster than anaerobic degradation; it is thus of great importance to keep the soil aerated (e.g., by plowing).

b. Fertilizer
Depending on the type of waste and on the composition of the soil, it may be necessary to add nutrients in the form of artificial fertilizer or animal manure.

c. Moisture Content
A high moisture content hinders oxygen transport and a low moisture content inhibits the biological activity.

d. pH
A pH value of 6 to 8 is normally required; the pH can be adjusted by addition of, for example, $Ca(OH)_2$ or $(NH_4)_2SO_4$.

e. Temperature
An increase in temperature (up to 30°C) generally increases the rate of the biodegradation. In landfarming, however, the temperature depends on the climate, and can hardly be influenced.

f. Seeding with Microorganisms
Addition of microorganisms, which are adapted to the biodegradation of the contaminants, can fasten the onset and possibly the decontamination process.

2. Implementation
A site to be decontaminated by landfarming must first be stripped of trees, bushes, large stones, etc. using, for example, a bulldozer. Further treatment consists mainly of regularly plowing (e.g., once a month) and fertilizing (if needed). The necessity of fertilizing is determined by analysis. Solid fertilizer (artificial fertilizer pellets or manure) can be applied using the usual agricultural machinery. Liquid fertilizer or aqueous suspensions of microorganisms can be applied using equipment developed by the "Instituut voor Mechanisatie van Arbeid en Gebouwen" (IMAG, Institute for Mechanization of Labor and Buildings), see Figure 13.[123]

An advantage of the use of this equipment is the fact that fertilizer or microorganisms can be introduced into the ground without being spread into the atmosphere.

3. State of Development
At this moment no cases of the use of landfarming for soil decontamination are known.

4. Costs
Costs for *in situ* landfarming only comprise costs for clearing of the site, plowing, and fertilizing.

As clearing of the site may include removal of vegetation (shrubs and trees) and structures, it is impossible to give an exact figure for the costs of this operation. For a site free of building structures, however, the costs will be in the order of Dfl. 4,000.-/ha.

Labor costs for monthly plowing, will amount to about Dfl. 3,000.-/ha/year. The costs of fertilizing are highly dependent on soil composition, etc., according to Bonnier[11] the

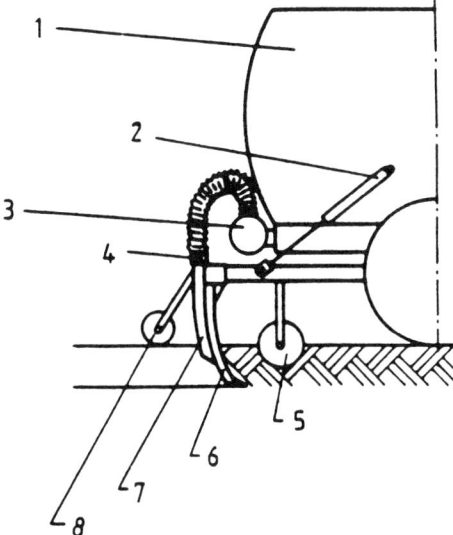

FIGURE 13. Apparatus for the introduction of fluids into soil coupled to a tanker. (1) The tanker, (2) lift (lower) cylinder, (3) distribution tube, (4) flow indicator, (5) coulter, (6) apparatus for opening the soil, (7) supply line, and (8) apparatus for closing the soil. (From Bosma, A. H. et al., Ontwikkeling van een onderzoek aan een zelfrijdende mestinjecteur, IMAG-publikatie 87, 1977. With permission.)

processing of oil-contaminated soil requires about 2500 kg of artificial fertilizer per hectare per year. This will cost about Dfl. 5,000.-.

The total costs for treatment over a 2-year decontamination period will amount to Dfl. 5.00l/ton for a treatment depth of 0.3 m. This does not include costs for environmental protection, replanting, etc.

B. Bioextraction
1. Principle

In the bioextraction process, biodegradation of compounds in contaminated soil is stimulated by circulating water to which nutrients, oxygen, and/or adapted microorganisms have been added through the soil (Figure 14). There are three possibilities for recirculating the water:

1. The water is infiltrated into the soil and drained from under the contaminated layer.
2. The water is injected under the contaminated layer and withdrawn from the surface.
3. The water is passed horizontally through the contaminated layer.

Part of the water extracted is drained off to prevent the accumulation of salts, etc. If necessary, the soil to be processed is contained using protective techniques described in Section VI and Appendix C.

The recirculated water, which may contain contaminants, is purified biologically after infiltration in the first layer of soil it is being transported through. The decontamination rate may be increased by warming the water.

The remarks in Section II.A concerning the parameters pH, temperature, and seeding with microorganisms also apply for bioextraction.

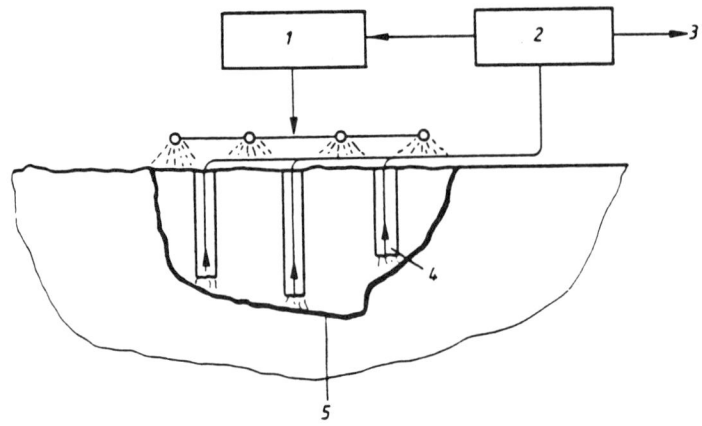

FIGURE 14. Principle of the bioextraction process. (1) Supply of water with nutrients, etc.; (2) separation of solids; (3) to waste; (4) wells; and (5) contaminated
soils.

The decontamination process will proceed at the same rate at which oxygen can be
introduced. Since the solubility of molecular oxygen in water is low, hydrogen peroxide or
nitrate may be added as a supplementary electron acceptor.

2. Implementation

Obstacles preventing a proper distribution of the recirculating water must be removed
from the site.

A spraying installation, such as an overhead irrigation installation of the type used in
agriculture, must be installed on the site.

Clogging of pipes and nozzles must be prevented. Pumps to recirculate the water must
be installed in wells. As colloidal and suspended particles will also be pumped up with the
recirculating water, it may be necessary to include a (biological) purification step before the
water is returned to the soil.

It will be clear that the application of bioextraction will be restricted to the decontamination
of relatively permeable soils such as sandy soils. It is also of great importance that the
permeability of the soil is retained during the process.

3. Use in Soil Decontamination

At this moment no cases of the use of full scale bioextraction for soil decontamination
are known in the Netherlands. However, several pilot plant projects are in progress.

4. Costs

The literature[126] contains calculations of costs for soil decontamination by leaching which,
in terms of implementation, is comparable with bioextraction. The size of the site, the
drainage system, and the location of the wells used in the calculation are given in Figure
15.

For this system based on infiltration channels the material and labor costs amount to about
Dfl. 130,000.- (based on U.S. price levels).

If it is assumed that the contamination is to be removed down to a maximum depth of
5 m, the annual costs will amount to Dfl. 2.-/ton of soil, exclusive of the costs for containment, water decontamination, etc.

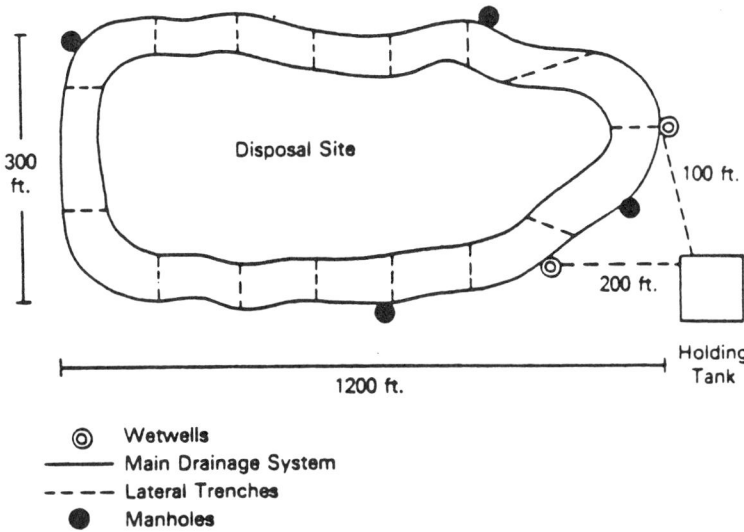

FIGURE 15. Example of a drained waste disposal site to which bioextraction could be applied.

III. DECONTAMINATION AFTER EXCAVATION

As mentioned in Section III a number of factors can make it impossible to decontaminate the soil *in situ*. The soil must then be excavated, after which it can be treated on or off site at a special location. Excavated soil can be decontaminated biologically in the following ways:

● Landfarming can be applied by spreading the contaminated soil on special plots (see also Section II.A).
● Composting-like techniques such as are used for the "upgrading" of domestic refuse may also be applied to the biotreatment of contaminated soil.
● Reactors (as found in chemical industries) can also be used for biotreatment if the residence periods required for the biodegradation are not too long.

A. The Excavation Process

Various machines for the excavation of contaminated soil are available. The costs of excavation by a dragline amount are[126,135]

	Volume capacity of the bucket	
	0.5 m³	1 m³
Max excavation depth (m)	4	5
Costs (Dfls/ton)	1.5	1

If the contaminated soil is not to be treated on site, it must be transported to the processing site by lorry; transport costs then amount to about Dfl. 0.60/km/ton.

If central processing locations are constructed in each province of the Netherlands, the average distance from a contaminated site to the processing location will be about 50 km. The costs for excavation and then transport amount to approximately Dfl. 65.-/ton.

B. Landfarming

1. Principle

Landfarming has already been described in Section II.A. The only difference between *in situ* application and application for excavated soil is that in the latter application contaminated soil is spread in a layer of 50 cm on a drainage bed on an impermeable foil.

2. Implementation

Various authors[11,121,122,127,128,130] have described the application of landfarming to oil-containing sludges; this process is called sludge farming. Table 20 summarizes a number of applications of sludge farming in the U.S.

The influence of aeration and fertilization is illustrated in Figures 16 and 17.[11]

Bölsing[122] has described the biodegradation of oil-containing refuse by sludge farming using lime. The oil was bound by the lime to a powder. As a result the oil could be better distributed in the soil, and the active biomass/oil interface thus enlarged.

A reduction of 50% of the initial concentration of 40 kg of oil per cubic meter was obtained in 30 days.

Lektomaki[129] has described investigations into the influences of aeration, irrigation, fertilization, and seeding with microorganisms on the results of sludge farming. It appeared, however, that adding of brewer's yeast was the most effective method of increasing biodegradation.

3. Use in Soil Decontamination

It appears that various oil-containing wastes can be biodegraded in about 2 years by means of sludge farming. Landfarming should therefore offer a good possibility for decontaminating soil polluted with inherently biodegradable compounds.

4. Costs

Costs of central processing of contaminated soil by landfarming amount to the sum of the costs for excavation, transport, and processing.

The costs for excavation and transport have been given as Dfl. 65/ton in Section III.A, and those for processing as about Dfl. 5.00/ton/year in section II.A. The costs of landfarming with a 50 km transport distance, amount to about Dfl. 70.- for the first year, exclusive of any other costs, e.g., purchase of soil and containment.

The total costs, including costs for soil use (rent or interest) for a 3-year period amount to approximately Dfl. 80.-/ton. Backfill of the decontaminated soil onto a "new" polluted site where soil has been excavated could reduce total hauling costs (including those of site restoration procedures).

C. Composting-Like Techniques

1. Principle

Composting is a process that is already used for the processing of, for example, domestic refuse (since 1932 at Wijster), surplus activated sludge, manure, and garden refuse; organic components of the refuse are degraded by aerobic microorganisms into stable, humus-like products. Important process parameters are as follows.

Oxygen content — Composting is a largely aerobic process, so that sufficient oxygen must be supplied.

Nutrients — For biodegradation to run smoothly the ratios C to N and C to P should not exceed 10 and 100, respectively. Addition of fertilizer may be necessary to achieve this. If nutrients are present in abundance then straw or peat may be mixed with the waste to facilitate the composting process by increasing the porosity of the contaminated soil.

Moisture content — The moisture content of the soil to be "composted" must not be

Table 20
SOME APPLICATIONS OF THE SLUDGE FARMING TECHNIQUE IN THE U.S.

Location	Type of sludge	Pretreatment	Farming technique	Rate of application	Fertilizer	Soil testing	Effects on land	Landfarming experience	Soil type	Land area	Remarks
Refinery V	Sludge from API oil-water separators; typical analysis: 20% hydrocarbon, 25% solids, and 55% water	Dewatering in storage pits for up to 2 years	Spread by bulldozer on grassland; disking to incorporate into soil after drying for 1 month	10—12 cm depth of sludge equivalent to 100 ton/acre of oil	2 ton agricultural lime; 22 kg/acre of N; 22 kg/acre of P as an initial application; no further additions	Initial pH 4.4—4.8; N content "medium"; P content "low"; no further testing	Normal grass cover established within a year	More than 20 years	Clay	75—100 acres	Slight odor detected for a few hours from freshly spread sludge
Refinery W	Oil-water separators; tank bottoms; stable emulsions	—	Area graded to gentle slope with 50 cm high terraced ridges; drainage from lowest point routed to oil separator; retrieved oil returned to land; four sections worked in rotation	15 cm depth of sludge mixed with 15 cm of soil using bulldozer; mixing repeated 2—4 times per month	—	*Pseudomonas stutzeri* is species which showed much rapid growth after application of the oily waste	3—9 months to decompose sludge; wet soil inhibits decomposition	Since 1961	Clay, naturally poorly drained	7 acres	—
Refinery X	Oily sludge from lagoons; typical analysis: 35% oil, 25% ash, and 40% water	—	Single application of 10-year accumulation from lagoons used as clarifiers	10—12 cm of sludge mixed with soil to depth of 45 cm by disking	—	"Acid" treatment showed microbial population of *Pseudomonas* not present in untreated soil	—	Since 1969	Silty loam surface impermeable; clay subsoil poorly drained	?	Heavy rainfall prior to disking contaminated lake with runoff which killed fish in the lake
Refinery Y	Desalter and tank cleanings, separator bottoms, spill cleanings, biosludge, and filter clays	—	Site divided into quadrants; earth dykes controlled water runoff; land spreading during only 9 months of the year; after spreading sludge disking to depth of 15—20 cm at a frequency of once per week	Equivalent of 100 ton/acre of oil per year spread to depth of 8—15 cm	Lime applied to keep soil pH at 7.0—7.5; commercial fertilizer applied to give 20—30 mg/kg nitrates and phosphates	After application, top 15 cm of soil contains 8—9% oil	—	Experimental scale 1972; full scale since mid 1973	—	8 acres	Lead and zinc accumulation has been noted
Refinery Z	Separator sludge and biosludge	—	Spreading followed by disk harrowing when partially dried	8—10 cm in depth	Crushed limestone added	—	—	—	—	—	—

From Bonner, P. E. et al., Sludgefarming: A Technique for the Disposal of Oily Refinery Wastes, Concawe report no. 3/80, Concawe, The Hague, 1980. With permission.

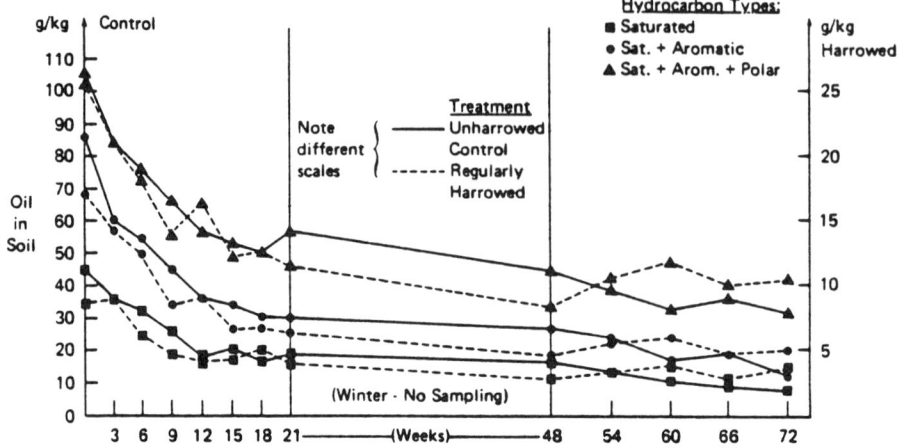

FIGURE 16.   The influence of aeration by harrowing on oil degradation in soil.

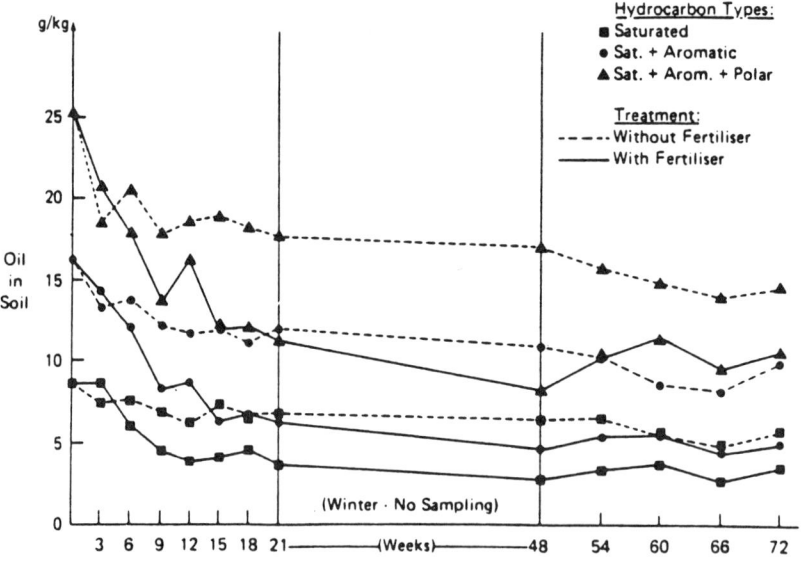

FIGURE 17.   The influence of fertilizing on oil degradation in soil.

below about 15%, otherwise the composting process stops. If the moisture content is too high, however, the permeability for air decreases.

Temperature — During composting of, for example, domestic waste, a temperature rise takes place inside the accumulated material due to the activity of microorganisms. If soil is treated, however, such as rise in temperature should not be expected. If an increase is advantageous this will have to be obtained artificially.

Porosity — The porosity of a material to be composted is a measure of the presence of cavities in the mass of material. Sufficient porosity is required for the necessary access of air to the material.

Inoculation — Biodegradation of a contaminant may be accelerated by inoculating the material with adapted microorganisms.

2. *Implementation*

A great number of systems for the implementation of composting-like processes is available; only the most important systems will be discussed here.[133,134]

Composting systems can be divided into systems in which the material is composted in an open space (open system) and those in which it is composted in an enclosed system.

a. Open Systems

Windrow — In this system the material to be treated is piled up in long mounds (about 5 m wide and about 2 m high) called windrows. Aeration takes place during periodical turning of the material. Additives such as special microorganisms and nutrients can also be mixed with the waste in this way. The composting period for activated sludge amounts to 3 to 4 weeks. The period required for the decontamination of soil will depend on the nature of the contaminant and is expected to exceed this period.

Beltsville (Figure 18) — This method is named after the place in the U.S. where it has been developed. The material to be composted is piled up in long mounds as with the windrow method. The aeration of the material to be composted is, however, achieved by the use of perforated tubes installed under the mounds. A fan is connected to the tubes, so that turning of the material is not necessary. Additives, if any, are mixed with the material to be composted prior to the composting. The composting period for activated sludge is about 21 days; for other material composting is completed in about 30 days.

b. Closed Systems

Dano (Figure 19) — Named after the producer, this system consists of a horizontal drum, which rotates around the longitudinal axis, so that the contents are continuously mixed. As a result of the intensive mixing, in which aeration and moisture content can be adjusted accurately, the composting process proceeds rapidly. The average residence time of the waste in the installation is several days, after which the material is heaped up to allow completion of the composting process. The Dano system is applied in the Netherlands for the processing of domestic refuse; abroad mixtures of domestic refuse and surplus activated sludge are also treated in this way.

Kneer (Figure 20) — This bioreactor, named after the inventor, consists of a vertical cylindrical vessel. Waste is fed in at the top of the reactor and an equal volume of composted material is withdrawn at the bottom. Air and water are introduced at the bottom. Additives, if necessary, are mixed with the waste. Aeration can be adjusted in response to changes in the temperature, so that the process may be influenced to a certain extent. However, the extent to which this is possible is restrictive, since, as a result of the continuous nature of the process, various composting phases occur simultaneously in one reactor. This makes the system sensitive to disturbances of the process. The average residence time is easily adjustable and may amount to several weeks.

Schnorr (Figure 21) — This reactor consists of a rectangular tower, which is divided into a number of stories by means of hinged floors. Periodically, the hinged floor of a story is opened, to allow the material to drop to a lower level. The composted material is removed at the bottom of the reactor. Forced aeration takes place by means of a ventilator. One existing plant for activated sludge consists of a 9-m high silo with about 18 stories. The total residence time in the plant is about 1 month. As a result of falling the mass is intensively mixed, thus producing a homogeneous end product. A disadvantage of this system is the high costs of investment. At present limitation of these costs by installing only two stories through which materials will be recirculated, is under investigation.

Triga (Figure 22) — The structure and function of this system are similar to those of the Kneer system. In the Triga system the cylindrical vessel is divided into four equal compartments by vertical partition walls. As a result of this setup the process can be continued even though one of the compartments is disturbed.

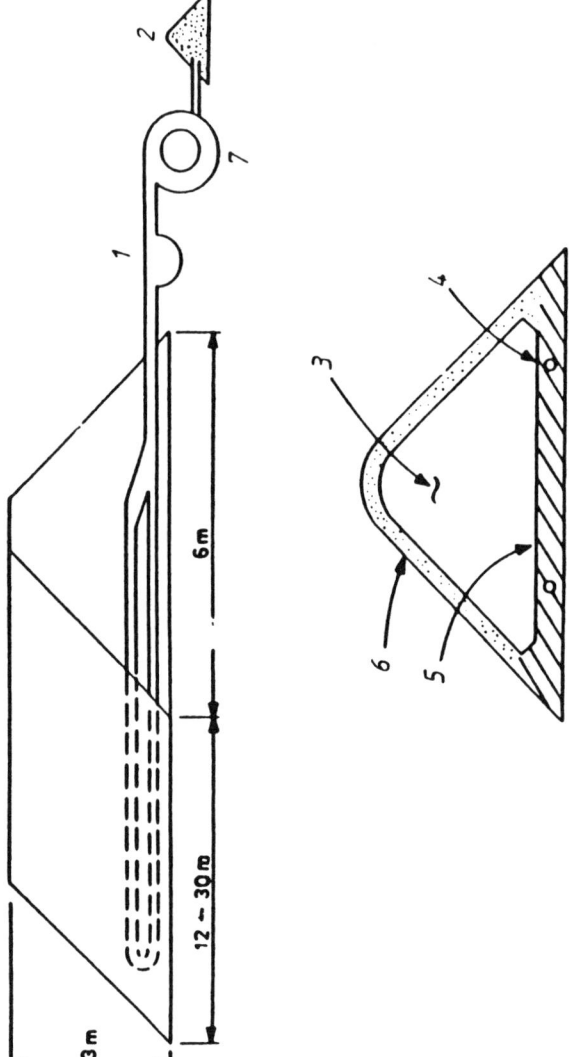

FIGURE 18. Composting according to an open system (Beltsville), with forced aeration. (1) Water separator, (2) fan, (3) additive or sieved compost, (4) additive or course compost, (5) additive and/or microorganisms, and (6) perforated pipe.

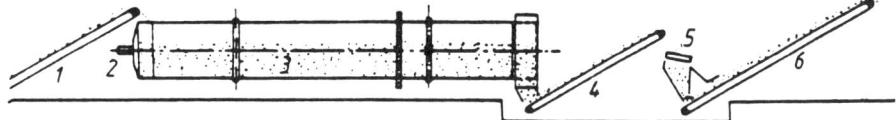

1	Conveyor belt 1	5	Shaking sieve
2	Inlet water and air	6	Conveyor belt 3
3	Drum		
4	Conveyor belt 2		

FIGURE 19.   Dano horizontal rotating bioreactor.

c. Mixing

Depending on the properties of the soil to be treated and on the nature of the contaminants, additives such as straw, fertilizers, and microorganisms may need to be mixed with the soil prior to decontamination with the aid of a composting-like technique. A mixing plant as used for preparing garden mold can be used.

This plant consists of stocks of the components to be mixed and conveyor belts which deliver the different constituents. Adjustment of the speed of the belts allows the mixing ratio to be adjusted to any desired value. The constituents are further mixed by grinding.

d. Current Uses

According to de Renzo[124] pharmaceutical refuse (not further defined) was completely degraded in about 30 days by application of the Windrow system.

A patent[120] indicates the possibility of processing oil-containing waste by composting. The material to be composted is sprayed with effluent from the biological water purification plant of an oil refinery.

A compost filter, applied for deodorizing air, practically completely degraded compounds such as toluene or butanol.[119]

Doyle et al.[125] described the decomposition of trinitrotoluene (TNT) and trinitrotrimethylenetriamine (RDX) by means of composting within 6 weeks.

3. Use in Soil Decontamination

At this moment no cases of the use of composting-like techniques for soil decontamination are known.

4. Costs

Ponsen[131] gives the following costs for composting activated sludge by means of the open Beltsville system and the closed Schnorr system.

- Open system: Dfl. 30.-/ton of dehydrated sludge (dry substance content 25%)
- Closed system: Dfl. 50.-/ton of dehydrated sludge (dry substance content 25%)

The costs of excavation, transport (50 km), and processing of contaminated soil should therefore amount to:

- Open system: Dfl. 95.-/ton
- Closed system: Dfl. 115.-/ton

The costs of treatment of soil contaminated with slowly degradable compounds will be higher than those given above; however, lack of experimental data prevents a reliable estimation of the costs in this case.

1. contaminated
   soil, addi-
   tives, etc.
2. control system
3. spent air
4. water
5. heat
6. fresh air

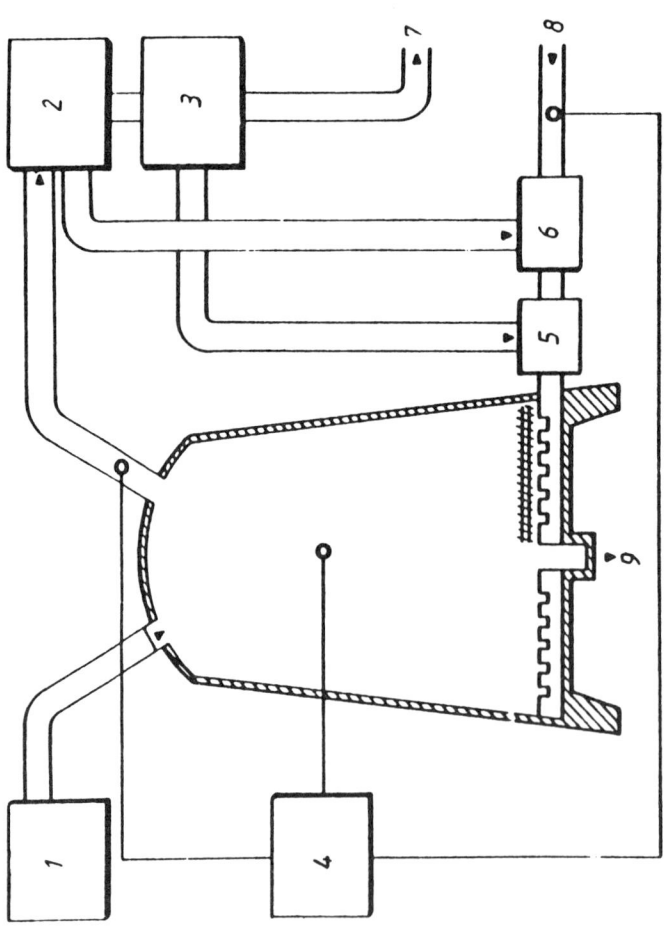

FIGURE 20. Kneer bioreactor.

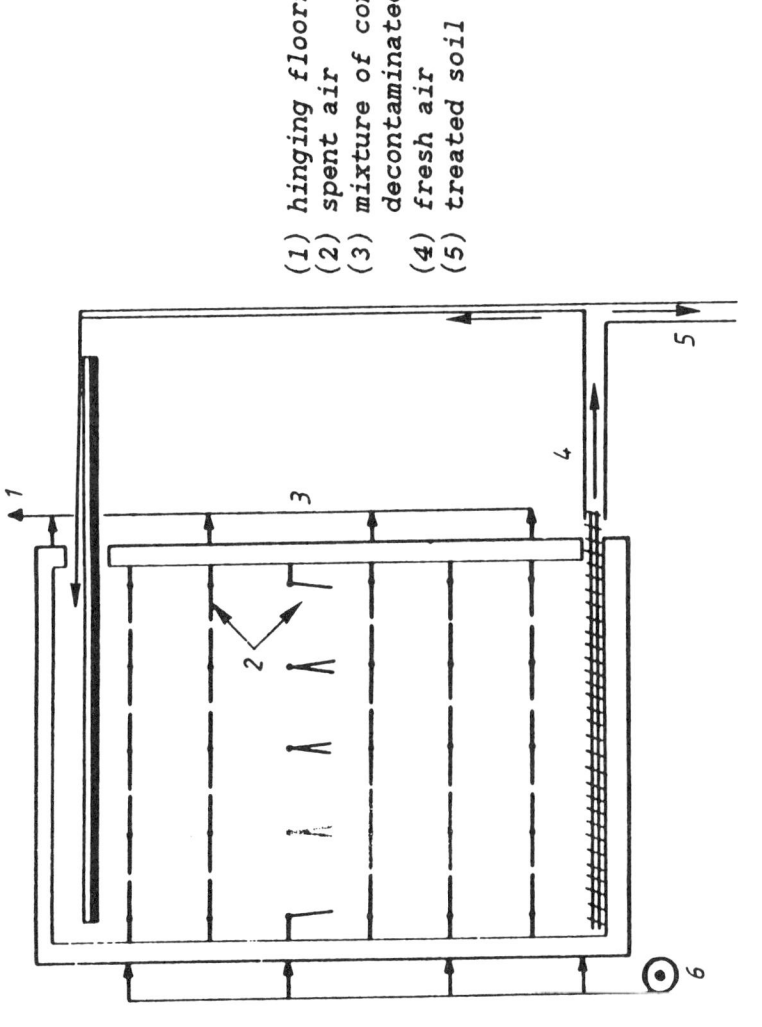

(1) *hinging floors*
(2) *spent air*
(3) *mixture of contaminated and decontaminated soil*
(4) *fresh air*
(5) *treated soil*

FIGURE 21. Schnorr bioreactor.

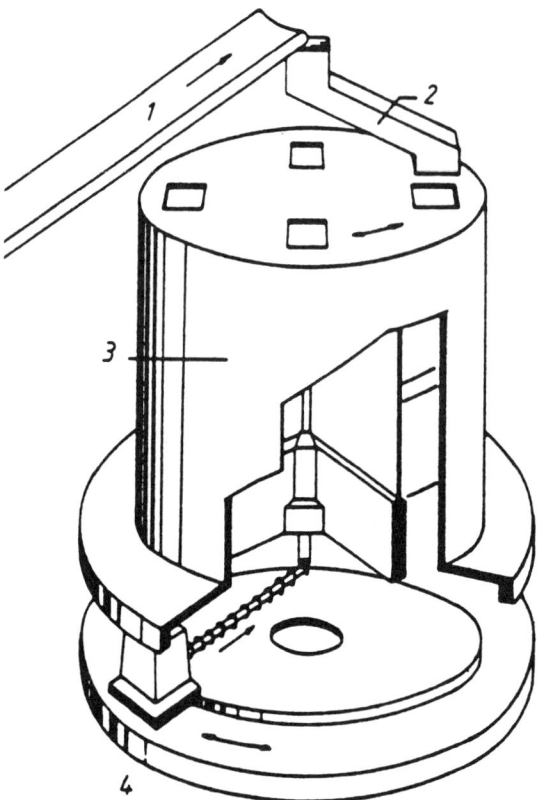

(1) conveyor belt
(2) distributor
(3) compartimentalized
    reactor
(4) orbital hopper screw

FIGURE 22.   Triga system reactor.

D. Industrial Processing Systems

1. Principle

In cases where the biological decontamination of soil would proceed within a few hours, the bioreactors of the types used in the processing industry could be used. Basically, the reactor must consist of a system in which a microbiological inoculum, various additives (nutrients), and oxygen are mixed with the soil. Various types of mixers for solid-substance mixers used in the processing industry may be suitable.[132]

The following parameters play an important role in selecting a suitable mixer:

- Composition of the soil (sand, clay, or humus)
- Water content of the soil (in relation to rheological properties)
- Formation of dust
- Safety, e.g., hazard of dust explosion by static loading
- Sensitivity to wear
- Loading and unloading facilities
- Mixing quality required
- Residence time required in the mixer

2. Implementation

Mixers for solids (in slurry or clay) of which an example is shown in Figure 23 may be suitable.

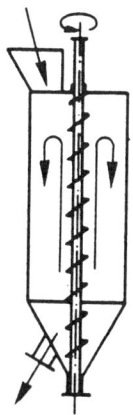

FIGURE 23.   Vertical screw mixer.

3. Use in Soil Decontamination

At this moment no cases of the use of industrial mixers for soil decontamination are known.

4. Costs

It is expected that the costs will roughly be comparable to the composting costs mentioned in Section III.C.4, because the significantly higher energy costs will be compensated by the short residence times (a few hours). The applicability of this technique will, therefore, probably depend on the availability of a potent culture of microorganisms biodegrading the contaminants involved.

APPENDIX C: EMISSION CONTROL MEASURES ACCOMPANYING MICROBIOLOGICAL SOIL DECONTAMINATION*

I. INTRODUCTION

When microbiological techniques are used for decontamination of soil, care must be taken not to pollute the surroundings of the treatment site. This appendix describes measures for containment of contaminated soil, containment and treatment of leachates and gaseous products from the contaminated soil or the decontamination technique, and techniques for environmental monitoring (particularly of ground water via sampling wells) around the treatment site.

II. CONTAINMENT TECHNIQUES

A number of techniques to isolate a decontaminated soil packet from the surroundings are available.[146,147] The isolation may be achieved using either natural or artificial means. Moreover, both vertical and horizontal isolation are possible.

Vertically placed means for containment include the following (the permeabilities (K) given in this section are these for water).

Steel sheet piling — The costs of installation are between Dfl. 70.00/m² of piling for lighter profiles (at slight depth) and Dfl. 200.00/m² of piling for heavier profiles (at greater depth). If the chemical resistance is increased by application of a coating or by hot zinc dipping, the total costs will rise considerably: $K \cong 10^{-7}$ to 10^{-8} m/sec.

---

*   This appendix was prepared by W. J. Th. van Gemert and H. J. van Veen.

Plastic sheet piling — The costs of installation are Dfl. 60.00 to Dfl. 90.00/m². The extent of permeability of the plastic sheet piling is mainly determined by the quality of the joints.

Jet grouting — A "cement grout" is injected at high pressure into the ground. The technique has not been applied in the Netherlands; the costs can therefore only be estimated with difficulty: $K \cong 10^{-8}$ m/sec.

Panel sheet piling — Bentonite-cement or concrete panels are formed in steel molds. The costs amount to Dfl. 70.00 to Dfl. 100.00/m², exclusive of costs for construction and removal of the molding plant required: $K \leq 10^{-8}$ m/sec.

Clay with or without plastic film in a trench — The costs are Dfl. 50.00 to Dfl. 100.00/m²: $K \cong 10^{-7}$ to 10^{-9} m/sec, depending on the type of clay applied and the necessity to use film.

Slurry trench technique — A trench is dug, and the sand removed is mixed with bentonite and poured back into the trench. The costs are highly dependent on local conditions and amount to at least Dfl. 30/m²: $K \cong 10^{-8}$ m/s.

Diaphragm wall — A narrow trench is dug and a fluid supporting the sides is simultaneously introduced. This fluid is either "self hardening" or is replaced by, e.g., bentonite or grout. The technique can be applied for practically all types of soil. The costs amount to at least Dfl. 175.00 to Dfl. 200.00/m²: $K \leq 10^{-8}$ m/sec.

Plastic film — The film (up to a thickness of 200 µm) is applied by means of a chain mortise machine. The costs are Dfl. 5.00 to Dfl. 25.00/m (maximum depth 4.85 m). The film is impermeable to water.

Bored piles wall — Overlapping piles of either bentonite cement or concrete are cast in prebored holes in the ground. The costs are Dfl. 125.00 to Dfl. 275.00/m², exclusive of transport costs of the plant required: $K \cong 10^{-8}$ m/sec.

Horizontally placed means for containment include the following.

Injection into permeable soil — Agents based on one of the following are injected into permeable soil:

1. Water glass
2. Synthetic resin
3. Bentonite, etc.

The costs are Dfl. 250.00 to Dfl. 750.00/m² and are highly dependent on the conditions: $K \leq 10^{-8}$ m/s.

Covering by asphalt-concrete layer — The cost amount to about Dfl. 50.00/m²: $K \leq 10^{-10}$ m/sec.

Application of a bentonite-sand layer — This layer is applicable as a sealing above or below the contaminated soil. Upon absorbing moisture, the bentonite swells. The costs are Dfl. 15.00 to Dfl. 40.00/m²: $K \cong 10^{-9}$ m/sec.

The choice of containment technique to be used in a particular case of soil contamination will have to be made after consideration of the permeability of the material used and the chemical stability for the relevant contaminants and the costs.

III. REMOVAL OF GASES

The gaseous products that can emerge from the soil or from the refuse during the microbiological decontamination are divided into two categories:

1. Volatile contaminants: in particular some compounds may volatilize when the temperature is increased to 30 to 40°C to stimulate biodegradation.

2. Volatile compounds that are produced during the process: these are by-products, intermediates, or end products of biological degradation. When the biodegradation is anaerobic these gases may for the greater part consist of methane and some H_2S. In aerobic degradation processes much carbon dioxide will be formed.

If gaseous compounds, which may inhibit biodegradation, accumulate during decontamination, these gaseous compounds must be dissipated and processed; this is also the case if emission of the particular gaseous compounds to the atmosphere is inadvisable.

The following information is required to establish the optimum ventilating system in any particular case:

1. The nature of the wastes or soil
2. The permeability of the waste or the soil
3. The depth of the contaminated area
4. Geological characteristics (type of stratification)

Items 3 and 4 are directly related to *in situ* treatment.

A number of systems for the control of volatile compounds are available;[142] these are discussed below.

A. Ventilation Through Pipes
1. Principle

Slotted pipes are placed near or in the contaminated area. These pipes are surrounded by coarse gravel to prevent blockage and they vent naturally or mechanically to the atmosphere or to treatment facilities (Figure 24). PVC pipes (10- to 15-cm diameter) are normally used, although galvanized steel can be used when PVC is attacked by the compounds concerned.

2. Applicability

The applicability of atmospheric vertical pipes is restricted to those cases where horizontal migration into the soil takes place within impermeable areas, or where the gases migrate under the surface towards central areas. Control of horizontal migration of gases by installing a number of atmospheric pipes around the contaminated area has little success, unless the pipes are placed very closely to each other.[140]

If dangerous gases are produced, a closed system with forced ventilation is required to prevent toxic fumes from migrating horizontally and vertically to the atmosphere.

The required ventilation capacity for volatile compounds in soil depends on the diffusion behavior of the gas in that soil. An investigation into this behavior was carried out by Farmer et al.[139] and by Shen et al.[141] The ventilation capacity must be somewhat above the calculated gas mass flux.

3. Costs

The installation costs for the ventilation pipes mentioned amount to about Dfl. 180.00 to 150.00/m.

PVC pipe costs Dfl. 40.00 (Ø 10 cm) to Dfl. 60.00 (Ø [diameter] 15 cm)/m. Additional costs for a "fan" for forced ventilation amount to about Dfl. 1800.

The total installation costs for a vertical ventilation pipe of 10 m, inclusive of exhaust, then amount to about Dfl. 3000.

The number of pipes per unit of surface area will depend on the action radius per pipe. Farmer et al.[139] and Shen et al. describe a procedure to calculate this. In general, an average distance of 15 m between pipes is required if forced ventilation is used.

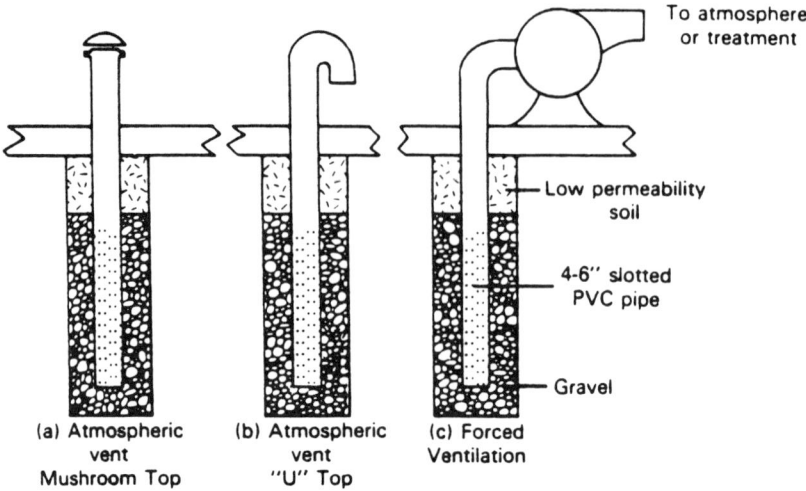

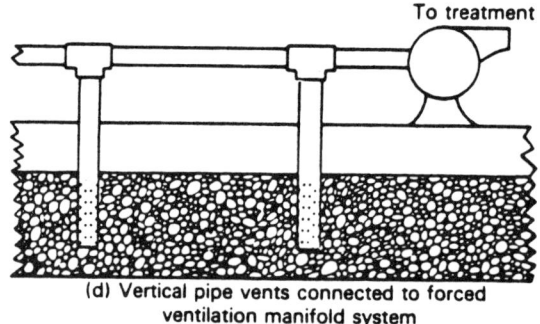

FIGURE 24.   Schematic diagram of the ventilation by pipes.

B. Ventilation Through Trenches

1. Principle Description

Narrow, deep trenches are dug around the contaminated area and filled with gravel. Due to the low resistance of gases in comparison to that of the surrounding soil, they escape through the trenches, and are therefore prevented from polluting the surroundings. The trenches may be either open, or covered by, for example, clay. If the trenches are covered it is possible to collect the gases for further processing (Figure 25).

Open trenches may be infiltrated by rain water and blocked by fouling. Open trenches are, therefore, not fit for use in flat terrain. As a remedy the edges may be heightened to facilitate rain water runoff and prevent too much infiltration (see Figure 25a and b) or a clay capping may be applied (see Figure 25c and d).

The operation of passive trenches can be improved by connecting the trenches by means of perforated pipes (Ø 30 cm) at a mutual distance of about 15 m.[137] The calculations indicated under Section III.A.3 apply also to closed trenches with forced ventilation.

2. Applicability

Trenches can particularly be applied where the migration below the contaminated area is limited by, for example, ground water or an impermeable formation layer. The trench, then, has to extend to a sufficient depth to make a "seal" with that layer.

Ventilation trenches with passive ventilation only are generally not capable of preventing

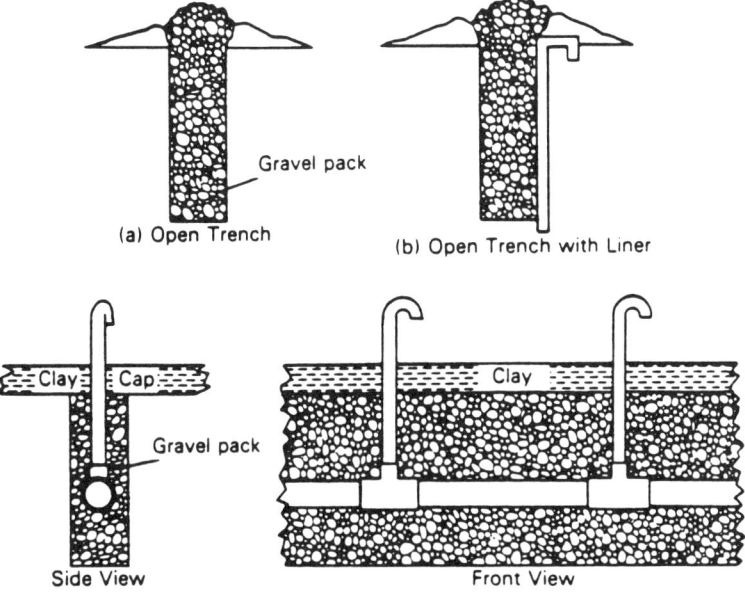

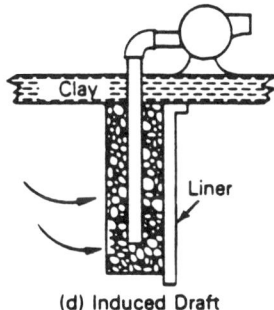

FIGURE 25. Schematic diagram of ventilation through trenches.

migration of gases to the surroundings. Combination with "sheet-pilings" (see Section II) are, however, also possible.[143,144]

3. Costs

The installation costs of a ventilation trench are composed of costs for digging, dehydration, gravel filling, and installation of impermeable screens and perforated pipes. The number of trenches needed can be estimated as indicated under Section III.A.3. For example, the installation costs of a ventilation trench 7 m deep, 1.5 m wide, and 170 m long, inclusive of perforated pipes and screens, are estimated at Dfl. 800,000.00.

C. Gas Barriers

1. Description

Barriers against migration of gases consist of materials with a low permeability for gases, such as clay, cement, or plastics. Two materials with a low gas permeability are commercially available: Hypalon (chlorosulfonated polyethylene) and Neoprene. Hypalon is preferred to Neoprene, because of lower costs and better properties.

It is important that the material is not damaged by stones, etc. For this reason it is recommended that two layers be applied, with a buffering layer between them.

Table 21
CHEMICAL RESISTANCE CHART

Vapor type	Hypalon[a]	Neoprene[a]
Acetic acid (glacial)	X	C
Acetone	T	B
Benzene	C	C
Butane	A	A
Butyraldehyde	X	C
Carbon tetrachloride	C	C
Cyclohexane	C	C
Dioctyl phthalate	C	C
Ethyl acetate	C	C
Ethyl alcohol	A (70°C)	A (70°C)
Formic acid	A	A
Gasoline	B	B
Hydrochloric acid (conc)	—	—
Hydrocyanic acid	B	A
Hydrogen sulfide	B	A
Kerosene	X	B
Methyl alcohol	A (70°C)	A (70°C)
Methylene chloride	C	C (37°C)
Methylethyl ketone	C	C
Naphtha	B	C
Perchloroethylene	X	C
Toluene	C	C
Trichloroethylene	C	C
Xylene	C	C

Note: All ratings are at room temperature, unless specified. A: chemical has little or no effect, B: chemical has minor to moderate effect, C: chemical has severe effect, and X: no data — not likely to be compatible.

[a] Cured sheet.

The material used should be resistant to the organic compounds present in the soil to be treated. Table 21 shows the resistance of Hypalon and Neoprene to a number of organic compounds.[126]

2. Applicability

Compact clay can be applied to check vertical migration of gases.[145] The fact that clay cracks when it desiccates is, however, a serious problem. Cement, or similar materials, have not yet been used to check gas migration.

Synthetic materials have been applied to prevent both horizontal and vertical migration of gases, although satisfactory results are only obtained in combination with ventilation pipes or with trenches.[126]

3. Costs

The total costs are determined by material costs and installation costs. For covering ventilation trenches, the use of two layers with a material called "geotextile" between them are recommended. An estimate of the costs is given in Table 22.[126]

Table 22
COSTS FOR GAS BARRIERS

Material	Notes	Total costs
Hypalon	36 mm thick	Dfl. 20.00/m²
Teflon®	10 mm thick	Dfl. 70.00/m²
Geotextile intermediate material		Dfl. 45.00 to 60.00/m²

D. Treatment of Gases

This topic is outside the terms of reference of the present study, and will therefore, not be considered here. Suffice it to state that a number of techniques (such as after-burning and adsorption to activated charcoal) are available.

IV. MONITORING SYSTEMS

Once the decontamination process to be used has been selected, a selection of procedures for monitoring both the progress of the decontamination process and the possible environmental pollution resulting from it, must be made (surface water, ground water, and air). Thus, the objectives of the monitoring system are

- To follow the process of decontamination by chemical analysis
- To detect possible sudden changes in the process (calamities)
- To protect the surrounding environment: surface water, ground water (particularly with *in situ* decontamination), and air, and any consumers of these

Brief descriptions of the possible principles of these monitoring activities are given below.

A. Process Monitoring

Soil samples are taken at regular intervals, depending on the biodegradation rate, for analysis of relevant compounds.

Soil samples must be taken at various depths and places for the final evaluation of *in situ* decontamination. During treatment it suffices to analyze samples from drainage wells (see Figure 14).

B. Ground Water Monitoring

Any ground water monitoring program will consist of the following steps:

- Determination of background concentrations of relevant contaminants in ground water.
- Choice of locations for "monitoring wells": Figure 26 shows an example of this.[138] In well A background concentrations can be determined. Well B is localized in the area of the contamination and functions as the first indicator. Well C is placed "downstream", outside the area of the contamination, in such a way that variations in ground water composition are recorded as early as possible. The design of the wells depends on the number of aquifers. If there is only one aquifer the design shown in Figure 27[138] can be applied. If there are several aquifers, the design as shown in Figure 28[138] may be applied. The costs for the installation of wells vary greatly, depending on the local conditions, but can be roughly estimated as between Dfl. 6,000.- (for a 3-m deep well with one sampling point) and Dfl. 20,000.- (for a 30-m deep well with several sampling points).[126]
- Sampling program: samples can be taken from the wells in three ways:

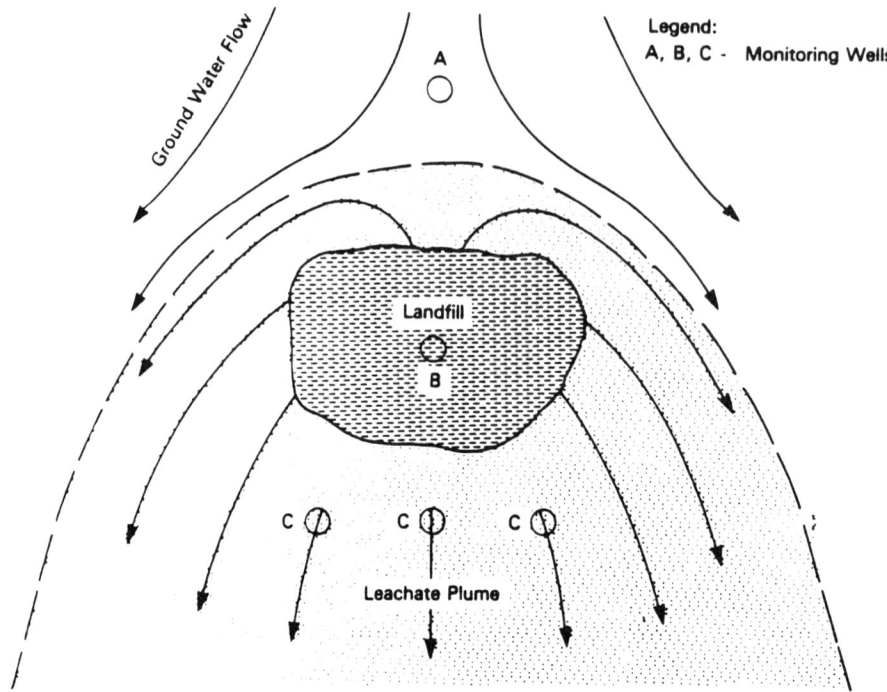

FIGURE 26.   Typical monitor well network, areal view.

1. By means of a permanent pump; this is recommended if checking over a longer period is required.
2. By means of mobile pumps.
3. By means of an air lift, with which compressed gases bring the water to the surface. With this technique contamination of samples taken at different times and from different wells (as might be found when mobile pumps are used) can be excluded. Care must be taken not to lose volatile compounds from samples.

The final choice is determined by frequency of sampling, the number of wells, and the local conditions.
- Laboratory analysis.
- Processing of the data.

C. Monitoring of Gaseous Emissions (*In Situ*)
If emission of (toxic) gases from contaminated soil is possible, it is necessary to monitor this (for both nature and amount of the gases). The location of the sampling points is dependent on the local conditions. It is recommended, however, that at least two monitoring pits be located at places where emissions of gases are the most probable or involve most risks.[136] Samples must be taken at about 1 m below the soil surface. An example of a suitable system is shown in Figure 29. A system for sampling at different depths is shown in Figure 30.

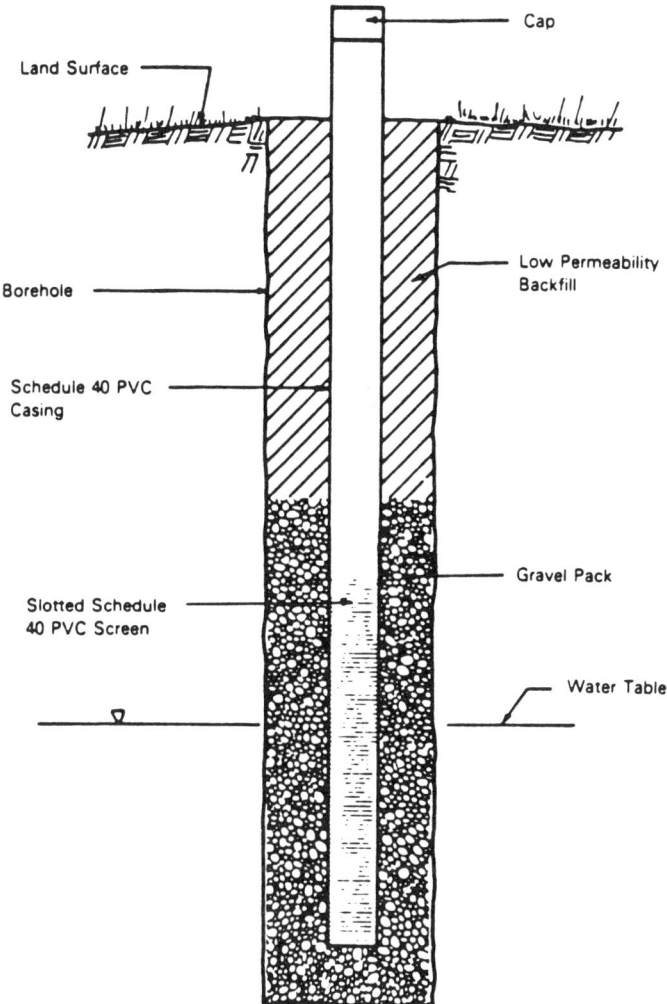

FIGURE 27. Typical monitoring well screened over a single vertical interval.

APPENDIX D: NONMICROBIOLOGICAL SOIL DECONTAMINATION TECHNIQUES*

I. INTRODUCTION

This appendix summarizes part of the report entitled *Study on Contaminated Land* by Rulkens et al.[140a]

A number of methods are available for the rehabilitation of contaminated soil, the most important of which are

1. Excavation of the contaminated soil, followed by transport to a permanent or temporary tip
2. Restriction of the spreading of the contamination by means of physical barriers (Appendix C)

---

* This appendix was prepared by W. H. Rulkens, W. J. Th. van Gemert, and H. J. van Veen.

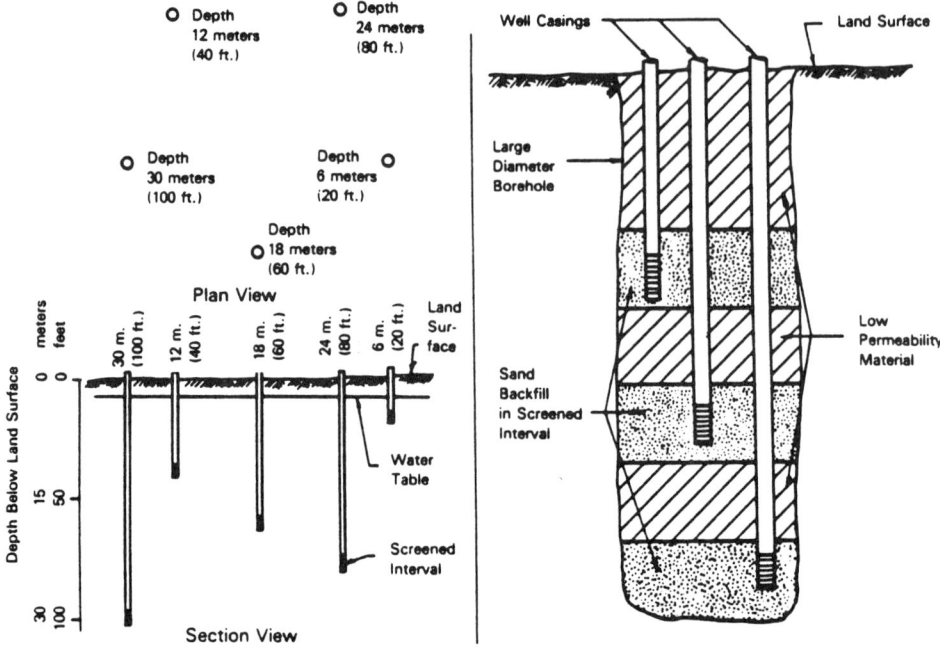

FIGURE 28. Typical well cluster configuration.

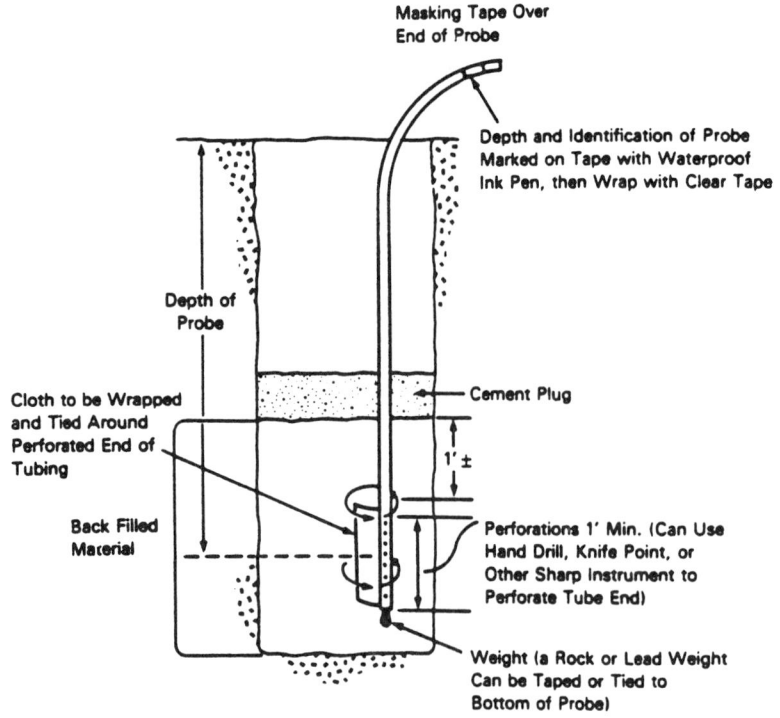

FIGURE 29. Typical gas probe placement.

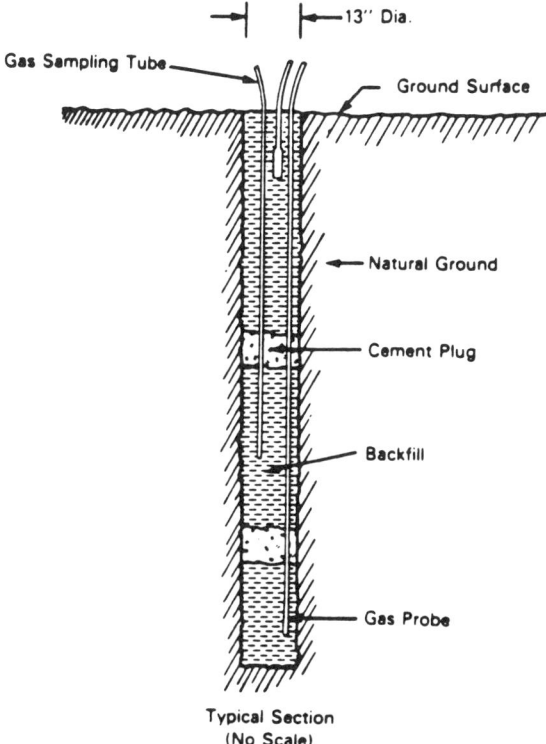

FIGURE 30. Typical multilevel gas sampling probe installation.

3. Restriction of the spreading of the contamination by means of drainage systems
4. Restriction of the spreading of the contamination by means of physical or chemical immobilization techniques
5. Excavation and decontamination of the soil
6. Decontamination of the soil *in situ*

Methods 1 to 4 are exclusively aimed at preventing the spread of contamination to the surrounding environment, while methods 5 and 6 result in decontamination of the polluted soil.

The method to be chosen in a specific case will depend on a great number of factors, such as nature and concentration of the contamination, the type of soil (peat, clay, or sand), the location, the intended use of the site, rehabilitation methods available, costs, and environmental hygiene and social factors.

This appendix will only discuss the rehabilitation methods that are aimed at decontamination of the soil (5 and 6), and, in particular, the application of these methods to cases of more or lesser diffusely distributed contaminants.

The following facets of the methods will be discussed:

● The principle and background
● Potential applications
● The state of the art

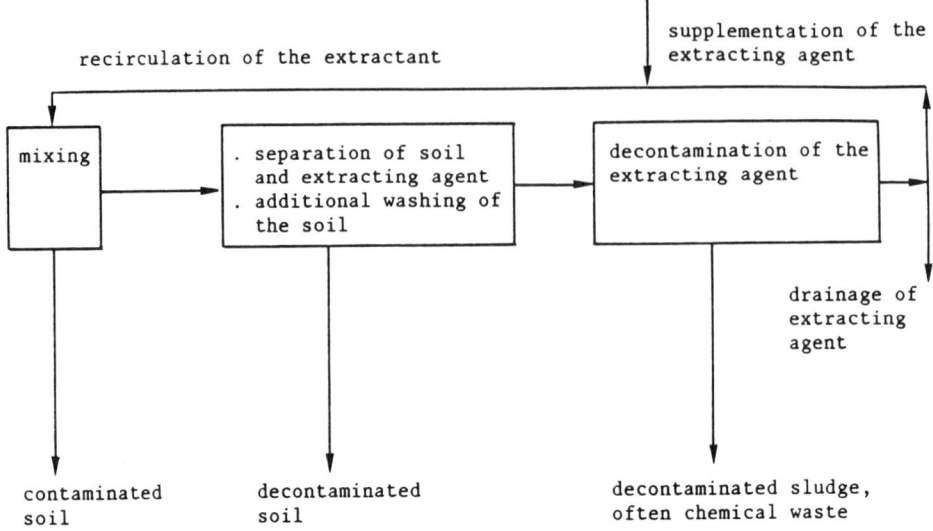

FIGURE 31. The extraction process.

II. TREATMENT OF EXCAVATED SOIL

Most of these decontamination techniques are still being developed and are in general based on processes that are already applied in other fields. The following techniques can be mentioned:

● Unit operations that are applied in the processing industry
● Treatment processes for chemical waste
● Processes that are applied to the physicochemical decontamination of waste water
● Techniques that are applied to soil handling, such as transport, storage, and decontamination of soil

The following decontamination techniques can then be distinguished:

● Extraction (leaching)
● Thermal treatment
● Steam stripping
● Chemical treatment

A. Extraction
1. Principle and Background
The principle of the extraction process is shown schematically in Figure 31. The contaminated soil is intensively mixed with an extracting agent such as cold or hot water to which chemicals such as acid, alkali, surface-active compounds, or an organic extracting agent have been added. The contaminants are transferred to the extraction agent.

Equipment used in the chemical processing industry for waste water purification or for soil processing that can be used are extracting equipment, such as mixers, upflow columns, and screw-extractors; apparatus for waste water purification, such as hydrocyclones, settlers, thickeners, centrifuges, and adsorption columns with activated charcoal; and specific equipment for classifying and/or washing gravel and sand.

2. Potential Applications

The extraction process is particularly suitable for the decontamination of soil that contains few clay particles, because these particles can only be separated from the extracting agent with much difficulty, and in general they highly adsorb contaminants. Some special groups of compounds that cannot be easily treated in other ways are discussed below.

a. Heavy Metals

Removal of heavy metals or of compounds of heavy metals can be carried out by extraction with diluted acids (e.g., HCl) and/or diluted alkaline solutions (e.g., NaOH solution) in which many heavy metals dissolve. It is also possible to use alkaline solutions to separate compounds of heavy metals from soil particles suspended in the extracting agent.

b. Cyanides

A distinction must be made between free and complex-bound cyanides. Free cyanide can generally be rinsed out by means of water or dilute alkali. Bound cyanide (generally iron cyanide complexes found as waste in former gasworks) can be dissolved in alkali; these complex-bound iron cyanides can be only decomposed with much difficulty, but they can easily be precipitated from the extracting agent, and separated from it as residual sludge.

c. Hydrocarbons (Including Halogenated Hydrocarbons)

Hydrophilic hydrocarbons (e.g., phenol) can generally be rinsed out with water. The contaminated extracting agent can then be decontaminated by means of adsorption, chemical, or biological treatment or other existing techniques for waste water decontamination.

Hydrophobic hydrocarbons dissolve hardly or not at all in water. Decontamination with an aqueous extracting agent is only possible if the extracting agent possesses dispersing properties, i.e., solutions containing (alkaline) or surface-active agents. These extracting agents bring the contaminants, together with humus-like components and the very fine mineral particles, into colloidal "solution".

The most important "bottleneck" in the extracting process is the presence of humus-like substances and clay particles. These substances, which strongly adsorb the various types of contaminants, generally end up in the residual sludge.

3. State of the Art

Extraction has been investigated on a laboratory scale for various types of soil samples contaminated with various compounds (heavy metals such as Cr, Pb, Cd, and As, oil, cyanides, aliphatic bromine compounds, and lindane). In many cases decontamination with an aqueous extracting agent was possible with an extraction efficiency of over 95%.

In investigation on a pilot-plant scale (about 500 kg of soil per hour), two soil samples contaminated with iron-cyanides (ferrous or ferric) and aliphatic bromine compounds indicated that the cyanide content in the soil could be reduced from about 1000 ppm to about 15 ppm and the bromine content from about 100 ppm to <1 ppm.

Extraction has been applied on a practical scale for decontamination of sandy soils contaminated with oil products.

For further development of the extracting process, however, additional research is needed to minimize the amount of residual sludge (chemical waste) and to develop an installation in which various types of soil can be treated without much adaptation (multipurpose unit).

B. Thermal Treatment

Three processes for thermal decontamination of soil can be considered.

1. Evaporation of the contaminants by direct heating with hot gases

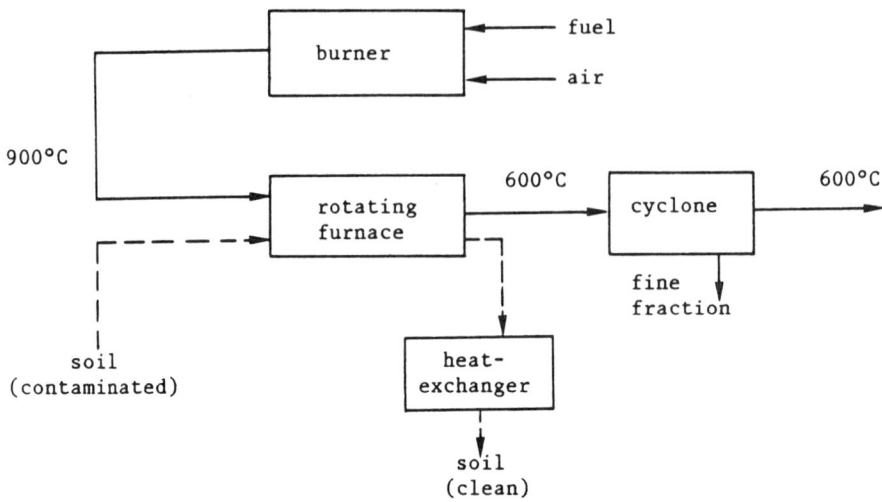

FIGURE 32.   Soil decontamination by direct heating.

2.   Evaporation of the contaminants by indirect heating via a heat exchanger
3.   Thermal destruction of the contaminants

The gases emitted by the first two processes will nearly always have to be treated to destroy or collect the contaminants.

1. Evaporation of the Contaminations by Means of Direct Heating with Hot Gases
a. Principle and Background

The process is shown schematically in Figure 32. Contact between contaminated soil and the hot gases, generated by a burner, takes place in a rotating furnace. The temperature in the furnace can be above 600°C, depending on the construction material of the furnace. The contaminants evaporate together with water and are transported by the gas flow through a cyclone in which fine soil particles are collected.

The emitted gases can be treated in one of three ways:

1.   Treatment at high temperature in an after-burner, causing complete oxidation of the (organic) contaminants. (In certain cases a temperature of about 1400°C is required, e.g., for burning PCBs.)
2.   Destruction of the (organic) contaminants by catalytic oxidation, which generally takes place at considerably lower temperatures than those required for noncatalytic after-burning. Metals such as Ni, Zi/Cu, Fe/Cu, and Al/Cu can be used as catalysts.
3.   Treatment by cooling, followed by scrubbing. The contaminants are collected in a suitable liquid phase (e.g., water) to which compounds to promote colloidal "dissolution" of the contaminants have been added.

In a direct heating process the heat required for the evaporation of water and contaminants must be supplied entirely by the hot gases, which means that a relatively great gas throughput is required (the specific heat of the gas is low) and consequently a relatively great gas treatment system is needed.

b. Potential Applications

Heavy metals — In general it is not possible to remove heavy metals, with the exception of mercury or compounds of mercury, by hot gas treatment.

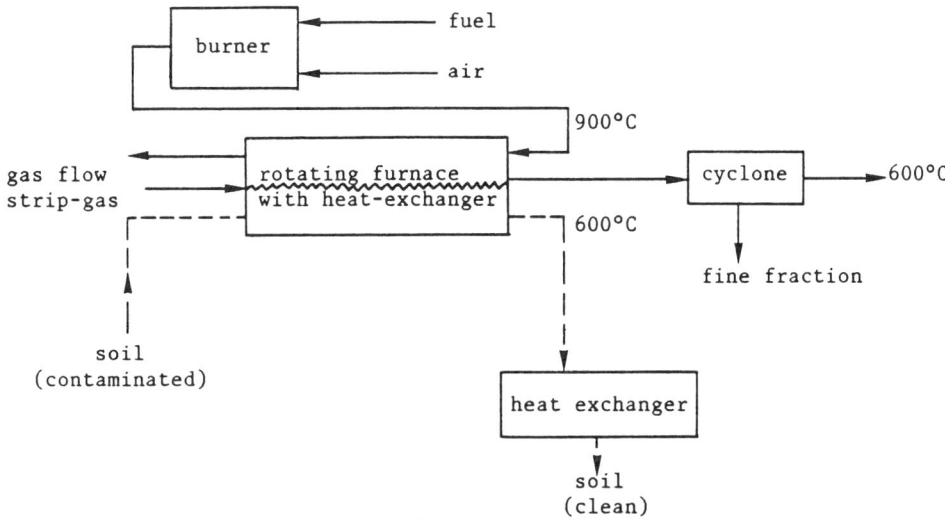

FIGURE 33.   Soil decontamination by indirect heating.

Cyanides — Complex-bound cyanides and, in particular, iron cyanides (ferrous or ferric) can hardly or not be removed from contaminated soil by evaporation, certainly not at relatively low temperatures.

Hydrocarbons — Evaporation of those hydrocarbons with a relatively high vapor pressure at the temperature applied is possible using this technique. These include many aliphatic compounds and compounds such as benzene and toluene. Treatment of the emitted gases may take place by after-burning at 800°C or at 500°C in the presence of a catalyst.

Halogenated hydrocarbons — Very volatile compounds such as trichloroethylene and perchloroethylene can be quite easily evaporated with treatment of the emitted gases by after-burning, or passage through an air condensor, followed by scrubbing. Less volatile compounds such as PCBs, insecticides, and pesticides can only be evaporated with much more difficulty and a stringent after-burning treatment is required: high temperatures (over 1000°C) and relatively long residence times (seconds) in the after-burner (over 1000°C) to prevent the formation of very hazardous components.

The most important bottlenecks in the hot gas evaporation process are

● The relatively large flow of gas required to supply the heat needed (and consequently a large installation for after-treatment)
● The high energy requirement of the process
● Explosion hazard

2. Evaporation of the Contaminants by Indirect Heating Via a Heat Exchanger

This process is shown schematically in Figure 33. The essential difference from the direct heating process is that the heat is transferred to the soil via a heat exchanger. A great advantage of this method is that only a small gas flow is needed to remove the volatilized contaminants, so that the installation for gas treatment can be of a much smaller size.

The evaporation plant is, however, of considerably greater size and complexity. The applications are about the same as those of the direct heating process (Section II.B.1.b).

3. Thermal Destruction of the Contaminants

If the vapor pressure of contaminants is so low that they cannot be removed from the soil by evaporation, then thermal destruction is necessary. This can take place in a rotating furnace at temperatures between 700 and 1200°C. The temperature, however, must not be so high that sintering or glazing of the soil particles takes place. After-treatment of the waste gases can be carried out as indicated in Section II.B.1.a.

4. State of the Art of the Various Thermal Techniques

At the moment only the evaporation method via direct heating with hot gases is operational for the large-scale thermal treatment of the contaminated soil (tens of tons of soil per hour). A mobile plant, developed by "Ecotechniek", has been applied for the decontamination of a former gasworks site, which had been contaminated with a great number of volatile hydrocarbons. The hydrocarbons were evaporated at 200 to 300°C and then after-burnt at about 800°C. It is expected that adaptation of the rotating furnace will allow evaporation temperatures up to 600°C.

A mobile treatment plant has also been developed by the Environmental Protection Agency in the U.S. In this plant it is possible to heat soil in a rotating furnace at 1000°C. After-burning of the emitted gases can take place at temperatures of more than 1200°C with a residence time in the after-burner of a few seconds. The method seems suitable to remove, for instance, PCBs. However, the capacity of the plant so far is low (a few tons of soil per hour).

The most important developments that can be expected in the field of the thermal treatment in the future are

- Plants that are suitable for the treatment of soil at temperatures above 1000°C
- Plants with indirect heat transfer
- Plants for catalytic after-burning

C. Steam Stripping

In addition to the evaporation of volatile contaminants by means of hot gases, as described in Section II.B.1, it is in principle also possible to use steam as stripping gas. The emitted gases (steam and contaminants) from a rotating furnace are condensed and the contaminants are removed from the condensate using one of a number of waste water treatment methods.

The process seems suitable for the removal of organic components having a high relative volatility as compared with water, such as perchloroethylene and trichloroethylene. The most important advantage of the steam stripping process over the evaporation process is the lower gas throughput to be after-treated. On the other hand the energy costs are higher.

The technique of volatile compound stripping is applied on an extensive scale in the processing industry. Use of the process for decontamination of soil is being studied by TNO.

D. Chemical Treatment

1. Principle and Background

Decontamination of soil via a chemical treatment is based on the reaction of the contaminant present with an agent added to the soil. This process can only be called decontamination if the reaction products are not harmful for the environment.

Two basic methods can be distinguished: (1) a dry method in which the (dry) soil is mixed with the agent and (2) a wet method in which the agent is added together with water. The use of water may in certain cases increase the reaction rate and the decontamination efficiency.

Various types of chemicals can be used for the chemical treatment of the soil:

- Chlorine
- Acid/alkali (particularly for neutralization)
- Ozone and hydrogen peroxide (for oxidation)
- Calcium hypochlorite and sodium hypochlorite (for the oxidation of cyanides)
- Potassium permanganate

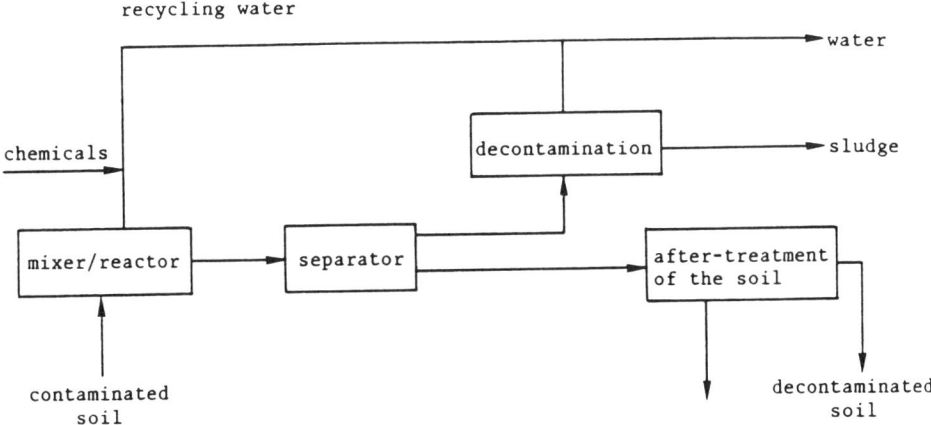

FIGURE 34.   Chemical decontamination of soil.

The wet chemical process is shown schematically in Figure 34. Contaminated soil, water, and the amount of chemicals needed are fed into a mixer/reactor such as a rotating drum, mixing vessel, screw washer, or scrubber. The emerging suspension is separated (e.g., in a laminated separator). After purification the water phase is partially drained and partially reused. The soil separated is dehydrated and further decontaminated if desired.

A simple mixing system is generally sufficient for dry treatment.

2. Potential Application
a. Heavy Metals
The method is not suitable for heavy metals.

b. Cyanides
Chemical treatment is probably suitable for the degradation of certain types of cyanide in soil. Suitable reagents include sodium hypochlorite and calcium hypochlorite. The effectiveness of the treatment can be promoted by adding water. Complex-bound iron (ferric or ferrous) cyanides cannot, however, be effectively treated with these chemicals.

c. Hydrocarbons and Halogenated Hydrocarbons
It is not clear whether (chlorinated) hydrocarbons can be transformed with, for example, hydrogen peroxide in a soil matrix. If such a process is to be used the ectoxicological properties of the reaction products should be determined in advance.

3. State of the Art
The chemical treatments mentioned above were generally developed for processing chemical wastes and consequently seem to offer prospects for decontamination of soil. As far as is known these methods are not operational. It is expected, however, that in particular the use of oxidizing agents (ozone, hydrogen peroxide, sodium hypochlorite, and calum hypochlorite) will enable future developments of techniques for decontamination of soil contaminated with easily oxidizable organic compounds.

The most important bottlenecks of the chemical treatment are

●   The presence of natural organic compounds in soil (humus acid-like compounds)

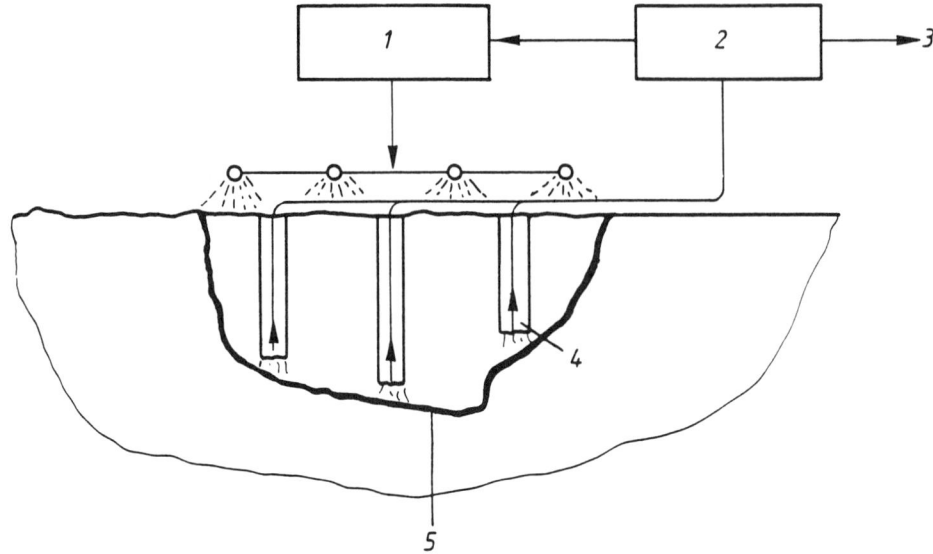

FIGURE 35.   *In situ* extraction of soil. (1) Addition of chemicals, (2) purification of the extractant, (3) sludge, (4) wells, and (5) contaminated soil.

- The long reaction times that are often necessary
- The necessity of using an excess of chemicals
- Safety and environmental consequences

III. *IN SITU* TREATMENT OF SOIL

A. Extraction (Leaching)

The principle of the extraction process is shown schematically in Figure 35. The soil to be treated is first isolated from the surroundings if necessary. Wells are then dug, and pumps installed. A suitable extracting agent is sprayed onto the surface of the soil. The extracting agent percolates through the contaminated soil and extracts the contaminants present. The extracting agent is pumped to a purification plant from the collecting wells. The purified extracting agent then can be reused.

From an environmental viewpoint an extracting agent based on water is to be preferred. The extracting agent listed in Section II.A.1 may be used. Water is to be preferred to organic solvents because a certain amount of water is already present in soil and because an organic solvent has to be removed after the purification process has been completed.

It is clear that the use of the extracting process is restricted to the decontamination of soils which are relatively permeable to the extracting agent (generally sandy soils).

The most important potential field of application of the method is the removal of cyanides, certain heavy metals, and organic compounds having a relative high solubility in water such as lower alcohols, acetone, phenol, etc.

The most important bottlenecks of the method are

- The maintenance of a good permeability of the soil during the extracting process
- The removal of contaminants from spots with local high concentrations
- The control of the decontamination process

To our knowledge the method is not yet operational. Investigation on both laboratory and semitechnical scale is necessary. It is expected that the method will be less expensive than extractive decontamination in a mobile plant. The field of application is, however, restricted.

Table 23
A SURVEY OF NONBIOLOGICAL DECONTAMINATION METHODS

Technique	Developmental stage	Potential field of application	Amount of additives required	Amount of energy required	Amount of residues formed
With excavation					
Extraction	..	....	..	....	.
Thermal treatment	...	....	....	.	...
Steam stripping	.	.	....	.	...
Chemical treatment	.	.	.	....	....
Without excavation					
Extraction	.	.	..	...	...
Chemical treatment	.	.	.	...	....

Note: The number of dots is a measure of the favorability of the evaluation.

B. Chemical Treatment

In *in situ* chemical treatment, the chemicals required are added to the soil via a liquid phase (mostly water) either by spraying onto the surface or through wells. The water can be recirculated if desired. In this case a method partially analogous to the extractive decontamination method is applied. This method is attractive when the reaction between the chemical and the contaminants only proceeds very slowly.

The most important fields of application are said to be degradation of cyanides by means of sodium hypochlorite or calcium hypochlorite (complex-bound iron cyanides cannot be decomposed in this way), the oxidation of organic chemicals with, for example, hydrogen peroxide, and the neutralization of either acid or alkaline contaminants. To our knowledge this method is not yet operational.

The most important bottlenecks of this method are

● High concentrations of contaminants leading to a high consumption of chemicals
● The presence of organic compounds in the soil which may increase consumption of chemicals

IV. DISCUSSION

Various aspects of the decontamination techniques discussed are reviewed in Table 23. This review must be considered preliminary. It appears from the table that extractive or thermal treatments of excavated soil are the most developed techniques and have the broadest range of application.

An important factor in the determination of the decontamination method to be used is the cost per ton of soil to be treated. Due to the lack of practical data (only thermal decontamination by means of evaporation and after-burning has yet been applied in practice), quantitative determination and comparison of the costs is not possible. A rough indication and the buildup of the costs can, however, be given.

V. TREATMENT OF EXCAVATED SOIL

A. Extraction

Estimated costs are Dfl. 100.00 to Dfl. 200.00/ton of soil. The costs depend highly on the amount of soil to be treated, the amount of chemicals that has to be added to the extracting agent, and the amount and nature of residual sludge that has to be either discharged or processed.

B. Thermal Treatment

Estimated costs are Dfl. 150.00 to Dfl. 300.00/ton of soil. The costs depend highly on the amount of soil to be treated, the water content of the soil, the temperature level required for either evaporation or burning of the contaminants, and the method of after-treatment of the flue gases.

C. Steam Stripping

Estimated costs are Dfl. 100.00 to Dfl. 200.00/ton of soil. The costs depend highly on the amount of soil to be treated, the amount of steam required, and the method of after-treatment of the condensate.

D. Chemical Treatment

Estimated costs are Dfl. 30.00 to Dfl. 100.00/ton of soil. The costs depend highly on the amount of soil to be treated and the amount of chemicals required (depending on the nature and the concentration of the contaminations).

VI. *IN SITU* TREATMENT OF SOIL

In general the costs of *in situ* decontamination may be expected to be considerably lower than those of decontamination after excavation (in the range of Dfl. 10.00/ton of soil). The most important factors for the costs of *in situ* decontamination are

- The size of the location to be treated
- The total duration of the decontamination process

Insufficient evidence is yet available to allow clear-cut conclusions concerning the optimum method for nonmicrobiological soil decontamination to be reached.

REFERENCES

1. **Alexander, M.**, Biodegradation: problems of molecular recalcitrance and microbial fallibility, *Adv. Appl. Microbiol.*, 7, 35, 1965.
2. **Alexander, M.**, Non-biodegradable and other recalcitrant molecules, *Biotechnol. Bioeng.*, 15, 611, 1973.
3. **Alexander, M.**, Biodegradation of chemicals of environmental concern, *Science*, 211, 132, 1981.
4. **Anderson, J. P. E., Lichtenstein, E. P., and Wittingham, W. F.**, Effect of *Mucor alternans* on the persistence of DDT in culture and in soil, *J. Econ. Entomol.*, 63, 1595, 1970.
5. Anon., Bodemsanering, Programma 1982, Vols. 1 and 2, Province of Overijssel, February 1982.
6. **Anthony, R. M. and Breimhurst, L. H.**, Determining maximum influent concentrations of priority pollutants for treatment plants, *J. Water Pollut. Control Fed.*, 53, 1457, 1981.
7. **Atlas, R. M.**, Stimulated petroleum biodegradation, *Crit. Rev. Microbiol.*, 5, 371, 1977.
8. **Atlas, R. M.**, Microbial degradation of petroleum hydrocarbons: an environmental perspective, *Microbiol. Rev.*, 45, 180, 1981.
9. **Audus, L. J.**, Biological detoxication of 2,4-dichlorophenoxyacetic acid in soil: isolation of an effective organism, *Nature*, 166, 356, 1950.
10. **Boethling, R. S. and Alexander, M.**, Microbial degradation of organic compounds at trace levels, *Env. Sci. Technol.*, 13, 989, 1979.
11. **Bonnier, P. E. et al.**, Sludgefarming: A Technique for the Disposal of Oily Refinery Wastes, Concawe Report no. 3/80, Concawe, The Hague, The Netherlands, 1980.
12. **Bourquin, A. W. and Pritchard, P. H., Eds.**, Proceedings of the Workshop: Microbial Degradation of Pollutants in Marine Environments, U.S. Governmental Protection Agency, Gulf Breeze, Fla., 1979.
13. **Chatterjee, D. K., Kilbane, J. J., and Chakrabarty, A. M.**, Biodegradation of 2,4,5-trichlorophenoxyacetic acid in soil by a pure culture of *Pseudomonas cepacia*, *Appl. Env. Microbiol.*, 44, 514, 1982.

14. **Chou, T. W. and Bohonos, N.,** Diauxic and co-metabolic phenomena in biodegradation evaluations, in *Microbial Degradation of Pollutants in Marine Environments,* Bourquin, A. W. and Pritchard, P. H., Eds., U.S. Environmental Protection Agency, Gulf Breeze, Fla., 1979, 76.

15. **Chu, J. P. and Kirsch, E. J.,** Metabolism of pentachlorophenol by an axenic bacterial culture, *Appl. Microbiol.,* 23, 1033, 1972.

16. **Clark, C. G. and Wright, S. J. L.,** *Soil Biol. Biochem.,* 2, 19, 1970.

17. **Clark, R. R., Chian, E. S. K., and Griffin, R. A.,** Degradation of polychlorinated biphenyls by mixed microbial cultures, *Appl. Environ. Microbiol.,* 37, 680, 1979.

18. **Clarke, P. H.,** Experiments in microbial evaluation: new enzymes, new metabolic activities, *Proc. R. Soc. London, Ser. B,* 207, 385, 1980.

19. **Colwell, R. R. and Walker, J. D.,** Ecological aspects of microbial degradation of petroleum in the marine environments, *Crit. Rev. Microbiol.,* 5, 423, 1977.

20. **Daughton, C. G. and Hsieh, D. P. H.,** Accelerated parathion degradation in soil by inoculation with parathion-utilizing bacteria, *Bull. Environ. Contam. Toxicol.,* 18, 48, 1977.

21. **Davis, J. B.,** *Petroleum Microbiology,* Elsevier, Amsterdam, 1967.

22. **Edgehill, R. U. and Finn, R. K.,** Microbial treatment of soil to remove pentachlorophenol, *Appl. Environ. Microbiol.,* 45, 1122, 1983.

23. **Evans, W. C.,** Biochemistry of the bacterial catabolism of aromatic compounds in anaerobic environments, *Nature,* 270, 17, 1977.

24. Federal Ministry of the Interior, Bonn, 1970, personal communication.

25. **Franklin, F. C. H., Bagdasarian, M., and Timmis, K. N.,** Manipulation of degradative genes of soil bacteria, in *Microbial Degradation of Xenobiotics and Recalcitrant Compounds,* Leisinger, Th., Cook, A. M., Hütter, R., and Nüesch, J., Eds., Academic Press, London, 1981, 109.

26. **Fussell, D. R. et al.,** Revised Inland Oil Spill Clean-up Manual, Concawe Report no. 7/81, Concawe, The Hague, The Netherlands, 1981.

27. **Grondmij,** Soil rehabilitation, Gouderak ''Zellingweg e.o.''. Further investigation on contaminant levels, Grondmij, PWS South Holland, October, 1982.

28. **Haller, H. D. and Finn, R. K.,** Kinetics of biodegradation of p-nitrobenzoate and inhibition by benzoate in a Pseudomonad, *Appl. Environ. Microbiol.,* 35, 890, 1978.

29. **Hill, R. and Wright, S. J. L., Eds.,** *Pesticide Microbiology,* Academic Press, New York, 1978.

30. **Hopper, D. J.,** Microbial degradation of aromatic hydrocarbons, in *Developments in Biodegradation of Hydrocarbons,* Vol. 1, Watkinson, J. R., Ed., Applied Science, London, 1978.

31. **Horvath, R. S.,** Microbial co-metabolism and the degradation of organic compounds in nature, *Bacteriol. Rev.,* 36, 146, 1972.

32. **Hsu, T. S. and Bartha, R.,** Accelerated mineralization of two organophosphate insecticides in the rhizosphere, *Appl. Environ. Microbiol.,* 37, 36, 1979.

33. **Jamison, V. M., Raymond, R. L., Hudson, J. O.,** Biodegradation of high-octane gasoline, in *Proc. 3rd Int. Biodegradation Symposium,* Sharpley, J. M., and Kaplan, A. M., Eds., Applied Science, London, 1976, 187.

34. **Josephson, J.,** Hazardous waste landfills, *Environ. Sci. Technol.,* 15, 250, 1981.

35. **Klug, M. J. and Markovetz, A. J.,** Utilization of aliphatic hydrocarbons by micro-organisms, *Adv. Microbiol. Physiol.,* 5, 1971.

36. **Knackmuss, H. J.,** Degradation of halogenated and sulfonated hydrocarbons, in *Microbial Degradation of Xenobiotics and Recalcitrant Compounds,* Leisinger, Th., Cook, A. M., Hütter, R., and Nüesch, J., Eds., Academic Press, London, 1981, 189.

37. **Kobayashi, H. and Rittmann, B. E.,** Microbial removal of hazardous organic-compounds, *Environ. Sci. Technol.,* 16, 170, 1982.

38. **Lal, R. and Saxena, D. M.,** Accumulation, metabolism and effects of organochloride insecticides on microorganisms, *Microbiol. Rev.,* 46, 95, 1982.

39. **Leisinger, Th., Cook, A. M., Hütter, R., and Nüesch, J., Eds.,** *Microbial Degradation of Xenobiotics and Recalcitrant Compounds,* Academic Press, London, 1981.

40. **Lindsay, W. L.,** *Chemical Equilibria in Soils,* Wiley Interscience, New York, 1979.

41. **McClure, G. W.,** Accelerated degradation of herbicides in soil by the application of microbial nutrient broths, *Contrib. Boyce Thompson Inst.,* 24, 235, 1970.

42. **McConell, G., Ferguson, D. M., Pearson, C. R.,** Chlorinated hydrocarbons and the environment, *Endeavour,* 34, 13, 1975.

43. **Munnecke, D. M.,** The use of microbial enzymes for pesticide detoxification, in *Microbial Degradation of Xenobiotics and Recalcitrant Compounds,* Leisinger, Th., Cook, A. M., Hütter, R., and Nüesch, J., Eds., Academic Press, London, 1981, 215.

44. Information from officials of the provinces, Rijnmond and Amsterdam, personal communication.

45. **Patterson, J. W. and Kodukala, P. S.,** Biodegradation of hazardous organic pollutants, *Chem. Eng. Progr.,* 77, 48, 1981.

46. **Peakall, D. B.,** PCBs and their environmental effects, *Crit. Rev. Environ. Control,* 5, 469, 1975.

47. **Sijmons, R.,** Kleine gifatlas van Nederland, Annexe to "Vrij Nederland", No. 11, 1983.

48. **Vonk, J. W. et al.,** Kwaliteitskenmerken ten behoeve van der bodembescherming, Eindrapport, Netherlands Organization for Applied Scientific Research, March 23, 1983.

49. **Wise, H. E. and Fahrenthold, P. D.,** Predicting priority pollutants from petrochemical processes, *Environ. Sci. Technol.,* 15, 1292, 1981.

50. **Wood, J. M.,** Chlorinated hydrocarbons: oxidation in the biosphere, *Environ. Sci. Technol.,* 16, 291, 1982.

51. "Volgermeerpolder", Results of investigations in the Volgermeerpolder, Amsterdam, 1981.

52. **Ahmed, M. and Focht, D. D.,** Degradation of polychlorinated biphenyls by two species of *Achromobacter, Can. J. Microbiol.,* 19, 47, 1973.

53. **Aftring, R. P., Chalker, B. E., and Tayler, B. F.,** Degradation of phthalic acids by denitrifying, mixed cultures of bacteria, *Appl. Environ. Microbiol.,* 41, 1177, 1981.

54. **Allan, J.,** Loss of biological efficiency of cattle dipping wash containing benzene hexachloride, *Nature,* 175, 1131, 1955.

55. **Baker, M. D. and Mayfield, C. I.,** Microbial and non-biological decomposition of chlorophenols and phenol in soil, *Water Air Soil Pollut.,* 13, 411, 1980.

56. **Bakker, G.,** Anaerobic degradation of aromatic compounds in the presence of nitrate, *FEMS Microbiol. Lett.,* 1, 103, 1977.

57. **Baxter, R. A., Gilbert, P. E., Lidgett, R. A., Mainprize, J. A., and Vodden, H. A.,** The degradation of polychlorinated biphenyls by microorganisms, *Sci. Total Environ.,* 4, 53, 1975.

58. **Bixby, M. W., Boush, G. M., and Matsumura, F.,** Degradation of dieldrin to carbon dioxide by soil fungus *Trichoderma koningii, Bull. Environ. Contam. Toxicol.,* 6, 491, 1971.

59. **Bouwer, E. J., Rittman, B. E., and McCarty, P. L.,** Anaerobic degradation of halogenated 1-carbon and 2-carbon compounds, *Environ. Sci. Technol.,* 15, 596, 1981.

60. **Bouwer, E. J. and McCarty, P. L.,** Removal of trace chlorinated organic compounds by activated carbon and fixed-film bacteria, *Environ. Sci. Technol.,* 16, 836, 1982.

61. **Bouwer, E. J. and McCarty, P. L.,** Transformation of 1- and 2-carbon halogenated aliphatic compounds under methanogenic conditions, *Appl. Environ. Microbiol.,* 45, 1286, 1983.

62. **Bouwer, E. J. and McCarty, P. L.,** Transformations of halogenated organic compounds under denitrification conditions, *Appl. Environ. Microbiol.,* 45, 1295, 1983.

63. **Brunner, W., Staub, D., and Leisinger, T.,** Bacterial degradation of dichloromethane, *Appl. Environ. Microbiol.,* 40, 950, 1980.

64. **Chapman, P. J.,** Degradation mechanisms, in Proceedings of the Workshop: Microbial Degradation of Pollutants in Marine Environments, Bourquin, A. W. and Pritchard, P. H., Eds., U.S. Environmental Protection Agency, Gulf Breeze, Fla., 1979, 28.

65. **Chatterjee, D. K. and Chakrabarty, A. M.,** Plasmids in the biodegradation of PCBs and chlorobenzoates, in *Microbial Degradation of Xenobiotics and Recalcitrant Compounds,* Leisinger, Th., Cook, A. M., Hütter, R., and Nüesch, J., Eds., Academic Press, London, 1981, 213.

66. **Clarke, P. H. and Ornston, L.,** Metabolic pathways and regulation, in *Biochemistry and Genetics of Pseudomonas,* Vol. 1, Richmond, M. H. and Clarke, P. H., Eds., John Wiley & Sons, London, 1975, 191.

67. **Dagley, S.,** New perspectives in aromatic catabolism, in *Microbial Degradation of Xenobiotics and Recalcitrant Compounds,* Leisinger, T., Cook, A. M., Hütter, H., and Nüesch, J., Eds., Academic Press, London, 1981, 181.

68. **DiGeronimo, M. J., Nikaido, M., and Alexander, M.,** Utilization of chlorobenzoates by microbial populations in sewage, *Appl. Environ. Microbiol.,* 37, 619, 1979.

69. **Doelle, H. W.,** *Bacterial Metabolism,* Academic Press, London, 1975.

70. **Duxbury, J. M., Tiedje, J. M., Alexander, M., and Dawson, J. W.,** 2,4-D Metabolism: enzymatic conversion of chloromaleylacetic acid to succinic acid, *J. Agric. Food Chem.,* 18, 199, 1970.

71. **Eaton, R. W. and Ribbons, D. W.,** Utilization of phtalate esters by *Micrococci, Arch. Microbiol.,* 132, 185, 1982.

72. **Edgehill, R. U. and Finn, R. K.,** Isolation, characterization and growth kinetics of bacteria metabolizing pentachlorophenol, *Eur. J. Appl. Microbiol. Biotechnol.,* 16, 179, 1982.

73. **Engesser, K. J., Schmidt, E., and Knackmuss, H. J.,** Adaptation of *Alcaligenes euthropus* B9 and *Pseudomonas* sp. B13 to 2-fluorbenzoate as growth substrate, *Appl. Environ. Microbiol.,* 39, 68, 1980.

74. **Evans, W. C., Smith, B. S. W., Moss, P., Fernley, H. N., and Davis, J. I.,** Bacterial metabolism of 4-chlorophenoxyacetate, *Biochem. J.,* 122, 509, 1971.

75. **Fewson, C. A.,** Biodegradation of aromatics with industrial relevance, in *Microbial Degradation of Xenobiotics and Recalcitrant Compounds,* Leisinger, T., Cook, A. M., Hütter, R., and Nüesch, H., Eds., Academic Press, London, 1981, 141.

76. **Fischer, P. R., Appleton, J., Pemberton, J. M.,** Isolation and characterization of the pesticide-degrading plasmid pJP1 from *Alcaligenes paradoxus, J. Bacteriol.,* 135, 798, 1978.

77. **Furukawa, K., Tonomu, K., and Kamibayashi, K.,** Effect of chlorine substitution on the biodegradability of polychlorinated biphenyls, *Appl. Environ. Microbiol.,* 35, 223, 1978.

78. **Furukawa, K., Tomizuka, N., and Kamibayashi, A.,** Effect of chlorine substitution on the bacterial metabolism of various polychlorinated biphenyls, *Appl. Environ. Microbiol.,* 38, 301, 1979.

79. **Furukawa, K. and Chakrabarty, A. M.,** Involvement of plasmids in total degradation of chlorinated biphenyls, *Appl. Environ. Microbiol.,* 44, 619, 1982.

80. **Furukawa, K.,** Microbial degradation of polychlorinated biphenyls (PCBs), in *Biodegradation and Detoxification of Environmental Pollutants,* Chakrabarty, A. M., Ed., CRC Press, Boca Raton, Fla., 1982, 33.

81. **Gaunt, J. K. and Evans, W. C.,** Metabolism of 4-chloro-2-methylphenoxyacetate by a soil pseudomonad. Preliminary evidence for the metabolic pathway, *Biochem. J.,* 122, 519, 1971.

82. **Gibson, D. T.,** Microbial degradation of aromatic compounds, *Science,* 161, 1093, 1968.

83. **Gibson, D. T.,** The microbial oxidation of aromatic hydrocarbons, *Crit. Rev. Microbiol.,* 1, 199, 1971.

84. **Gibson, D. T., Gschwendt, B., Yeh, W. K., and Kobal, V. M.,** Initial reactions in the oxidation of ethylbenzene by *Pseudomonas putida, Biochemistry,* 12, 1520, 1973.

85. **Harper, D. B. and Blakley, E. R.,** The metabolism of *p*-fluorophenylacetic acid by a *Pseudomonas* sp. II. The degradative pathway, *Can. J. Microbiol.,* 17, 645, 1971.

86. **Hooftman, R. N. and de Kreuk, F. J.,** Investigation on the environmental load of chlorinated benzenes (a literature study), CL 81/153a, MT-TNO, Delft, The Netherlands, 1981.

87. **Jagnow, G. and Haider, K.,** Evolution of $^{14}CO_2$ from soil incubated with dieldrin-^{14}C, *Soil Biol. Biochem.,* 4, 43, 1972.

88. **Jagnow, G,. Haider, K., and Ellwardt, P. C.,** Anaerobic dechlorination and degradation of hexachlorocyclohexane by anaerobic and facultative anaerobic bacteria, *Arch. Microbiol.,* 115, 285, 1977.

89. **Janssen, D. B., Scheper, A., and Witholt, B.,** Biodegradation of 2-chloroethanol and 1,2-dichloroethane by pure bacterial cultures, *Prog. Ind. Microbiol.,* 20, 169, 1984.

89a. **Janssen, D. B., Schepen, A., Dijkhuizen, L., and Witholt, B.,** Degradation of halogenated aliphatic compounds by *Xanthobacter autotrophicus* GJ10, *Appl. Environ. Microbiol.,* 49(3), 673, 1985.

90. **Kamp, P. F. and Chakrabarty, A. M.,** Plasmids specifying *p*-chloribiphenyl degradation in enteric bacteria, in *Plasmids of Medical, Environmental and Commercial Importance,* Timmis, K. N. and Puhler, A., Eds., Elsevier/North-Holland, Amsterdam, 1979, 275.

91. **Klecka, G. M.,** Fate and effects of methylene chloride in activated sludge, *Appl. Environ. Microbiol.,* 44, 701, 1982.

92. **Korte, F. and Stiasni, M.,** Umwandlung von Telodrin durch Mikroorganismen und Moskitolargen, *Justus Liebigs Ann. Chem.,* 673, 146, 1964.

93. **Kujawa, M., Haertig, M., Macholz, R. M., and Engst, R.,** Der Abbau von ^{14}C-Lindane durch eine Schimmelpilzkultur, *Nahrung,* 20, 181, 1976.

94. **Kurane, R., Suzuki, T., and Takahara, Y.,** Isolation of microorganisms growing on phthalate esters and degradation of phthalate esters by *Pseudomonas acidovorans* 256-1, *Agric. Biol. Chem.,* 41, 2119, 1977.

95. **Lichtenstein, E. P. and Schulz, K. R.,** Epoxidation of aldrin and heptachlor in soils as influenced by autoclaving, moisture and soil types, *J. Econ. Entomol.,* 53, 192, 1960.

96. **MacRae, I. C., Raghu, K., and Castro, T. F.,** Persistence and biodegradation of four common isomers of benzene hexachloride in submerged soils, *J. Agric. Food Chem.,* 15, 911, 1967.

97. **MacRae, I. C., Raghu, K., and Bautis, E. M.,** Anaerobic degradation of insecticide lindane by *Clostridium* sp., *Nature,* 221, 859, 1969.

98. **Marinucci, A. C. and Bartha, R.,** Biodegradation of 1,2,3- and 1,2,4-trichlorobenzene in soil and in liquid enrichment culture, *Appl. Environ. Microbiol.,* 38, 811, 1979.

99. **Matsumura, F. and Benezet, H.,** Microbial degradation of insecticides, in *Pesticide Microbiology,* Hill, R. and Wright, S. L. J., Eds., Academic Press, New York, 1978, 623.

100. **Matsumura, F., Patil, K. C., and Boush, G. M.,** Formation of photodieldrin by microorganisms, *Science,* 170, 1206, 1970.

101. **Miller, S.,** Drinking water and its treatment, *Environ. Sci. Technol.,* 14, 510, 1980.

102. **Ohisa, M. and Yamaguchi, M.,** Gamma-BHC degradation accompanied by the growth of *Clostridium* isolated from paddy field, *Agric. Biol. Chem.,* 42, 1819, 1978.

103. **Pearson, C. R. and McConell, G.,** Chlorinated C1 and C2 hydrocarbons in the marine environment, *Proc. R. Soc. London, Ser. B,* 189, 305, 1975.

104. **Reineke, W. and Knackmuss, H. J.,** Microbial metabolism of haloaromatics: isolation and properties of a chlorobenzene-degrading bacterium, *Appl. Environ. Microbiol.,* 47, 395, 1984.

105. **Rittmann, B. E. and McCarty, P. L.,** Utilization of dichloromethane by suspensed and fixed-film bacteria, *Appl. Environ. Microbiol.,* 39, 1225, 1980.

106. **Sethunathan, N., Bautista, E. M., and Yoshida, T.,** Degradation of benzene hexachloride by soil bacterium, *Can. J. Microbiol.,* 15, 1349, 1969.

107. **Shelat, Y. A. and Patel, K. S.,** Utilization of bromobenzene as a sole source of carbon by *Bacillus polymyxa, Curr. Sci.,* 42, 368, 1973.

108. **Stucki, G., Brunner, W., Staub, D., and Leisinger, T.,** Microbial degradation of chlorinated C1 and C2 hydrocarbons, in *Microbial Degradation of Xenobiotics and Recalcitrant Compounds,* Leisinger, T., Cook, A. M., Hütter, R., and Nüesch, H., Eds., Academic Press, London, 1981, 131.

109. **Stucki, G., Gälli, R., Ebersold, H.-R., and Leisinger, T.,** Dehalogenation of dichloromethane by cell extracts of Hyphomicrobium DMz, *Arch. Microbiol.,* 130, 366, 1981.

110. **Stucki, G., Krebser, U., and Leisinger, T.,** Bacterial growth on 1,2-dichloroethane, *Experientia,* 39, 1271, 1983.

111. **Suflita, J. M., Horowitz, A., Shelton, D. R, and Tiedje, J. M.,** Dehalogenation: a novel pathway for the anaerobic degradation of halo-aromatic compounds, *Science,* 218, 1115, 1982.

112. **Symons, J. M., Stevens, A. A., Clark, R. M., Geldreich, E. E., Love, O. T., and DeMarco, J.,** Removing trihalomethanes from drinking water, *Water Eng. Manage.,* 128, 50, 1981.

113. **Tu, C. M., Miles, J. R. W., and Harris, C. R.,** Soil microbial degradation of aldrin, *Life Sci.,* 7, 311, 1968.

114. **Tucker, E. S., Saeger, V. W., and Hicks, O.,** Activated sludge primary biodegradation of polychlorinated biphenyls, *Bull. Environ. Contam. Toxicol.,* 14, 705, 1975.

115. **Verscheuren, K.,** *Handbook of Environmental Data on Organic Chemicals,* Van Nostrand Reinhold, New York, 1977.

116. **Vockel, D. and Korte, F.,** Beitrage zur ökologischen Chemie. LXXX. Versuche zum Mikrobiellen Abbau von dieldrin und 2,2-dichlorobiphenyl, *Chemie (Prague),* 20, 412, 1974.

117. **Vonk, J. W., Hanstveit, A. O., Barug, D., and Adema, D. M. M.,** Milieutoxicologisch onderzoek van gechloreerde fenolen, Rapport CL 81/34 (MT-TNO), Netherlands Organization for Applied Scientific Research, 1981.

118. **Williams, P. A.,** Genetics of biodegradation, in *Microbial Degradation of Xenobiotics and Recalcitrant Compounds,* Leisinger, T., Cook, A. M., Hütter, R., and Nüesch, J., Eds., Academic Press, London, 1981, 97.

119. **Adema, E.,** Compostfiltratie, Report no. 81-05572 (MT-TNO), Netherlands Organization for Applied Scientific Research, 1981.

120. **Anon.,** Verfahren zur Herstellung von Bodenverbesserungsmittel, East Germany Patent 147, 663, 1981.

121. **Atlas, R. M. et al.,** Abundance, distribution and Oil Biodegradation Potential of Microorganisms in Raritan Bay, *Environ. Pollut.,* 4, 291, 1973.

122. **Bölsing, F.,** Mikroben gegen Ölschäden, in Das technische Umweltmagazin, June 1975.

123. **Bosma, A. H. et al.,** Ontwikkeling van een onderzoek aan een zelfrijdende mestinjecteur, IMAG-publikatie 87, 1977.

124. **de Renzo, A. J.,** Unit Operations for Treatment of Hazardous Industrial Wastes, Noyes Data Corporation, Park Ridge, N.J., 1978.

125. **Doyle, R. C. et al.,** Treatment of TNT and RDX contaminated soils by composting, paper presented at the Natl. Conf. Manage. of Uncontrolled Hazardous Waste Sites, Washington, D.C., 1982.

126. EPA, EPA Handbook Remedial Action at Waste Disposal Sites, EPA — 625/6-82-006, U.S. Environmental Protection Agency, Washington, D.C., 1982.

127. **Gudin, C. et al.,** Biological aspects of land rehabilitation following hydrocarbon contamination, *Environ. Pollut.,* 8, 107, 1975.

128. **Grove, G. W.,** Use landfarming for oily waste disposal, *Hydrocarbon Process.,* 57(5), 130, 1978.

129. **Lektomäki, M. et al.,** Improving microbial degradation of oil in soil, *Ambio,* 4(Suppl. 3), 126, 1975.

130. **van Oudenhoven, J. A. C. M.** Disposal Techniques for Split Oil, Concawe Report No. 9/80, Concawe, The Hague, The Netherlands, 1980.

131. **Ponsen, R. A.,** Slib-Compostering, een nieuwe vorm van natuurlijke slibverwerking, H_2O, 14(Suppl. 18), 401, 1981.

132. **Schaake, P.,** Het mengen van vaste stoffen, *Procestechniek,* 16 (Suppl. 5), 313, 1973.

133. Compost en zwarte grond uit zuiveringsslib. I. Systemen, technologie en ervaring (inventarisatie) STORA rapport, September, 1982.

134. Slibontwatering tot meer dan 40% droge stof, STORA rapport, May 1979.

135. WEBCI prijzenboekje, 10e druk, 1982.

136. **Anon.,** Classifying Solid Waste Disposal Facilities: a Guidance Manual, SW-828, Office of Solid Waste, U.S. Environmental Protection Agency, Washington, D.C., 1980.

137. **Constable, T. et al.,** *Gas Migration and Modelling,* Proc. 5th Annu. Res. Symp. on Municipal Solid Waste: Land Disposal, EPA-600-9-79-023a, U.S. Environmental Protection Agency, Washington,D.C., 1979.

138. EPA, U.S. EPA Procedure Manual for Groundwater Monitoring at Solid Waste Disposal Facilities, EPA/530/SW-616, Office of Solid Waste, U.S. Environmental Protection Agency, Washington, D.C., 1979.

139. **Farmer, W. et al.,** Land disposal of hexachlorobenzene wastes, controlling vapor movements in soils. Land disposal of hazardous wastes, EPA-600/9-78-016, Municipal Environmental Research Laboratory, U.S. Environmental Protection Agency, Cincinnati, Ohio, 1978.

140. **Moore, C. A.,** Theoretical approach to gas movement through soils. Gas and leachate from landfills, formation, collection, EPA-600-19-76-004, U.S. Environmental Protection Agency, Cincinnati, Ohio, 1976.

140a. **Rulkens, W. H. et al.,** Study on contaminated land, in Project B: On-Site Processing of Contaminated Soil, NATO/CcMS, November 1982.

141. **Shen, T. T. et al.,** Air pollution aspects of land disposal of toxic waste. Hazardous materials risk assessment, disposal and management, *J. Environ. Eng. Div., Am. Soc. Civ. Eng.,* 106 (EEI), 211, 1980.

142. **Shen, T. T.,** Control techniques for gas emissions from hazardous waste landfills, *J. Air Pollut. Control Assoc.,* 31 (Suppl. 2), 132, 1981.

143. **Stone, R.,** Reclamation of landfill methane and control of off-site migration hazards, *Solid Waste Manage./Refuse Removal J.,* 1978.

144. **Stone, R.,** Preventing the underground movement of methane from sanitary landfills, *Civ. Eng.,* 1978.

145. **Thibodeaux, L. et al.,** Estimating the air emissions of chemicals from hazardous waste landfills, in American Institute of Chemical Engineers Annual Meeting, San Francisco, 1979.

146. **VROM,** Natuurlijke afdichtingsmaterialen, in Reeks Bodembescherming, Vol. 8, Staatsuitgeverij, The Hague, The Netherlands, 1983.

147. **VROM,** *Handboek Bodemsaneringstechnieken,* Staatsuitgeverij, The Hague, The Netherlands, 1983.

148. **Raymond, R. L., Hudson, J. O., and Jamison, V. W.,** Oil degradation in soil, *Appl. Environ. Microbiol.,* 31, 522, 1976.

149. **Dibble, J. T. and Bartha, R.,** Rehabilitation of oil-inundated agricultural land: a case history, *Soil Sci.,* 128(1), 56, 1979.

150. **Motosugi, K. and Soda, K.,** Microbial degradation of synthetic organochlorine compounds, *Experientia,* 39(11), 1214, 1983.

151. **Bollag, J. M. and Loll, M. J.,** Incorporation of xenobiotics into soil humus, *Experientia,* 39(11), 1221, 1983.

152. **Ghisalba, O.,** Chemical wastes and their biodegradation — an overview, *Experientia,* 39(11), 1247, 1983.

Chapter 3

TREATMENT OF HAZARDOUS WASTES IN A SEQUENCING BATCH REACTOR

Philip A. Herzbrun, Robert L. Irvine, Kenneth C. Malinowski, and Michael J. Hanchak

TABLE OF CONTENTS

I.   Abstract 158

II.   Introduction 158

III.   Bench-Scale Studies 160
     A.   Operating Strategies 161
     B.   Results 162
         1.   Organic Removal 162
         2.   Oxygen Uptake-Rate Studies 162
         3.   Additional Studies 165

IV.   Full-Scale Sequencing Batch Reactor Demonstration Facility 166

V.   Conclusions 166

Acknowledgments 167

References 167

I. ABSTRACT

The primary means for removing organic carbon compounds at the Chemical and Environmental Conservation Systems (CECOS) International Niagara Falls Wastewater Treatment Plant (NFWTP) is activated carbon. Because of the cost and energy demands associated with the use of activated carbon, CECOS investigated the sequencing batch reactor (SBR), a biological treatment system, as an alternate method for the removal of organic carbon from waste water. Coincidentally, the New York State Energy Research and Development Authority (NYSERDA) was seeking candidates for the cofunding of a full-scale demonstration of an advanced biological treatment system which would result in energy savings. A proposal submitted by CECOS to NYSERDA was accepted. Bench-scale studies were initiated by CECOS in August 1983. The bench-scale reactors were operated at several loadings and four detention times. A description of the CECOS hazardous waste disposal site is presented along with SBR bench-scale results on organic carbon and phenol removal, sludge-settling characteristics, and oxygen-uptake rates. Based on the results reported herein, a full-scale SBR system was designed, with construction beginning in December 1983. The full-scale SBR was placed into operation in June 1984.

II. INTRODUCTION

The CECOS International Wastewater Treatment Plant in Niagara Falls is one of only two fully permitted treatment, storage, and disposal facilities in New York. It treats both leachates from industrial landfills and industrial wastes shipped either in bulk or drums.

Located on 14.9 km² in Niagara County, CECOS presently manages a number of operations, including a secure chemical management facility, a secure sludge management facility, a sanitary landfill, a copper recovery unit for decontamination of PCB transformers, a fuels blending program, and a multifaceted waste water treatment system. The waste water treatment plant located in Niagara Falls is a two-phase operation with the first phase dedicated to storage, equalization, chemical precipitation, oxidation, neutralization, and sludge handling and the second phase to storage, neutralization, and organic carbon removal.

The waste water treatment system typically receives 9460 to 11,260 m³/month. Approximately 50% of this is leachate pumped from both active and inactive landfills on the CECOS site, approximately 30% is pumped as part of a remedial ground water program from an unrelated facility bordering the CECOS site, and the remaining 20% is received in bulk or drums from various industries in a 640-km radius.

The first phase of the Niagara Falls waste water treatment facility is shown in Figure 1. This phase involves the chemical oxidation of organics using waste oxidizers such as dichromate in Tank L-1, reduction of inorganic heavy metal oxidizers in Tank L-R, lime addition for acid neutralization and metal-hydroxide precipitation, and sludge dewatering via filter press.

The second phase, shown in Figure 2, involves pH adjustment in Tank L-8 using hydrochloric acid, sand filtration, organic removal via three granular activated carbon columns (a fourth column is used as an operating standby), and biological treatment in an aerated lagoon (Tank L-10). No provisions are made for biological-solids recovery in the aerated lagoon. The effectiveness of the biological degradation is not controlled and performance is directly related to seasonal fluctuations in temperature. The treated effluent is discharged to the NFWTP.

As can be seen from Figure 2, all discharging to the NFWTP is done on a batch basis utilizing either Tanks L-10, -11, or -12. All batches are sampled and analyzed, permission is verbally given by the operators of the NFWTP, and the appropriate tank is discharged. A monitoring station automatically takes a flow-proportioned sample which is subsequently

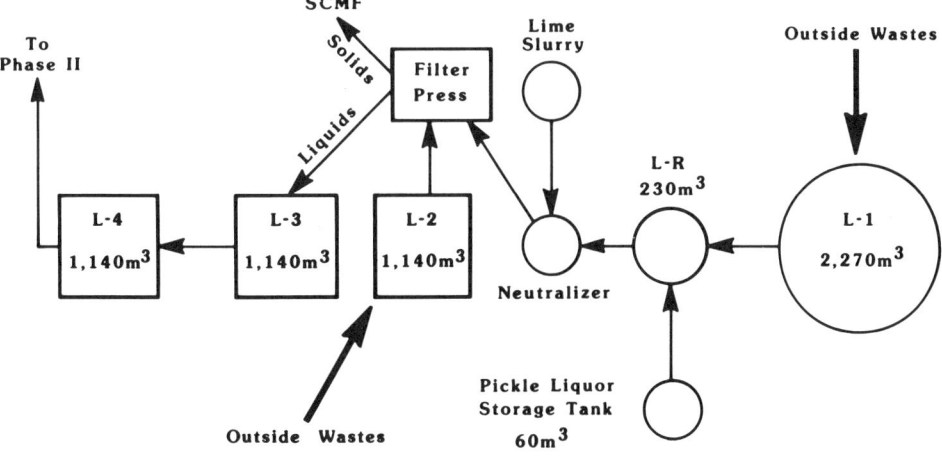

FIGURE 1.   Phase I waste water treatment system.

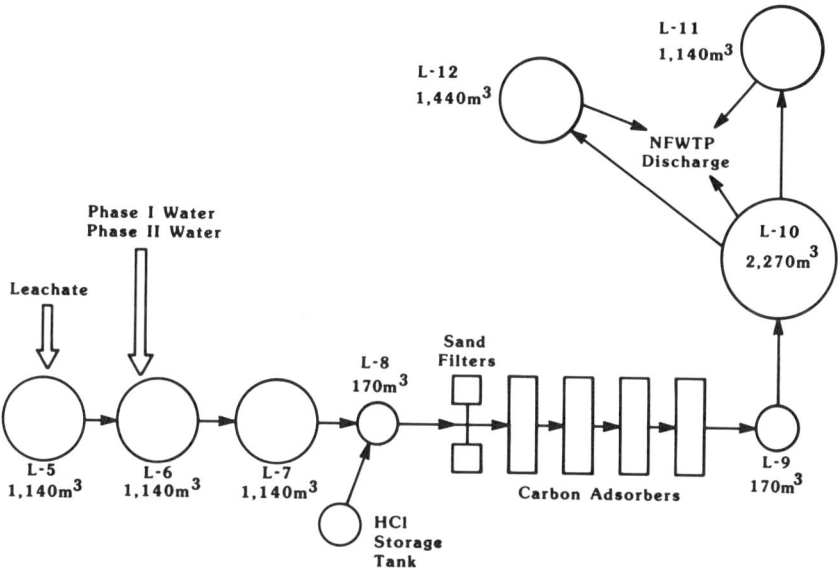

FIGURE 2.   Phase II waste water treatment system.

split with NFWTP personnel and reanalyzed for appropriate parameters. The effluent limitations which must be met are summarized in Table 1.

Due to the variability and toxic nature of the industrial waste waters being treated, the system was initially designed so that primary removal of toxic organics was accomplished by adsorption on granular activated carbon, followed by additional treatment through biological action. Replacement of activated carbon quickly became the single largest operating expense in the system. Each activated carbon column requires a bed change approximately once every 30 to 60 days at a cost per change of roughly $20,000.

In late 1982, CECOS decided to upgrade the biological system. Concurrently, a grant was offered by NYSERDA to cofund a new and innovative biological treatment system which could demonstrate significant energy conservation. CECOS applied for this grant in February 1983, based on use of an innovative technology, the SBR.[1] The benefit expected

Table 1
EFFLUENT LIMITATIONS

Total organic carbon	1000 g/m^3
Total halogenated organics	0.1 g/m^3
Volatile halogenated priority pollutants	1.0 g/m^3 each
Chemical oxygen demand	1040 kg/day
Phenol	0.8 kg/day
Total suspended solids	990 kg/day
Organic phosphorus	1.6 kg/day
pH	5.5—9.0

Table 2
RAW WASTE WATER
CHARACTERISTICS

pH	7.5—8.2
Phenol (g/m^3)	20.4—44.4
Total solids (g/m^3)	22,000—32,000
Total organic carbon (g/m^3)	1,100—1,500
Ammonia (g/m^3)	200—300
Total phosphorus (g/m^3)	1.8—3.8
Cu(g/m^3)	0.1—1.5
Ni(g/m^3)	0.7—1.2
Ba(g/m^3)	5—100
Zn(g/m^3)	0.1—0.3

from the use of biological degradation in the SBR was a dramatic drop in the use of activated carbon. Because carbon regeneration is extremely energy intensive, an energy savings of 180 million kcal/year was calculated. In July 1983, the NYSERDA awarded CECOS a contract to build a full-scale SBR demonstration plant for the treatment of hazardous wastes.

An earlier SBR demonstration for the treatment of municipal waste waters was highly successful[2] as were bench-scale SBR studies for the treatment of leachates from the Hyde Park Landfill Disposal Site.[3] Nevertheless, a bench-scale study was initiated in August 1983. The objectives of this study were to prove the treatability of CECOS waste water, provide data for proper sizing of the full-scale tank and aeration equipment, and become familiar with SBR operation and monitoring.

III. BENCH-SCALE STUDIES

The results presented herein were obtained from four SBRs which were operated and monitored over a period of 2 months. A major goal of the bench-scale studies was to determine as quickly as possible approximate information on reactor sizing. The first systems were operated at a detention time of 10 days because of experience obtained from previous studies.[3] As is described below, a range of detention times was tested by progressively increasing the volume of waste water treated each day. Typical variations in the characteristics of selected influent parameters are shown in Table 2.

Each of the four bench-scale reactors was a 10-ℓ polycarbonate container which was drilled and ported to allow gravity discharge of preset volumes through automatic solenoids. Air was supplied by an air compressor and controlled through automatic solenoids. Additional mixing was provided by magnetic stirrers. Feed was supplied to the reactors through peristaltic pumps and adjusted for appropriate volumes depending on which of four treatment strategies was employed. All pumping, aeration, agitation, and discharge functions were automatically sequenced through the use of laboratory timers.

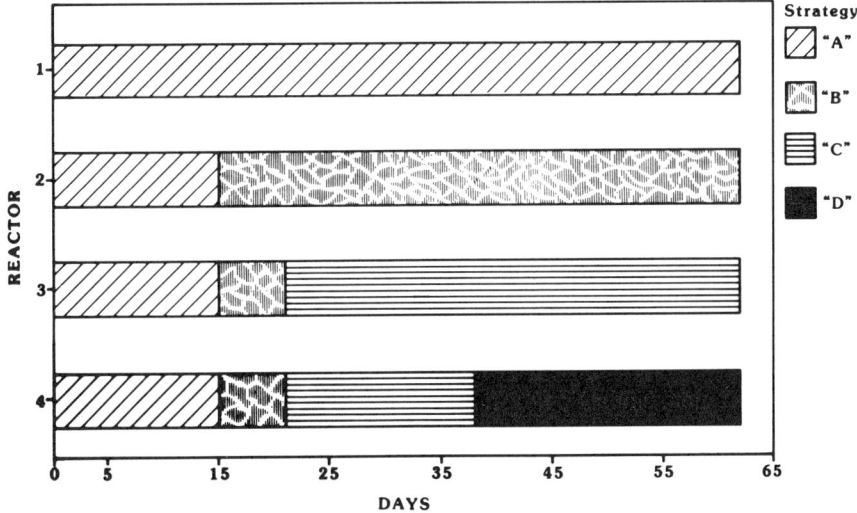

FIGURE 3. Reactor operating strategies.

Table 3
REACTOR OPERATING STRATEGIES

	Strategy			
	A	B	C	D
Detention time (days)	10	5	2.5	1.25
Maximum reactor liquid vol (ℓ)	10	10	10	10
Feed vol per cycle (ℓ)	1	2	2	4
Cycles per day	1	1	2	2
Total vol fed per day (ℓ)	1	2	4	8
Time for (hr)				
Fill	10	10	5	5
React	10	10	5	5
Settle	2	2	1	1
Draw/idle	2	2	1	1
Total cycle time (hr)	24	24	12	12

All four reactors were operated at room temperature with reactor temperatures ranging between 24 to 26°C. Feed was obtained from the precarbon pH adjustment tank (Tank L-8, see Figure 2), initially in 2-week supplies, later on a daily basis.

A. Operating Strategies

All reactors (arbitrarily labeled 1, 2, 3, and 4) were seeded with activated sludge obtained from the Niagara County Sewer District #1 Treatment Plant. Four operating strategies, designated A, B, C, and D, were tested. As can be seen in Figure 3, all four reactors were operated under strategy A for the first 15 days and switched, as shown, to the other operating strategies.

As can be seen from Table 3, strategy A involved the daily addition of 1 ℓ of waste water over a 10-hr period (designated "fill") to a reactor with an initial volume of 9 ℓ. The fill period, which was fully aerated, was followed by a 10-hr period of aeration without substrate addition (this period is designated "react"). In strategy B, the feed volume was doubled;

Table 4
PERFORMANCE DATA

Reactor number	1	2	3	4
Final treatment strategy	A	B	C	D
Weeks at final strategy	7	7	6	3.5
Influent TOC (g/m³)	1440(14)	1440(14)	1480(13)	1570(11)
Range (g/m³)	1200—1890	1200—1890	1300—1890	1370—1890
Effluent TOC (g/m³)	280(16)	300(15)	360(13)	710(7)
Range (g/m³)	200—340	230—410	270—440	480—940
TOC degraded (%)	81	79	76	55
Influent phenol (g/m³)	39.5(13)	39.5(13)	42.9(12)	52.8(10)
Range (g/m³)	20.6—57	20.6—57	31.0—57	43.8—64.6
Effluent phenol (g/m³)	0.3(10)	0.3(12)	1.0(16)	1.7(8)
Range (g/m³)	0.2—0.5	0.2—0.6	0.5—1.4	1.5—2.0
Phenol degraded (%)	99.2	99.2	97.7	96.8
MLSS (g/m³)	3200(14)	4300(14)	5100(14)	6900(7)
MLVSS (g/m³)	2300 (6)	3200 (6)	3400 (7)	4000(4)
Effluent SS (g/m³)	120 (9)	150 (9)	120 (9)	90 (5)
Range (gm/³)	60—200	50—260	40—180	70—120
SVI (mℓ/g)	30 (6)	25 (6)	30 (6)	25 (3)
Range (mℓ/g)	15—50	20—40	15—50	15—35

Note: Number of data points appear in parentheses.

in strategy C, the doubled feed volume was added twice each day; in strategy D, the feed volume per cycle was double again. The obvious intent of this experimental program was to systematically increase the waste loading but only after providing some time for acclimation.

B. Results

1. Organic Removal

Average performance data for each of the four reactors are summarized in Table 4 for percent removal of total organic carbon (TOC) and phenol. Effluent suspended solids (SS) concentrations as well as reactor mixed liquor suspended solids (MLSS), mixed liquor volatile suspended solids (MLVSS), and sludge volume index (SVI) are also reported in this table. With the exception of reactor 1, data for each reactor are only presented for the period during which a new operating strategy was tested. In the case of reactor 1, results from the initial 2-week period are not included so that the impact of the initial inoculum on performance could be minimized.

The percent degradation of both phenol and TOC is shown to decrease with decreasing detention time in Table 4, with the major impact being on TOC removal. This is shown graphically in Figure 4. Strategy D resulted in a clear system overload. As a result, the decision was made to design the full-scale system to operate with detention times between 2.5 and 10 days.

2. Oxygen Uptake-Rate Studies

In order to assess the overall reaction potential of organism systems developed for the different operating strategies, oxygen uptake rates were determined throughout fill and react periods for strategies A, C, and D (B was not run) by removing mixed liquor samples from the reactor and placing them in a closed 135-mℓ vessel and monitoring the change in dissolved oxygen (DO) with time. Unit oxygen uptake rates were determined by dividing the oxygen uptake rates by the MLVSS concentration in kilograms per cubic meter. Results from these studies are summarized in Tables 5, 6, and 7.

All three studies demonstrated a rapid increase in the oxygen uptake rates at the beginning of the feed cycle, indicating that the influent is readily biodegradable in the SBR. In addition,

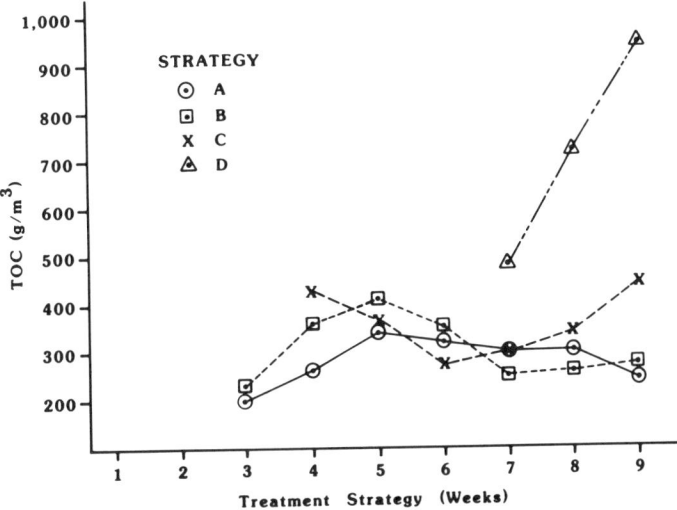

FIGURE 4. Weekly averages of effluent TOC for each operating strategy.

Table 5
OXYGEN UPTAKE-RATE STUDY: STRATEGY A

Cycle time (hr)	Reactor DO (g/m³)	Oxygen uptake rate (g/m³/hr)	Unit oxygen uptake (g/kg/hr)	TOC (g/m³)
Fill period				
0.0	—	—	—	250
0.2	5.5	14	4.8	280
1.2	6.0	15	5.0	—
2.5	6.5	15	5.0	—
4.6	6.6	15	4.9	270
7.0	6.4	14	4.7	—
9.8	6.1	15	4.9	—
React period				
10.2	6.3	13	4.4	230
11.2	6.5	11	3.6	—
12.3	6.4	11	3.6	—
13.3	6.4	9	2.9	—
14.3	6.5	8	2.7	260
23.0[a]	6.8	6	2.0	270

[a] React extended 3 hr to determine impact of additional aeration.

the two treatment strategies (A and C) likely to be utilized most in the full-scale demonstration study had unit oxygen uptake rates that were at or near endogenous rates before the end of the react period, and gave peak unit oxygen uptake rates well within the capabilities of the full-scale aeration equipment that was ultimately designed. The oxygen uptake rates for strategy D, however, were greater than those which can be easily met with conventional aeration equipment and did not reach endogenous levels during the react period.

Additional oxygen uptake-rate studies were conducted to better determine the impact of substrate concentration on the rate of oxygen consumption. In these studies, samples of the mixed liquor from reactors operating under strategies A and C were removed and aerated.

Table 6
OXYGEN UPTAKE-RATE STUDY: STRATEGY C

Cycle time (hr)	Reactor DO (g/m³)	Oxygen uptake rate (g/m³/hr)	Unit oxygen uptake (g/kg/hr)	TOC (g/m³)
Fill period				
0.3	1.8	38	8.5	500
1.3	1.5	65	14.3	330
2.3	1.6	67	14.8	475
3.6	1.1	64	14.3	390
4.9	0.9	60	13.3	310
React period				
5.9	4.7	31	6.8	360
7.3	5.3	24	5.2	325
8.5	5.3	22	5.0	300
9.5	5.3	21	4.7	320

Table 7
OXYGEN UPTAKE-RATE STUDY: STRATEGY D

Cycle time (hr)	Reactor DO (g/m³)	Oxygen uptake rate (g/m³/hr)	Unit oxygen uptake (g/kg/hr)	TOC (g/m³)
Fill period				
0.0	2.0	—	—	800
0.3	0.8	150	35.7	—
1.1	0.8	200	47.9	—
2.5	0.5	240	56.0	830
3.8	0.6	230	54.0	—
4.9	0.7	220	51.4	870
React period				
5.3	0.4	150	35.5	—
6.1	0.7	140	34.3	—
7.3	0.6	130	30.0	860
8.5	0.7	90	21.2	—
9.5	1.1	90	21.0	770
28.6[a]	—	20	4.3	610

[a] Sample removed from reactor at end of the react period and aerated for an additional 18.6 hr.

Of each, 135-mℓ aliquots were spiked with varying amounts of SBR-feed waste water obtained on September 20, 1983. The TOC of this waste water was 1430 g/m³ and the MLVSS for strategies A and C were 2000 and 3200 g/m³, respectively. Results are illustrated in Table 8 and Figure 5.

These data also confirm the biodegradability of the feed waste water in the SBR. In addition, the unit oxygen uptake-rate data as expressed suggest that there may be a greater fraction of viable organism in reactor C than in reactor A. Perhaps most importantly, the spiking of reactor A indicated that larger volumes of water may be able to be processed during peak periods during warm weather.

Table 8
SPIKED OXYGEN UPTAKE-RATE STUDIES:
REACTOR STRATEGIES A AND C

Reactor strategy	Spike vol (mℓ)	Oxygen uptake (g/m³/hr)	Unit oxygen uptake (g/kg/hr)	Theoretical TOC (g/m³)	Change in TOC (g/m³)
A	0.0	5	2.7	320	—
A	0.5	22	11.1	324	4
A	1.0	48	24.0	328	8
A	4.0	73	36.8	352	32
A	8.0	76	38.2	382	62
C	0.0	14	4.4	340	—
C	0.5	61	18.9	344	4
C	1.0	110	32.4	348	8
C	4.0	160	49.4	371	31
C	8.0	180	56.5	401	61

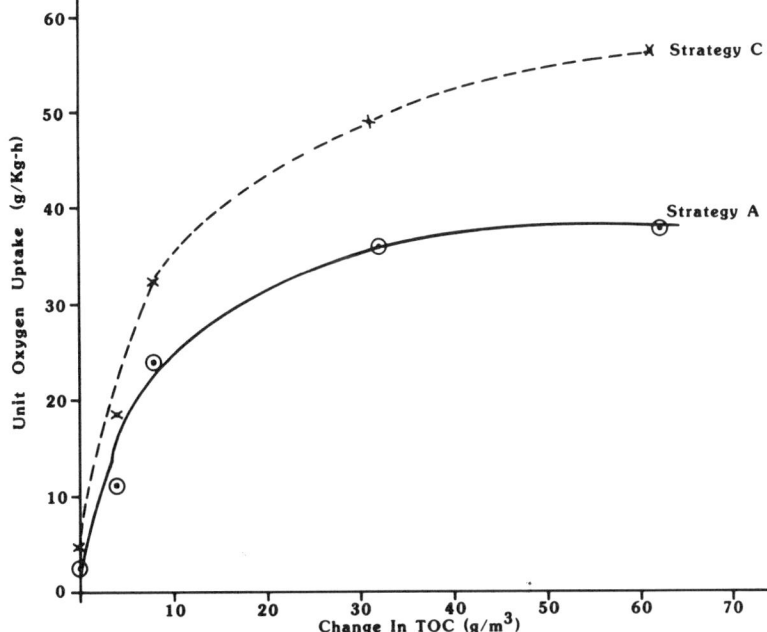

FIGURE 5.   Unit oxygen uptake vs. TOC.

3. Additional Studies

In order to check the need for phosphorus beyond that present in the raw waste water, supplemental phosphorus in the form of hexametaphosphate and/or phosphonate was added for a period of 3 weeks without differences in performance noted.

During the 3-month period, excessive foaming occurred on occasion. As required, minor amounts of a commercially available bubble breaker that is normally used at the CECOS Wastewater Treatment Plant was added with immediate improvement without any loss in treatment performance.

In order to evaluate a power failure or significant mechanical problems, one reactor was

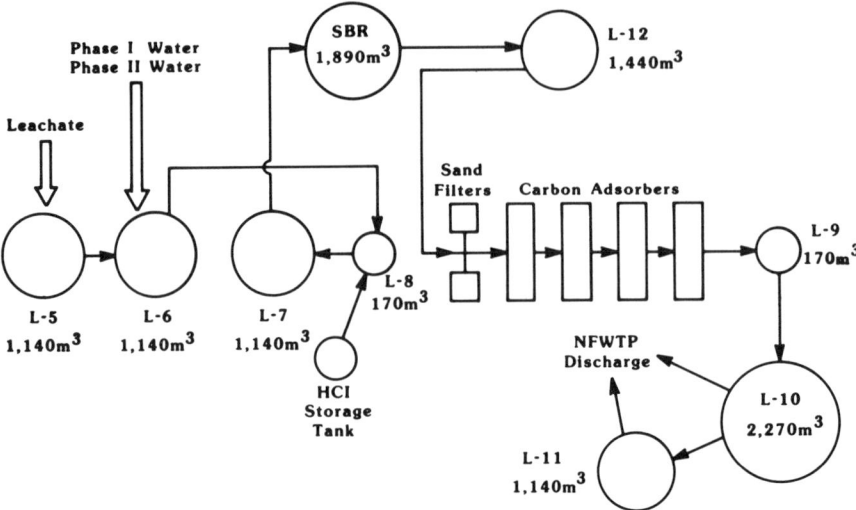

FIGURE 6.   Phase II waste water treatment system modified to include the SBR.

left totally quiescent for 48 hr. TOCs were taken immediately before and after this period, and monitoring continued on the usual schedule. No effects, such as elevated phenol or TOC, excessive odor, or flotation of the sludge blanket, were observed.

Studies to investigate the functioning of a reactor at temperatures as low as 5°C were performed after those reported herein and are presented elsewhere.[4]

IV. FULL-SCALE SEQUENCING BATCH REACTOR DEMONSTRATION FACILITY

As a result of the bench-scale studies, a full-scale demonstration facility was designed. Construction was initiated in December 1983. Modifications to the existing treatment facility were made to the Phase II operation as can be seen in Figure 6. Major changes involved the construction of a 1890 m³ SBR, rerouting of waste water from L-6 to -8 for pH adjustment with hydrochloric acid, storage of this water in L-7 for ultimate treatment in the SBR, and use of L-12 for any additional sedimentation that may be necessary and as a feed tank to the activated carbon system. The full-scale SBR was placed into operation in June 1984.[4]

As a result of these modifications, the role of activated carbon in the treatment system will be shifted from that of being the primary means for removing organic carbon to one which involves the final polishing of the effluent. Because of the reduction in the mass of TOC that will be removed by the carbon columns, the times between regeneration of the activated carbon will be extended. This will result in both cost and energy savings.

V. CONCLUSIONS

Four SBR reactors were operated at room temperature for a 2-month period. Detention times varied from 10 down to 1.25 days. TOC degradation ranged from 55 to 81%, and phenol degradation ranged from 96.8 to 99.2%. Both ranges easily result in effluent concentrations that are below permit limits.

Oxygen uptake rates and spiking studies indicate that the waste water is readily biodegradable, and that larger volumes of water can be handled in the full-scale reactor during peak periods.

Sufficient phosphorus was present in the raw waste water, and foaming should be easily controlled with bubble-breaking compounds. A 1- or 2-day power failure should have no short- or long-term effect on system performance.

The function of activated carbon will be shifted from one of primary organic removal to a more cost-effective polishing role. Associated with this will be a significant energy savings based on fewer carbon loads being regenerated.

ACKNOWLEDGMENTS

This study was supported in part under NYSERDA contract number 601-CON-IC-84. The full-scale aeration equipment and tankage were supplied and installed by Jet-Tech, Inc., Industrial Airport, Kan. The process design and bench-scale studies were supervised by SBR Technologies, Inc., Mishawaka, Ind.

REFERENCES

1. **Irvine, R. L. and Busch, A. W.**, Sequencing batch reactors — an overview, *J. Water Pollut. Control Fed.*, 51, 235, 1979.
2. **Irvine, R. L., Ketchum, L. H., Jr., Breyfogle, R., and Barth, E. F.**, Municipal application of sequencing batch treatment at Culver, Indiana, *J. Water Pollut. Control Fed.*, 55, 484, 1983.
3. **Irvine, R. L., Sojka, S. A., and Colaruotolo, J. F.**, Enhanced biological treatment of leachates from industrial landfills, *Hazard. Waste*, 1, 123, 1984.
4. **Herzbrun, P. A., Irvine, R. L., and Malinowski, K. C.**, Biological treatment of hazardous waste in the SBR, *J. Water Pollut. Control Fed.*, 57(12), 1163, 1985.

Chapter 4

ANAEROBIC DEGRADATION OF PHENOLIC COMPOUNDS WITH APPLICATIONS TO TREATMENT OF INDUSTRIAL WASTE WATERS

Phillip M. Fedorak and Steve E. Hrudey

TABLE OF CONTENTS

I.   Introduction 170

II.   Mechanisms of Anaerobic Degradation of Aromatic Compounds 185
        A.   Photometabolism 185
        B.   Nitrate Respiration 186
        C.   Sulfate Reduction 190
        D.   Fermentation 190

III.   Susceptibility of Various Phenolics Toward Anaerobic Degradation 193

IV.   Phenolic Content of Industrial Waste Waters 205

V.   Treatment of Phenolic Waste Waters 211
        A.   Treatment of Synthetic Waste Waters 211
        B.   Treatment of Authentic Industrial Waste Waters 212
                1.   General 212
                2.   Coal Conversion Waste Waters 215
                3.   Denitrification Treatment of Coking Waste Waters 216
                4.   Pulp and Paper Industry Waste Waters 217

VI.   Summary 218

Acknowledgments 219

References 219

I. INTRODUCTION

Phenolic compounds contain one or more hydroxyl groups attached to an aromatic nucleus. They may also contain other functional groups such as alkyl, amino, carboxyl, halo, methoxy, nitro, etc. attached to the aromatic nucleus.

At parts per billion levels, phenolics cause tastes and odors in drinking waters and may taint the flavor of fish taken from contaminated waters. At parts per million concentrations, these compounds can be toxic to aquatic life. The toxicity of phenol to aquatic biota has been reviewed by Babich and Davis.[1]

Phenolic compounds found in the environment may originate from a variety of sources. Many industrial waste waters contain phenolics[2] which may find ways into ground waters, rivers, lakes, and oceans. Plants synthesize a wide variety of phenolics ranging from simple compounds such as catechol, hydroxyquinone, and *p*-hydroxybenzoic acid, to complex polymers including lignins, catechol melanins, and flavolans.[3] These must ultimately undergo microbial degradation to allow the carbon to be recycled. Phenolics are often intermediates in the aerobic degradation of nonphenolic compounds. For example, the pesticide Silvex© yields a trichlorophenol and a dichlorocatechol.[4] Similarly, the aerobic degradation of many aromatic hydrocarbons (e.g., benzene, toluene, and naphthalene) produces intermediates which are phenols and catechols.[5]

A wide variety of phenolics are susceptible to aerobic microbial degradation and many types of microorganisms are capable of degrading these compounds. These processes have been studied extensively and have been reviewed by others.[1,2] However, much less information on the anaerobic degradation of phenolic compounds is available and this subject was reviewed in 1977 by Evans.[6]

Interest in the anaerobic degradation of phenolics has centered around (1) understanding the fate of these compounds in anaerobic environments,[7,8] (2) establishing biological treatment processes for purification of phenolic-containing waste waters,[9-12] and (3) converting hydrolysis products of lignin to methane.[13-15]

This chapter will review the mechanisms of anaerobic biodegradation of phenolic compounds with particular emphasis on the fermentation processes which ultimately yield methane. These mechanisms have been studied with mixed enrichment cultures,[14,16-18] pure cultures,[19-21] cocultures[21] and a partially purified enzyme.[22] Benzoic acid has been used as a model aromatic compound in numerous investigations[23-27] and more is known about the fate of benzoic acid under anaerobic conditions than is known about that of phenol. Since the degradative pathways of phenol and benzoate are similar,[6] and it is likely that the microbial interactions within methane-producing consortia fermenting these substrates are similar, discussion on the anaerobic decomposition of benzoic acid will also be included. Where information is available, the names or descriptions of the microorganisms involved will be given.

These microbial activities and capabilities will be considered in light of the use in the treatment of phenolic waste waters. The susceptibilities of a large number of phenolics to be degraded under anaerobic conditions will be summarized and the phenolic composition of various waste waters will also be surveyed. To date, only laboratory-scale studies on the anaerobic treatment of actual and synthetic phenolic waste waters have been reported and these will be reviewed.

Many of the phenolic compounds which are discussed in this chapter are known by several names. We have chosen a common name for each compound which is used throughout the text. Table 1 summarizes the common names and gives synonyms, molecular, and structural formulas, and the American Chemical Society registry number for each compound.

Table 1
SUBSTRATES TESTED AND INTERMEDIATES DETECTED IN STUDIES OF THE ANAEROBIC
DEGRADATION OF PHENOLICS

Common name	Synonyms or isomers	Molecular formula	A.C.S. registry number	Structural formula
Adipic acid	Hexanedioic acid 1,4-Butanedicarboxylic acid	$C_6H_{10}O_4$	[124-04-9]	$HOOC(CH_2)_4COOH$
o-Aminophenol	2-Aminophenol 2-Hydroxyaniline 2-Amino-1-hydroxybenzene	C_6H_7NO	[95-55-6]	
p-Aminophenol	4-Aminophenol 4-Hydroxyaniline 4-Amino-1-hydroxybenzene	C_6H_7NO	[123-30-8]	
m-Anisic acid	3-Methoxybenzoic acid	$C_8H_8O_3$	[586-38-9]	
Butyric acid	Butanoic acid *n*-Butyric acid Ethylacetic acid	$C_4H_8O_2$	[107-92-6]	$CH_3CH_2CH_2COOH$
Caffeic acid	3-(3,4-Dihydroxyphenyl)-pro-penoic acid 3,4-Dihydroxycinnamic acid	$C_9H_8O_4$	[331-39-5]	
n-Caproic acid	Hexanoic acid	$C_6H_{12}O_2$	[142-62-1]	$CH_3(CH_2)_4COOH$
Catechol	1,2-Dihydroxybenzene 1,2-Benzenediol Pyrocatechol	$C_6H_6O_2$	[120-80-9]	

Table 1 (continued)
SUBSTRATES TESTED AND INTERMEDIATES DETECTED IN STUDIES OF THE ANAEROBIC DEGRADATION OF PHENOLICS

Common name	Synonyms or isomers	Molecular formula	A.C.S. registry number	Structural formula
o-Chlorophenol	2-Chlorophenol	C_6H_5OCl	[95-57-8]	
m-Chlorophenol	3-Chlorophenol	C_6H_5OCl	[108-43-0]	
p-Chlorophenol	4-Chlorophenol	C_6H_5OCl	[106-48-9]	
Chlorohydroxybenzoic acid	3-Chloro-4-hydroxybenzoic acid	$C_7H_5O_3Cl$	[394-58-7]	
Chlorohydroxybenzoic acid	5-Chloro-2-hydroxybenzoic acid	$C_7H_5O_3Cl$	[321-14-2]	
Chlorocatechol	1,2-Dihydroxy-4-chlorobenzene 4-Chloro-1,2-benzenediol	$C_6H_5O_2Cl$	[2138-22-9]	

Name	Structure	Synonym	Formula	CAS
Chlororesorcinol		1,3-Dihydroxy-4-chlorobenzene	$C_6H_5O_2Cl$	[95-88-5]
Coniferyl alcohol		4-(3-Hydroxy-1-propenyl)-2-methoxyphenol; 3-(4-Hydroxy-3-methoxy-phenyl)-2-propen-1-ol	$C_{10}H_{12}O_3$	[458-35-5]
o-Cresol		2-Methylphenol	C_7H_8O	[95-48-7]
m-Cresol		3-Methylphenol	C_7H_8O	[108-39-4]
p-Cresol		4-Methylphenol	C_7H_8O	[106-44-5]
Cyclohexane-carboxylic acid		Hexahydrobenzoic acid	$C_7H_{12}O_2$	[98-89-5]
Cyclohex-1-ene-1-car-boxylic acid		1-Cyclohexene-1-carboxylic acid	$C_7H_{11}O_2$	[636-82-8]
Cyclohexanol		Hexalin; Hexahydrophenol	$C_6H_{12}O$	[108-93-0]

Table 1 (continued)
SUBSTRATES TESTED AND INTERMEDIATES DETECTED IN STUDIES OF THE ANAEROBIC DEGRADATION OF PHENOLICS

Common name	Synonyms or isomers	Molecular formula	A.C.S. registry number	Structural formula
Cyclohexanone	Ketohexamethylene Pimelic ketone Hytrol O Anone Nadone	$C_6H_{10}O$	[108-94-1]	
Dichlorophenols	2,3-Dichlorophenol	$C_6H_4OCl_2$	[576-24-9]	
Dichlorophenols	2,4-Dichlorophenol	$C_6H_4OCl_2$	[120-83-2]	
	2,5-Dichlorophenol	$C_6H_4OCl_2$	[583-78-8]	
	2,6-Dichlorophenol	$C_6H_4OCl_2$	[87-65-0]	
Dichlorophenols	3,4-Dichlorophenol	$C_6H_4OCl_2$	[95-77-2]	

	Name	Structure	CAS	Formula
	3,5-Dichlorophenol		[591-35-5]	$C_6H_4OCl_2$
Dichlorophenol	Unspecified		[25167-81-1]	$C_6H_4OCl_2$
Dihydrophloroglucinol	1,3-Cyclohexadiene-1,3,5-triol		[26932-12-7]	$C_6H_8O_3$
Dihydroxymethyl-benzenes	1,2-Dihydroxy-3-methylbenzene; 3-Methyl-1,2-benzenediol; 3-Methylcatechol; 2,3-Dihydroxytoluene		[488-17-5]	$C_7H_8O_2$
	1,3-Dihydroxy-2-methylbenzene; 2-Methyl-1,3-benzenediol; 2-Methylresorcinol; 2,5-Dihydroxytoluene		[608-25-3]	$C_7H_8O_2$
	1,3-Dihydroxy-6-methylbenzene; 4-Methylresorcinol; 2,4-Dihydroxytoluene		[496-73-1]	$C_7H_8O_2$
Dimethylphenols	2,3-Dimethylphenol; 2,3-Xylenol		[526-75-0]	$C_8H_{10}O$
	2,4-Dimethylphenol; 2,4-Xylenol		[105-67-9]	$C_8H_{10}O$

Table 1 (continued)

SUBSTRATES TESTED AND INTERMEDIATES DETECTED IN STUDIES OF THE ANAEROBIC DEGRADATION OF PHENOLICS

Common name	Synonyms or isomers	Molecular formula	A.C.S. registry number	Structural formula
	2,5-Dimethylphenol 2,5-Xylenol	$C_8H_{10}O$	[95-87-4]	
	2,6-Dimethylphenol 2,6-Xylenol	$C_8H_{10}O$	[576-26-1]	
	3,4-Dimethylphenol 3,4-Xylenol	$C_8H_{10}O$	[95-65-8]	
	3,5-Dimethylphenol 3,5-Xylenol	$C_8H_{10}O$	[108-68-9]	
Dimethylphenol	Unspecified Xylenol	$C_8H_{10}O$	[1300-71-6]	
Ethylphenols	2-Ethylphenol o-Ethylphenol Phlorol	$C_8H_{10}O$	[90-00-6]	

Compound	Synonyms	Formula	CAS
	3-Ethylphenol *m*-Ethylphenol	$C_8H_{10}O$	[620-17-7]
	4-Ethylphenol *p*-Ethylphenol	$C_8H_{10}O$	[123-07-9]
Ethylphenol	Unspecified	$C_8H_{10}O$	[25429-37-2]
Eugenol	2-Methoxy-4-(2-propenyl)-phenol 4-Allyl-2-methoxyphenol Allyl guaiacol Eugenic acid Caryophyllic acid	$C_{10}H_{12}O_2$	[97-53-0]
Ferulic acid	3-(4-Hydroxy-3-methoxy-phenyl)-2-propenoic acid 4-Hydroxy-3-methoxycinnamic acid 3-Methoxy-4-hydroxycinnamic acid Caffeic acid 3-methylether 3,4,5-Trihydroxybenzoic acid	$C_{10}H_{10}O_4$	[1135-24-6]
Gallic acid		$C_7H_6O_5$	[149-91-7]
Gentisic acid	2,5-Dihydroxybenzoic acid 5-Hydroxysalicylic acid	$C_7H_6O_4$	[490-79-9]

Table 1 (continued)

SUBSTRATES TESTED AND INTERMEDIATES DETECTED IN STUDIES OF THE ANAEROBIC DEGRADATION OF PHENOLICS

Common name	Synonyms or isomers	Molecular formula	A.C.S. registry number	Structural formula
Guaiacol	o-Methoxyphenol 2-Methoxyphenol Methylcatechol o-Hydroxyanisole 1-Hydroxy-2-methoxybenzene	$C_7H_8O_2$	[90-05-1]	
n-Heptanoic acid	Enanthic acid Oenanthic acid Oenanthylic acid n-Heptoic acid n-Hyptylic acid	$C_7H_{14}O_2$	[111-14-8]	$CH_3(CH_2)_5COOH$
Homopyrocatechol	1,2-Dihydroxy-4-methylbenzene 4-Methyl-1,2-benzenediol 4-Methylcatechol 3,4-Dihydroxytoluene	$C_7H_8O_2$	[452-86-8]	
Hydroquinone	1,4-Benzenediol p-Dihydroxybenzene Hydroquinol Quinol	$C_6H_6O_2$	[123-31-9]	
Hydroxybenzoic acids	o-Hydroxybenzoic acid 2-Hydroxybenzoic acid	$C_7H_6O_3$	[69-72-7]	
	m-Hydroxybenzoic acid 3-Hydroxybenzoic acid	$C_7H_6O_3$	[99-06-9]	

Name	Synonyms	Molecular formula	CAS number	Structure
Hydroxybenzoic acid	*p*-Hydroxybenzoic acid 4-Hydroxybenzoic acid	$C_7H_6O_3$	[99-96-7]	
	Unspecified	$C_7H_6O_3$	[29656-58-4]	
p-Hydroxycinnamic acid	3-(4-Hydroxyphenyl)-2-propenoic acid *p*-Coumaric acid	$C_9H_8O_3$	[7400-08-0]	
2-Hydroxycyclohexanecarboxylic acid		$C_7H_8O_3$	[28131-61-5]	
2-Hydroxycyclohexanone	Adipoin	$C_6H_{10}O_2$	[533-60-8]	
p-Hydroxyphenylacetic acid	Mandelic acid α-Hydroxybenzene acetic acid α-Hydroxy-α-toluic acid α-Hydroxyphenyl acetic acid Phenylhydroxyacetic acid Phenylglycolic acid Amygdalic acid Amygdalinic acid Paramendelic acid·	$C_8H_8O_3$	[90-64-2]	
α-Hydroxymuconic semialdehyde	2-Hydroxy-6-formylhexa-2,4-dienoic acid	$C_6H_6O_4$	[3272-98-2]	

Table 1 (continued)
SUBSTRATES TESTED AND INTERMEDIATES DETECTED IN STUDIES OF THE ANAEROBIC DEGRADATION OF PHENOLICS

Common name	Synonyms or isomers	Molecular formula	A.C.S. registry number	Structural formula
Methoxyphenols	o-Methoxyphenol 2-Methoxyphenol See guaiacol	$C_7H_8O_2$	[90-05-1]	
	m-Methoxyphenol 3-Methoxyphenol	$C_7H_8O_2$	[150-19-6]	
	p-Methoxyphenol 4-Methoxyphenol	$C_7H_8O_2$	[150-76-5]	
Methoxyphenol	Unspecified	$C_7H_8O_2$	[26638-02-9]	
cis,cis-Muconic acid	2,4-Hexadienedioic acid 1,3-Butadiene-1,4-dicarboxylic acid	$C_6H_6O_4$	[505-70-4]	HOOCCH:CHCH:CHCOOH
Naphthols	1-Naphthol α-Naphthol 1-Naphthalenol α-Hydroxynaphthalene	$C_{10}H_8O$	[90-15-3]	
	2-Naphthol β-Naphthol 2-Naphthalenol β-Hydroxynaphthalene Isonaphthol	$C_{10}H_8O$	[135-19-3]	

Name	Synonyms	Formula	CAS	Structure
Nitrophenols	o-Nitrophenol 2-Nitrophenol	$C_6H_5NO_3$	[88-75-5]	
	m-Nitrophenol 3-Nitrophenol	$C_6H_5NO_3$	[554-84-7]	
	p-Nitrophenol 4-Nitrophenol	$C_6H_5NO_3$	[100-02-7]	
Nitrophenol	Unspecified	$C_6H_5NO_3$	[25154-55-6]	
Orcinol	5-Methyl-1,3-benzenediol 5-Methylresorcinol 3,5-Dihydroxytoluene	$C_7H_8O_2$	[504-15-4]	
2-Oxocyclohexane Carboxylic acid		$C_7H_{10}O_3$	[18709-01-8]	
Pentachlorophenol	Penta PCP Penchlorol Santophen 20	C_6HOCl_5	[87-86-5]	
Phloroglucinol	1,3,5-Benzenetriol 1,3,5-Trihydroxybenzene Phloroglucin Dilospan S Spasfon-Lyoc	$C_6H_6O_3$	[108-73-6]	

Table 1 (continued)
SUBSTRATES TESTED AND INTERMEDIATES DETECTED IN STUDIES OF THE ANAEROBIC DEGRADATION OF PHENOLICS

Common name	Synonyms or isomers	Molecular formula	A.C.S. registry number	Structural formula
Phloroglucinol Carboxylic acid	2,4,6-Trihydroxybenzoic acid	$C_7H_6O_5$	[83-30-7]	
Pimelic acid	Heptanedioic acid 1,5-Pentanedicarboxylic acid	$C_7H_{12}O_4$	[111-16-0]	$HOOC(CH_2)_5COOH$
Propionic acid	Propanoic acid Methylacetic acid Ethylformic acid	$C_3H_6O_2$	[79-09-4]	CH_3CH_2COOH
Protocatechuic acid	3,4-Dihydroxybenzoic acid	$C_7H_6O_4$	[99-50-3]	
Pyrogallol	1,2,3-benzenetriol 1,2,3-Trihydroxybenzene Pyrogallic acid	$C_6H_6O_3$	[87-66-1]	
α-Resorcylic acid	3,5-Dihydroxybenzoic acid	$C_7H_6O_4$	[99-10-5]	
β-Resorcylic acid	2,4-Dihydroxybenzoic acid	$C_7H_6O_4$	[89-86-1]	

Salicylic acid	2-Hydroxybenzoic acid o-Hydroxybenzoic acid	[69-72-7]	$C_7H_6O_3$	
Sinapic acid	3-(4-Hydroxy-3,5-dimethoxy-phenyl)propenoic acid	[530-59-6]	$C_{11}H_{12}O_5$	
Succinic acid	Butanedoic acid Amber acid Ethylene succinic acid	[110-15-6]	$C_4H_6O_4$	
Syringic acid	4-Hydroxy-3,5-dimethoxyben-zoic acid	[530-57-4]	$C_9H_{10}O_5$	
Syringaldehyde	4-Hydroxy-3,5-dimethoxybenzaldehyde	[134-96-3]	$C_9H_{10}O_4$	
Syringol	2,6-Dimethoxyphenol	[91-10-1]	$C_8H_{10}O_3$	
Trimethoxybenzoic acids	2,3,4-Trimethoxybenzoic acid	[573-11-5]	$C_7H_{12}O_5$	
	2,4,6-Trimethoxybenzoic acid	[570-02-5]	$C_7H_{12}O_5$	

Table 1 (continued)

SUBSTRATES TESTED AND INTERMEDIATES DETECTED IN STUDIES OF THE ANAEROBIC DEGRADATION OF PHENOLICS

Common name	Synonyms or isomers	Molecular formula	A.C.S. registry number	Structural formula
	3,4,5-Trimethoxybenzoic acid	$C_7H_{12}O_5$	[118-41-2]	COOH, OCH$_3$, OCH$_3$, CH$_3$O
Tyrosine	β-(p-Hydroxyphenyl)-alanine α-Amino-p-hydroxyhydrocin-namic acid	$C_9H_{11}NO_3$	[60-18-4]	OH, CH$_2$CHNH$_2$COOH
Vanillic acid	4-Hydroxy-3-methoxybenzoic acid	$C_8H_8O_4$	[121-34-6]	OH, OCH$_3$, COOH
Vanillin	4-Hydroxy-3-methoxybenzaldehyde Methylprotocatechuic aldehyde Vanillic aldehyde 3-Methoxy-4-hydroxybenzaldehyde	$C_8H_8O_3$	[121-33-5]	OH, OCH$_3$, CHO
Vanillyl alcohol	4-Hydroxy-3-methoxybenzyl alcohol 4-Hydroxy-3-methoxybenzene methanol	$C_8H_{10}O_3$	[498-00-0]	OH, OCH$_3$, CH$_2$OH
Veratic acid	3,4-Dimethoxybenzoic acid Dimethylprotocatechuic acid	$C_9H_{10}O_4$	[93-07-2]	COOH, OCH$_3$, OCH$_3$

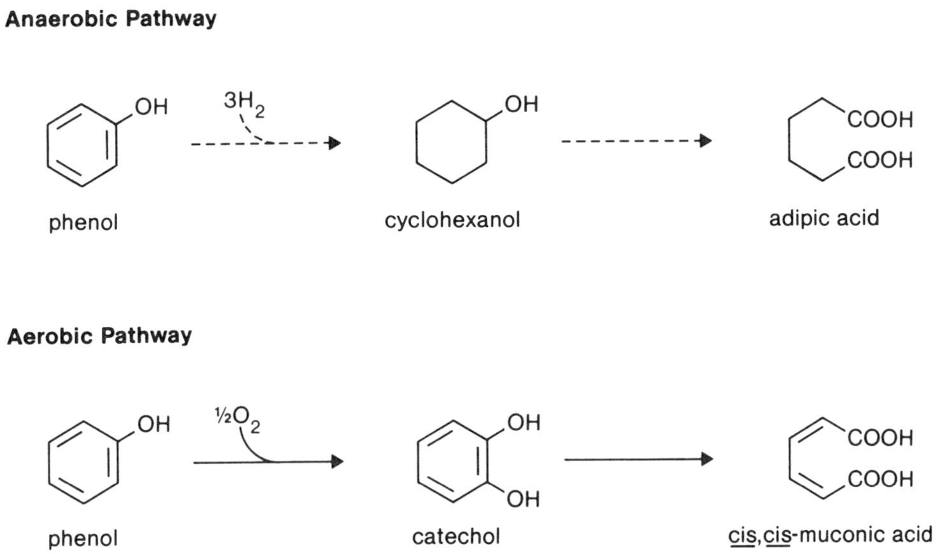

Anaerobic Pathway

phenol   cyclohexanol   adipic acid

Aerobic Pathway

phenol   catechol   cis,cis-muconic acid

FIGURE 1. Examples of intermediates formed during the anaerobic and aerobic degradation of phenol. In the latter case, the oxidation via ortho cleavage is shown.

II. MECHANISMS OF ANAEROBIC DEGRADATION OF AROMATIC COMPOUNDS

Bacterial degradation of aromatic compounds has been shown to involve a reductive pathway which differs markedly from the pathways used by microorganisms in the presence of molecular oxygen. Figure 1 illustrates some intermediates found in the metabolism of phenol under anaerobic and aerobic conditions. In the anaerobic pathways, the aromatic ring is reduced giving cyclohexanol and then ring cleavage occurs giving adipic acid. In contrast, the aromatic character of the ring is not altered and catechol is formed under aerobic conditions. The aromatic ring is then cleaved producing a diene (*cis,cis*-muconic acid). Figure 1 illustrates the aerobic ortho-cleavage pathway. Meta cleavage will give α-hydroxymuconic semialdehyde.[28]

Investigations of the anaerobic metabolism of aromatic compounds have used pure cultures, defined cocultures, and enriched consortia to elucidate biochemical pathways and to determine susceptibility of various compounds to degradation under the particular culture conditions.

Anaerobically, aromatic compounds can be degraded by four mechanisms: photometabolism, nitrate respiration, sulfate reduction, and fermentation. Each of these will be discussed separately. Benzoate has been shown to be degraded by all four mechanisms whereas phenol has not.

A. Photometabolism

Several species of the family *Rhodospirillaceae* (purple nonsulfur bacteria) have been found to metabolize aromatic compounds anaerobically in the presence of light. Indeed, the ability to photometabolize benzoate is also used as a taxonomic test.[29] In some cases, these photoheterotrophs can grow using the aromatic compound as the major carbon source while in other cases the organisms can photometabolize the compound but cannot utilize it for growth. In pure culture studies, these organisms are usually grown on benzoic acid and then the metabolic diversity of the culture is tested with other substrates. Only a few phenolics have been tested under these conditions.

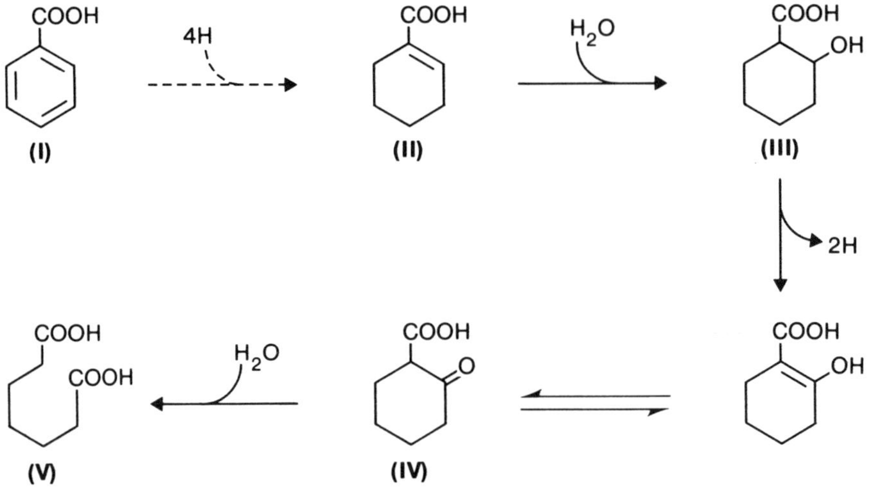

FIGURE 2.   Proposed pathway for the photometabolism of benzoic acid (I) by *Rhodopseudomonas palustris*. Identified intermediates include cyclohex-1-ene-1-carboxylic acid (II), 2-hydroxycyclohexanecarboxylic acid (III), 2-oxocyclohexanecarboxylic acid (IV), and pimelic acid (V). In some presentations of this pathway,[6] cyclohexanecarboxylic acid is considered as a precursor to compound (II). (Reprinted by permission from Dutton, P. L. and Evans, W. C., *Biochem. J.*, 113, 525, copyright ©1969. The Biochemical Society, London.)

Figure 2 shows the reductive pathway for the photometabolism of benzoic acid by *Rhodopseudomonas palustris*.[30,31] Using ^{14}C-labeled benzoic acid, Dutton and Evans[30] have identified compounds (II), (III), and (V) shown in Figure 2. Small amounts of cyclohexanecarboxylic acid were also found, and it is sometimes considered as a precursor to compound (II).[6] *R. palustris* would grow on the phenolics *m*- and *p*-hydroxybenzoate but not on *o*-hydroxybenzoate or protocatechuate under anaerobic photosynthetic conditions.[30] Dihydroxy compounds such as catechol and isomers of dihydroxybenzoate were not photometabolized by benzoate-grown cells during a 4-hr test period. This organism has also been shown to grow photosynthetically on phloroglucinol.[32]

Pfennig[33] reported the isolation and characterization of *Rhodocyclus purpureus* which photometabolizes benzoate. However, the ability of *R. purpureus* to grow on phenolics was not studied. Tanaka et at.[34] isolated a bacterium similar to *Rhodopseudomonas palustris* which, under anaerobic photosynthetic conditions, would grow on benzoate and *p*-hydroxybenzoate but not on phenol, *m*- or *o*-hydroxybenzoate, *p*-hydroxyphenylacetate, or protocatechuate. Benzoate-grown cells would photometabolize, but not grow on, *m*- and *o*-hydroxybenzoate, protocatechuate, and 2,4-dihydroxybenzoate. Phenol and 3,5-dihydroxybenzoate were not photometabolized.

Table 2 summarizes the phenolic compounds which have been shown to be photometabolized. The limited number of entries in this table suggests that treatment of waste waters by anaerobic photoheterophic bacteria would not be practical.

B. Nitrate Respiration

Nitrate respiration is used in the absence of O_2 by some bacteria which are facultative anaerobes. In this case, nitrate is used as the terminal electron acceptor. The process of reducing nitrate through nitrite to the gaseous forms of N_2O or N_2 is known as denitrification. The metabolism of aromatic compounds via nitrate respiration has been studied using mixed microbial population,[18] defined cocultures,[35] and pure cultures.[36-39] In all of the cases cited below, the evolution of N_2 was observed.

Liquid mineral medium containing 100 mg/ℓ phenol and 5 g/ℓ KNO_3 was inoculated with

Table 2
PHENOLICS WHICH
HAVE BEEN SHOWN
TO BE
PHOTOMETABOLIZED
UNDER ANAEROBIC
CONDITIONS BY
MEMBERS OF THE
FAMILY
RHODOSPIRILLACEAE

o-Hydroxybenzoate
m-Hydroxybenzoate
p-Hydroxybenzoate
2,4-Dihydroxybenzoate
Protocatechuate
Phloroglucinol

a mixture of soils, manure, and sludge from a phenol-degrading oxidation ditch.[18] Phenol degradation and evolution of gas were observed in this enrichment culture and three Gram-negative bacteria were isolated from the culture. Individually, these organisms degraded phenol very slowly; therefore, the more active original mixed population in the enrichment culture was used for further investigations. Bakker[18] screened the ability of this enrichment culture to anaerobically degrade eight aromatic compounds in the presence of nitrate. In order of decreasing rates of degradation, these were *m*-hydroxybenzoic acid, phenol, *p*-hydroxybenzoic and 3,4-dihydroxybenzoic acid, benzoic acid, *p*-, *o*-, and *m*-cresol. When uniformly labeled ^{14}C-phenol was supplied to the culture, 8% of the radioactivity was recovered in the acid-hydrolyzed cell material and 28% as CO_2. The author did not account for the remaining radioactivity.

From soil enrichments with protocatechuate, Oshima[35] isolated two Gram-negative rods. One was designated strain 5207 and the other was identified as a *Pseudomonas* strain 5233. In pure culture, both isolates grew using nitrate as a terminal electron acceptor but neither of these strains could degrade protocatechuate. However, in coculture protocatechuate, *p*-hydroxybenzoate, benzoate, and tyrosine (each at 0.05% concentration) served as sole carbon sources and supported anaerobic growth in the presence of nitrate. Catechol would not support growth. Strain 5207 could be coupled with any of ten other *Pseudomonas* species and the cocultures would degrade protocatechuate.

In a series of papers, Taylor and coworkers[36-38] have described the metabolic capabilities of *Pseudomonas* strain PN-1 which can grow on *p*-hydroxybenzoate, benzoate, and *m*-hydroxybenzoate aerobically or anaerobically by nitrate respiration. This organism could also grow anaerobically, but not aerobically, on protocatechuate and *o*-hydroxybenzoate. No growth occurred under either condition on phenol or catechol.[36] Studies with ^{14}C-labeled benzoate[37] showed that far more of the radioactivity was released as $^{14}CO_2$ during anaerobic metabolism under denitrifying conditions than under aerobic conditions. These results suggested that different biochemical pathways were involved in the degradation of benzoate by strain PN-1 under aerobic and anaerobic conditions.

More recent studies with this strain[38] have shown that it will grow anaerobically on vanillic acid, vanillin, and vanillyl alcohol in the presence of nitrate. Under these conditions, strain PN-1 could actively demethylate several lignoaromatic compounds such as guaiacol, ferulate, *m*-anisate, and 3,4,5-trimethoxybenzoate. As shown in Figure 3, this activity produced dihydric phenolics from monohydric compounds (e.g., from guaiacol and ferulate), and phenolic derivatives of benzoic acid from nonphenolic compounds (e.g., from *m*-anisate and

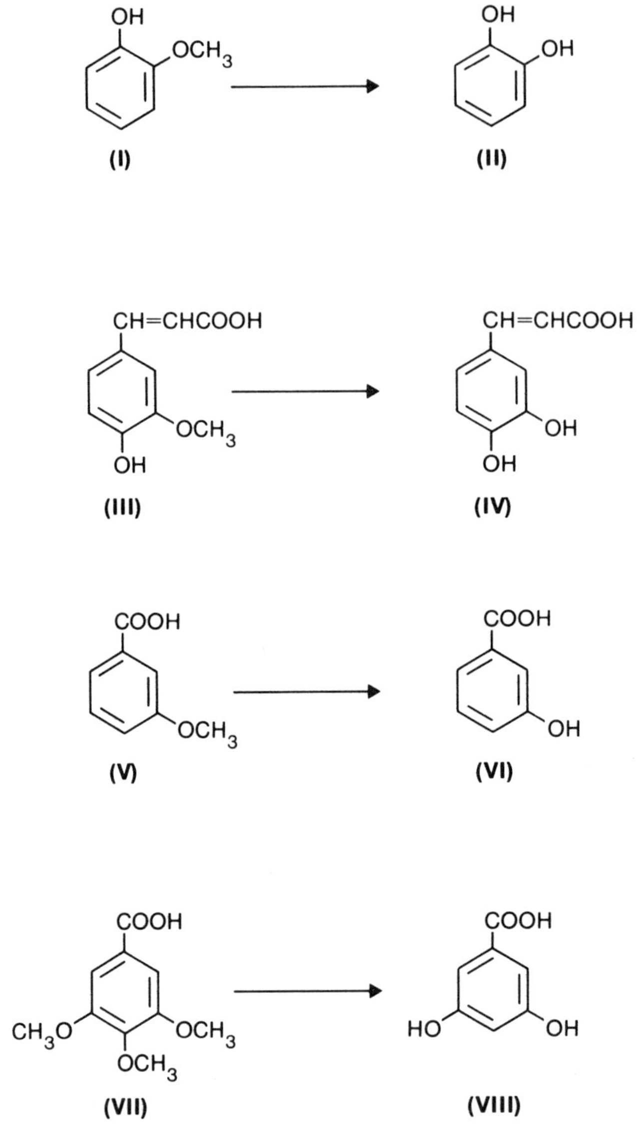

FIGURE 3.   Anaerobic demethylation reactions of *Pseudomonas* PN-1 in the presence of nitrate producing new phenolic compounds. Guaiacol (I) yields catechol (II); ferulate (III) yields caffeate (IV); *m*-anisate (V) yields *m*-hydroxybenzoate (VI); and 3,4,5-trimethoxybenzoate (VII) yields 3,5-dihydroxybenzoate (VIII).

3,4,5-trimethoxybenzoate). Bache and Pfennig[39] have also demonstrated demethylation of other aromatics under anaerobic conditions in the absence of nitrate. These transformations in an anaerobic environment yield new phenolics for further biodegradation.

Pure culture studies using a *Moraxella* sp. and ^{14}C-labeled substrate elucidated the reductive pathway for benzoate degradation under conditions of nitrate respiration.[40] Figure 4 shows the identified intermediates in this reductive pathway which are very similar to those found during photometabolism of benzoate (Figure 2). However, ring cleavage occurs differently in the two pathways. Pimelic acid results from the cleavage during photometabolism whereas adipic acid is produced during nitrate respiration. In the latter case, decar-

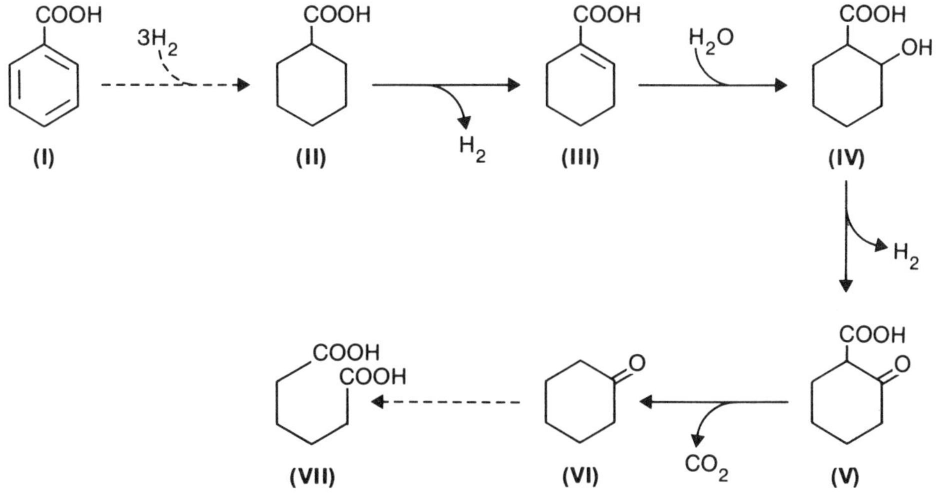

FIGURE 4. Proposed pathway for the anaerobic metabolism of benzoic acid (I) by *Moraxella* sp. through nitrate reduction. Identified intermediates include cyclohexane-carboxylic acid (II), cyclohex-1-ene-1-carboxylic acid (III), 2-hydroxycyclohexanecarboxylic acid (IV), 2-oxocyclohexanecarboxylic acid (V), cyclohexanone (VI) and adipic acid (VII). (Reprinted by permission from Williams, R. J. and Evans, W. C., *Biochem. J.,* 148, 1, copyright ©1975. The Biochemical Society, London.)

Table 3
PHENOLICS WHICH HAVE BEEN SHOWN TO BE
METABOLIZED THROUGH NITRATE RESPIRATION UNDER
ANAEROBIC CONDITIONS

By mixed enrichment cultures	By pure cultures or cocultures	Undergoes demethylation
Phenol	Phloroglucinol	Guaiacol
o-Cresol	*o*-Hydroxybenzoic acid	Ferulic acid
m-Cresol	*m*-Hydroxybenzoic acid	Syringic acid
p-Cresol	*p*-Hydroxybenzoic acid	*m*-Anisic acid[a]
m-Hydroxybenzoic acid	Protocatechuic acid	Veratric acid[a]
p-Hydroxybenzoic acid	Vanillic acid	3,5-Dimethoxybenzoic acid[a]
Protocatechuic acid	Vanillin	3,4,5-Trimethoxybenzoic acid[a]
	Vanillyl alcohol	3,4,5-Trimethoxycinnamic acid[a]
	p-Hydroxycinnamic acid	
	Caffeic acid	
	Tyrosine	

[a]   Demethylation yields phenolic product(s) from a nonphenolic substrate.

boxylation occurs while the ring is intact and cyclohexanone is formed which gives rise to the 6-carbon dicarboxylic acid upon ring cleavage.

Other phenolics which would support anaerobic growth of the *Moraxella* sp. through nitrate reduction were *m*- and *p*-hydroxybenzoic acids, protocatechuic acid, *p*-hydroxycinnamic acid, caffeic acid, and phloroglucinol.[40] Phenolics which would not support growth and were not utilized by benzoate grown cells were phenol, catechol, resorcinol, hydroquinone, *p*-cresol, salicylic acid, gentisic acid, and phloroglucinolcarboxylic acid.

Table 3 summarizes the phenolics which have been shown to be metabolized anaerobically through nitrate respiration. The number of compounds in this list is substantially greater than that in Table 2 which gave the phenolics which could be photometabolized. In both cases, the majority of the phenolics are derivatives of benzoic acid or other carboxylic acids which are typically found as hydrolysis products of lignin.

C. Sulfate Reduction

Sulfate can serve as a terminal electron acceptor under anaerobic conditions in a manner similar to that of nitrate. However, only a few microbial systems have been observed to degrade aromatic compounds by this means and only benzoate and *p*-hydroxybenzoic acid have been shown to be susceptible to this mode of metabolism.[41] Benzoate degradation coupled with sulfate reduction has been demonstrated in pure cultures[41] and in cocultures.[27,42]

Recently, Widdel and co-workers[41,43] described the isolation methods and characterization of three novel species of strict anaerobes (*Desulfococcus multivorans*, *Desulfonema magnum*, and *Desulfosarcina variabilis*) that use benzoate in pure culture as an electron donor and carbon source in the presence of sulfate. *Desulfonema magnum* will also use *p*-hydroxybenzoate but not *o*- or *m*-hydroxybenzoate nor cyclohexanecarboxylate.[41] Although obtained from freshwater and marine sediments, the abundance and distribution of these isolates in anaerobic environments are unknown and there have been no reports on the range of aromatics which they can degrade.

Balba and Evans[42] observed that neither *Pseudomonas aeruginosa* nor *Desulfovibrio vulgaris* could grow anaerobically in pure culture on benzoate. However, a coculture of these two organisms degraded benzoate with the concurrent production of sulfide from sulfate. Cyclohexanecarboxylic acid and cyclohexanol were detected in the coculture. Thus, benzoate metabolism was likely to occur via a reductive pathway similar to that found for nitrate respiration (Figure 4). It was postulated that the sulfate reducer was producing organic acids capable of acting as an electron acceptor for the facultative *P. aeruginosa* which attacked benzoate.

Using a H_2-utilizing sulfate-reducing *Desulfovibrio* sp. and coculture techniques, Mountfort and Bryant[27] isolated a benzoate-degrading anaerobic bacterium from domestic sewage sludge. These two organisms would grow on benzoate in syntrophic association. However, individually, neither could utilize this substrate. This coculture utilized only benzoate and failed to grow on any of the following phenolics: phenol, *o*- and *p*-hydroxybenzoate, ferulate, or vanillate.

Other investigations[44,45] have shown the importance of interspecies hydrogen transfer[46] in anaerobic systems. For example, the catabolism of propionate or butyrate are thermodynamically unfavorable unless the partial pressure of H_2 is kept low. A similar situation was found with the benzoate-degrading coculture.[27] If H_2 was present at 0.8 atm, benzoate degradation was strongly inhibited. Also, if sulfate (the terminal electron acceptor) was omitted from the medium, the coculture could not degrade benzoate. The sulfate reducer could be replaced by H_2-utilizing methanogens (*Methanospirillum hungatei*) and benzoate degradation occurred with the subsequent production of methane. When methane production was specifically inhibited with bromoethanesulfonic acid, 3 mol acetate per mole benzoate (and presumably CO_2 and H_2 or formate) were formed.

The mechanism of benzoate degradation described by Mountfort and Bryant[27] could also be considered to be fermentation-coupled to sulfate or carbon dioxide reduction and therefore included in the following section on fermentation. However, because of the importance of the sulfate reducer in the isolation procedure, their work has been considered an example of sulfate reduction.

D. Fermentation

Fermentation is strictly defined as metabolism in which energy is derived from the use of organic compounds as both the electron donor and the electron acceptor. That is, electron acceptors such as nitrate, sulfate, or carbon dioxide are not involved. In this sense, there have been very few examples of fermentation of aromatic compounds and these studies have been with pure cultures.[19-21] However, there are numerous reports of "fermentations" of phenolics coupled with the production of methane in mixed cultures.[13,14,16,17,47] In these methanogenic consortia, the nonmethanogenic microorganisms ferment the phenolics to substrates which are used by the methanogens and methane is produced from the reduction of carbon dioxide and/or the methyl group of acetate. This section will review the fermentation of phenolics in methanogenic consortia and in pure cultures. Again, benzoate will be included in the discussion as a model aromatic compound.

Tarvin and Buswell[48] were the first to report that phenol, benzoic acid, and four other aromatic compounds were transformed to methane and carbon dioxide by a mixed microbial population in sewage sludge. Phenol was not used as a test substrate. However, phenol was observed to be a transient intermediate in the degradation of tyrosine. Chmielowski et al.[47] screened 17 phenolics to determine whether they were amenable to anaerobic fermentation by methanogenic cultures. Eight of these (phenol, *p*-cresol, resorcinol, pyrogallol, phloroglucinol, *o*-, *m*-, and *p*-hydroxybenzoate) were converted to methane and carbon dioxide.

The results of other screening studies have also been reported. For example, Healy and Young[13] showed that 11 different aromatic compounds (including 9 phenolics) derived from lignin hydrolysis could be fermented to methane by mixed populations. Horowitz et al.[49] screened 78 aromatic compounds for susceptibilities to degradation by anaerobic consortia from a freshwater lake sediment and from municipal digested sludges. Among these were 28 benzoic acid derivatives including various isomers of halo, methyoxy, methyl, nitro, and amino compounds; 27 phenolics with similar substituents; 7 phthalates; and 12 anilines. Of these compounds, 42, including 12 phenolics, were metabolized.

Phenolics are somewhat persistent in anaerobic environments and characteristically, relatively long lag or acclimation times have been observed prior to the onset of degradation in the microbial consortia. This suggests that phenolic-degrading microorganisms likely occur in low numbers in anaerobic environments. Observed acclimation times for phenol in batch cultures have been 14,[13] 18,[50] and 20 days.[47] Acclimation times for catechol were reported to be 21[13] and 32 days.[50] Fedorak and Hrudey[51] have shown that acclimation times increase with increasing phenolic concentration. For example, methane production from batch cultures which received 100, 200, or 300 mg/ℓ phenol was observed after a 15-day incubation, whereas, in those which received 400 or 500 mg/ℓ, the acclimation times were 18 and 26 days, respectively.

Healy and Young[13] demonstrated that methanogenic consortia which were acclimated to one phenolic were often cross-acclimated (or coacclimated) to compounds with similar chemical structures. A culture which was fermenting vanillic acid was cross-acclimated to (that is, would give immediate gas production from) syringaldehyde, syringic acid, vanillin, benzoic acid, catechol, and protocatechuic acid.

The fate of benzoic acid during methane fermentation has been well documented.[23-26] Figure 5 shows a proposed anaerobic pathway for benzoate degradation in a methanogenic consortium as elucidated by Shlomi et al.[26] It differs from that reported by Keith et al.[25] in the intermediates observed after ring cleavage. For example, Figure 5 shows that the first product of ring cleavage was the dicarboxylic acid, pimelic, whereas Keith et al. found the monocarboxylic acid, heptanoate. In their work, Shlomi et al.[26] were unable to unequivocally identify cyclohex-1-ene-1-carboxylic acid (compound II, Figure 5). However, Keith et al.[25] identified this intermediate in the culture fluid of the methanogenic consortium. Nonetheless, the reductive pathway leading to ring cleavage shown in Figure 5 is very similar to those shown in Figures 2 and 4 in which anaerobic photometabolism and nitrate respiration, respectively, are the mechanisms of benzoate degradation.

To date, the anaerobic degradation of phenol has only been demonstrated in mixed en-

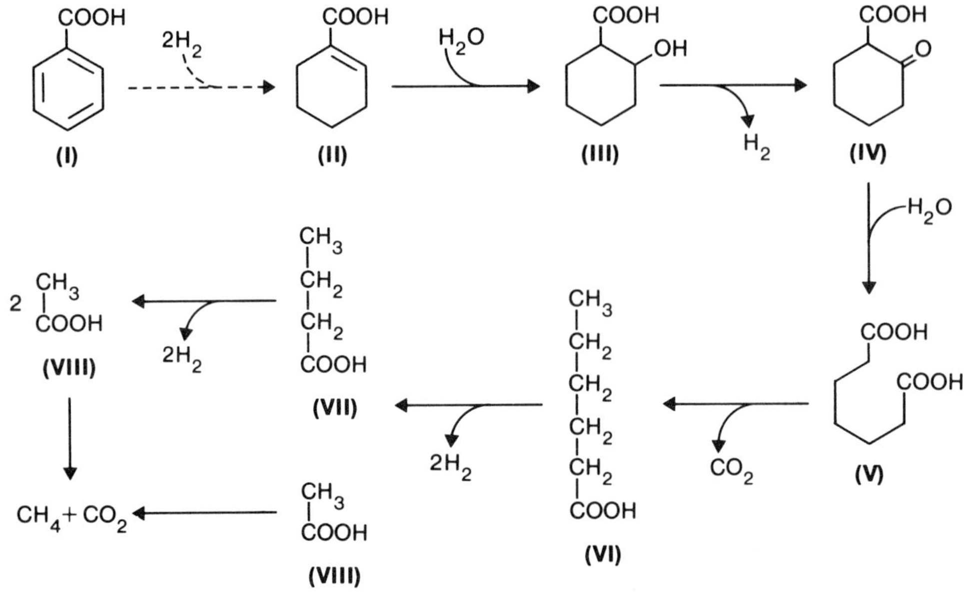

FIGURE 5.   Proposed pathway for the degradation of benzoic acid (I) in a methanogenic consortium.[26] Although cyclohex-1-ene-1-carboxylic acid (II) was not unequivocally identified, all other intermediates were identified: 2-hydroxycyclohexane carboxylic acid (III), 2-oxocyclohexanecarboxylic acid (IV), pimelic acid (V), caproic acid (VI), butyric acid (VII), and acetic acid (VIII).

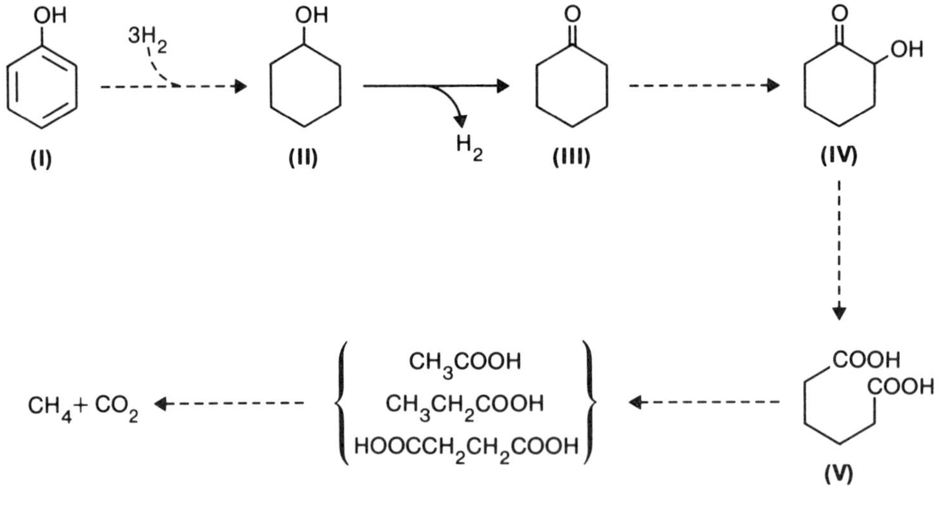

FIGURE 6.   Proposed pathway for the degradation of phenol (I) in a methanogenic consortium.[16] Intermediates found were: cyclohexanol (II), cyclohexanone (III), 2-hydroxycyclohexanone (IV), and adipic acid (V) leading to the formation of succinic acid, propionic acid, and acetic acid.

richment cultures under conditions of nitrate respiration[18] and under methanogenic conditions.[13,42,50] The proposed degradative pathway under the latter conditions is shown in Figure 6. Again, the aromatic ring is reduced to the level of cyclohexanol (II) before ring cleavage occurs. Balba and Evans[16] have detected cyclohexanone (III) and 2-hydroxycyclohexanone (IV) which leads to the ring cleavage product adipic acid (V). The short-chain organic acids which are formed undergo further metabolism leading to suitable substrates (e.g., acetate, formate, and H_2/CO_2) for methanogens.

Tsai and Jones[19] described the isolation of eight strains of rumen bacteria which can ferment phloroglucinol in pure culture. These could only be isolated from phloroglucinol enrichment cultures indicating that they were present in very low numbers. Five of these strains were facultative anaerobes and were identified as *Streptococcus bovis*. The other three were strict anaerobes and were assigned to the genus *Coprococcus*. In batch cultures containing 0.02 *M* phloroglucinol (2520 mg/ℓ) the *Coprococcus* isolates removed all of the substrate within 2 days. However, the degradation by the *S. bovis* strains was much slower. Two strains removed >90% of the substrate within 4 days while the others removed only 70 to 75% over the 5-day culture period. The metabolism of one of the *Coprococcus* isolates (strain Pe₁5) was studied in greater detail.[19] Resting cells degraded 1 mol of phloroglucinol to produce 2 mol of acetic acid and 2 mol of carbon dioxide. Strain Pe₁5 would not grow on any of the following phenolics: phenol, catechol, resorcinol, hydroquinone, pyrogallol, *o*-, *m*-, and *p*-hydroxybenzoic acids, 2,3-, 2,4-, 3,4-, 2,5-, and 3,5-dihydroxybenzoic acids, orcinol, gallic acid, caffeic acid, and ferulic acid.

Patel et al.[22] isolated a partially purified preparation of phloroglucinol reductase from strain Pe₁5 grown on phloroglucinol. Although the microorganism is a strict anaerobe, the enzyme was not sensitive to air. In the presence of NADPH₂ the enzyme converted phloroglucinol to dihydrophloroglucinol. Through a nonenzymatic elimination of H₂O, dihydrophloroglucinol can be transformed to resorcinol.

Schink and Pfennig[21] describe the isolation and metabolic studies of *Pelobacter acidigallici* gen. nov. sp. nov. This strictly anaerobic bacterium ferments the following trihydroxybenzenes: pyrogallol, phloroglucinol, 2,4,6-trihydroxybenzoic acid, and gallic acid. From each mole of the former two phenolics, 3 mol of acetate were produced, whereas from the benzoic acid derivatives, 3 mol of acetate and 1 mol carbon dioxide were formed per mole of substrate. Using a most probable number technique, the gallic acid degraders were found at a density of 23 cells per milliliter in a creek mud and 240 cells per milliliter in a river mud. Schink and Pfennig grew their isolates in cocultures with *Methanosarcina barkeri* and the trihydroxybenzenes were converted completely to methane and carbon dioxide.

Other biochemical transformations of phenolic compounds have also been observed in methanogenic consortia. As was observed in nitrate-reducing cultures,[38] demethylation also occurs in methane-producing cultures. Boyd et al.[52] observed the formation of dihydroxybenzene intermediates from methoxyphenolics. For example, *o*-methoxyphenol yielded catechol and *m*-methoxyphenol yielded resorcinol. Syringic acid (3,5-dimethoxy-4-hydroxybenzoic acid) was shown to demethylate sequentially to give 3,4-dihydroxy-5-methoxybenzoic acid and 3,4,5-trihydroxybenzoic acid.[53] During the period of demethylation, methane was produced in these anaerobic mixed cultures. Since acetate production and ring cleavage were not observed during this time, Kaiser and Hanselmann[53] concluded that methane was derived from the methoxyl carbon.

Microbially mediated dechlorinations of aromatic compounds have also been observed in methanogenic cultures.[17,52] The metabolism of *o*-chlorophenol produced phenol as a transient intermediate.[52] Boyd and Shelton[17] observed that, in cultures which had not previously been exposed to dichlorophenols, the ortho chlorine was preferentially removed. For example, 2,3- and 2,5-dichlorophenol yielded 3-chlorophenol. Similarly 2,4-dichlorophenol gave 4-chlorophenol and 2,6-dichlorophenol gave 2-chlorophenol. Studies using selected ¹⁴C-labeled substrates show that >90% of the radioactivity added as any of the three chlorophenol isomers or as 2,4-dichlorophenol was recovered as ¹⁴CH₄ and ¹⁴CO₂. Dehalogenation of benzoic acid derivatives has also been reported in methanogenic cultures.[54,55]

III. SUSCEPTIBILITY OF VARIOUS PHENOLICS TOWARD ANAEROBIC DEGRADATION

This section provides, in tabular form, a thorough survey of phenolic compounds which

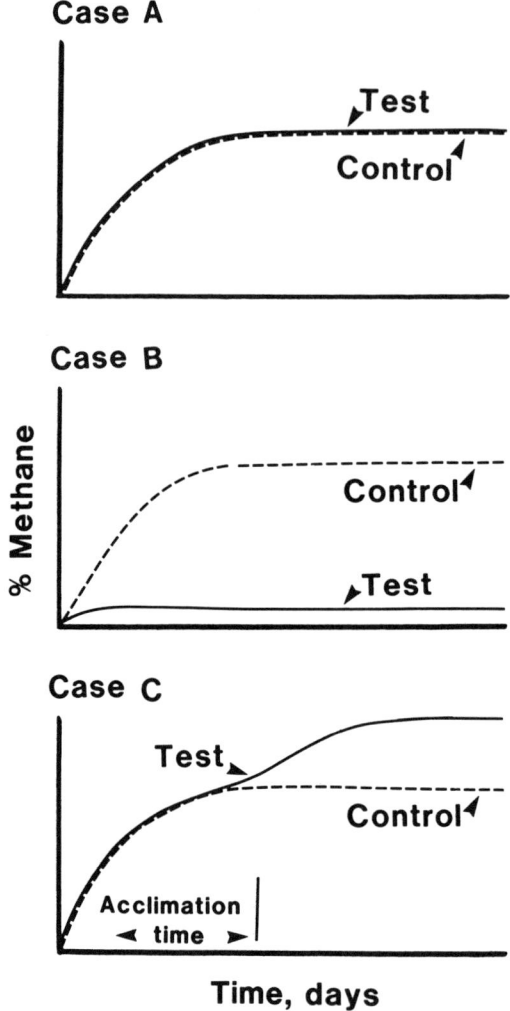

FIGURE 7. Typical trends in methane production observed in batch cultures.

have been tested in anaerobic cultures to determine biodegradability. The summary has focused on the methanogenic fermentation of these compounds, but has not excluded the other three mechanisms of biodegradation. Most of the data presented have been obtained from laboratory-scale batch cultures or occasionally from semicontinuous culture studies.

Investigations which consider the conversion of phenolics to methane are frequently done using the serum bottle modification of the Hungate technique.[56,57] Test culture volumes are typically small, ranging in size from 10 to approximately 100 mℓ. Using these small-scale systems, many compounds can readily be screened at various concentrations. Methane production is usually monitored by gas chromatographic methods and the total volume of gas production is measured using a variety of techniques.[13,56-59]

Figure 7 shows three distinct responses observed in methanogenic cultures which receive aliquots of a test substrate or waste water. If domestic anaerobic sludge or anaerobic mud is used as a source of microorganisms, each culture receives fermentable organic material from the inoculum. Thus, methane is produced in the control cultures.

In case A (Figure 7), the test solution contains no inhibitory compounds nor any compounds which can be fermented to methane. Thus, there is no difference between the methane in

Table 4
ANAEROBIC MICROBIAL METABOLISM OF PHENOL
AND NAPHTHOLS

Compound	Observations	Ref.
Phenol	Phenol found as transient intermediate in methane fermentation of tyrosine	48
	Methane produced after 20-day lag; fermentation continued when the concentration was gradually increased to 1500 mg/ℓ	47
	Unidentified alicyclic compounds formed as intermediates in methane fermentation; catechol, the intermediate in aerobic degradation of phenol, was not formed	63
	Phenol, cyclohexanol, and cyclohexanone were intermediates of catechol degradation in methanogenic cultures; adipic acid was the first product of ring cleavage	16
	Mixed culture in presence of nitrate produced cyclohexanol, cyclohexanone, and *n*-hexanoic acid; methane not produced	18
	Methane produced in cultures with 300 mg/ℓ phenol after 2.5 weeks acclimation time	50
	In batch cultures, acclimation times increase with increasing phenol concentrations; methane-producing organisms were found to be more tolerant to high phenol concentrations than were the phenol-degrading organisms	51, 64
	Cross-acclimation demonstrated; methanogenic cultures enriched on *p*-hydroxybenzoic acid could also ferment phenol	13
	Determined kinetic constants for the nonmethanogenic fermentation of phenol in a mixed reactor	65
	Anaerobic activated carbon filter operated for 735 days with phenol as the major carbon source; methane production observed	66
	Not photometabolized by an anaerobic photoheterotroph capable of growth on benzoic acid	34
	Degraded by pure cultures and mixed populations coupled to nitrate respiration	18
1-Naphthol	Not degraded to methane	47, 64
2-Naphthol	Not degraded to methane	47, 64

the control and that in the test culture. Case B indicates the presence of compounds which inhibit methane formation at the test concentration. Case C shows that the test solution contains no inhibitory compounds and that at least one fermentable substrate is present. After an acclimation period, methane production in excess of the control level is observed.

At elevated concentrations, many phenolics inhibit the methanogenic process. For example, at 300 mg/ℓ 3,4- or 3,5-dimethylphenol has shown no inhibition of methane production in batch cultures.[51] However, at 500 mg/ℓ of either compound, a reduction in methane evolution was observed. Other workers have challenged laboratory anaerobic cultures with a variety of phenolics specifically to determine whether these compounds were inhibitory to the methanogenic fermentation. Pearson et al.[60] used various concentrations of phenol and Johnson and Young[61,62] tested 2-chlorophenol, 2-nitrophenol, 4-nitrophenol, 2,4-dichlorophenol, 4-chloro-3-methylphenol, 2,4-dimethylphenol, and 2,4,6-trichlorophenol in their inhibition studies. Since these studies were testing for inhibition, rather than degradability, their results have not been included in Tables 4 to 9.

In Tables 4 to 9, 66 compounds have been listed. Of these, 46 have been shown to

Table 5
ANAEROBIC MICROBIAL METABOLISM OF
ALKYLPHENOLS

Compound	Observations	Ref.
2-Methylphenol (*o*-cresol)	Not degraded to methane	47, 49, 51, 52
	Degraded by a mixed culture via nitrate respiration	18
3-Methylphenol (*m*-cresol	Not degraded to methane	47
	Degraded by a mixed culture via nitrate respiration	18
	Acclimation time of about 4 weeks required in cultures containing 50 mg/ℓ; the substrate was completely degraded 3 weeks later with methane production reaching 90% of theoretical amount	52
	Acclimation time of 60 days required in cultures containing 80 mg/ℓ; acclimation time tends to be shorter if other fermentable phenolics such as phenol or *p*-cresol are present; methane production observed	64
4-Methylphenol (*p*-cresol)	Methane produced after 12-day lag; fermentation continued when *p*-cresol concentration was gradually increased to 1700 mg/ℓ	47
	In a reactor adapted to the fermentation of both *p*-cresol and resorcinol, *p*-cresol was degraded more slowly than resorcinol	63
	Methane produced with phenol and cyclohexanol as intermediates	67
	Degraded by a mixed culture via nitrate respiration	18
2,3-Dimethylphenol	Not degraded to methane	64
2,4-Dimethylphenol	Not degraded to methane	47, 64
2,5-Dimethylphenol	Not degraded to methane	51
2,6-Dimethylphenol	Not degraded to methane	51
3,4-Dimethylphenol	Not degraded to methane	47, 51
3,5-Dimethylphenol	Not degraded to methane	47, 51
3-Ethylphenol	Degraded in batch cultures with concomitant gas production from an initial concentration of 250 mg/ℓ	68
4-Ethylphenol	Not degraded to methane	64
1,2-Dihydroxy-3-methylbenzene (homopyrocatechol)	Not degraded to methane	64
1,2-Dihydroxy-4-methylbenzene	Not degraded to methane	49
	Degraded in batch cultures with concomitant gas production from an initial concentration of 150 mg/ℓ	68
1,3-Dihydroxy-2-methylbenzene	Not degraded to methane	49

Table 5 (continued)
ANAEROBIC MICROBIAL METABOLISM OF ALKYLPHENOLS

Compound	Observations	Ref.
1,3-Dihydroxy-5-methylbenzene (orcinol)	Not degraded to methane	49
1,3-Dihydroxy-6-methylbenzene	Not degraded to methane	49, 64

Table 6
ANAEROBIC MICROBIAL METABOLISM OF DI- AND TRIHYDRIC PHENOLS

Compound	Observations	Ref.
1,2-Dihydroxybenzene (catechol)	Not degraded to methane	47
	Acclimation times of 21 and 31 days required in cultures containing 300 mg/ℓ; methane production observed	13, 50
	In methanogenic consortia degradation intermediates included *cis*-benzenediol (*cis*-1,2-dihydroxy-3,5-cyclohexadiene), phenol, cyclohexanol, and cyclohexanone	16
	Anaerobic activated carbon filter operated for 490 and 617 days with catechol as the major carbon source; methane production observed	69, 70
	Methanogenic consortia enriched on syringic acid converted eight 3,4-substituted aromatics to catechol with little or no further degradation to ring cleavage products	71
	Not photometabolized by benzoate-grown culture of *Rhodopseudomonas palustris*	32
	Would not support growth of *Pseudomonas* strains growing anaerobic by nitrate respiration	35, 36
1,3-Dihydroxybenzene (resorcinol)	Methane produced after 15-day lag; fermentation continued when resorcinol concentration was gradually increased to 1600 mg/ℓ	47
	In reactor adapted to fermentation of both *p*-cresol and resorcinol, resorcinol was degraded more quickly than *p*-cresol	63
	Metabolized with production of gas from batch culture containing 500 mg/ℓ	72
	Readily converted to methane by the microbial population in one anaerobic municipal sludge but no methane produced by the population in a similar sludge from another municipality	49

Table 6 (continued)
ANAEROBIC MICROBIAL METABOLISM OF DI- AND TRIHYDRIC PHENOLS

Compound	Observations	Ref.
1,4-Dihydroxybenzene (hydroquinone)	Not degraded to methane	47
	Metabolized with production of gas from batch culture containing 500 mg/ℓ	72
1,2,3-Trihydroxybenzene (pyrogallol)	Methane produced after 9-day lag; fermentation continued when pyrogallol concentration was gradually increased to 1700 mg/ℓ	47
	Readily degraded to methane by microbial populations in anaerobic municipal sludge but not by that in a freshwater lake sediment	49
	Suggested intermediate in the methanogenic degradation of 2,6-dimethoxyphenol, 2,3,4-trihydroxybenzoic acid, and sinapic acid	71
	Pure culture of *Pelobacter acidigallici* produced 3 mol acetic acid from 1 mol of pyrogallol. Growth yield was 9.9 mg/mol substrate utilized; in coculture with *Methanosarcina barkeri*, the substrate was converted to methane and CO_2	21
1,3,5-Trihydroxybenzene (phloroglucinol)	Readily degraded to methane	47, 49
	Pure culture studies with a *Coprococcus* spp.; 1 mol substrate give 2 mol acetate and 2 mol CO_2	20
	Studies with partially purified enzyme from a *Coprococcus* spp.; resorcinol was an early intermediate of degradation	22
	In a methanogenic consortium containing 10 mM bromoethanesulfonic acid (to inhibit methane production), 1 mol phloroglucinol yielded 3 mol acetic acid	53
	Pure culture of *Pelobacter acidigalici* produced 3 mol acetic acid from each mole of phloroglucinol; growth yield was 9.9 mg/mol substrate utilized. In coculture with *Methanosarcina barkeri*, the substrate was converted to methane and CO_2	21
	Used anaerobically by *Moraxella* sp. by nitrate respiration	40
	Rhodopseudomonas gelatinosa photometabolized phloroglucinol with production of three unidentified acids	32

Table 7
ANAEROBIC METABOLISM OF CARBOXY- AND METHOXYPHENOLS

Compound	Observations	Ref.
2-Hydroxybenzoic acid (salicyclic acid)	Methane produced after 14-day lag	47
	Found as an intermediate of 2-methoxybenzoic acid degradation in methanogenic cultures	73
	Not photometabolized by *Rhodopseudomonas palustris*	30
	Photometabolized by benzoate-grown cells of an unidentified photoheterotroph; however, the bacterium could not grow on this compound	34
	Not utilized by *Moraxella* sp. incubated anaerobically with nitrate	40
	Supports anaerobic growth of *Pseudomonas* strain PN-1 in the presence of nitrate	36
3-Hydroxybenzoic acid	Degraded to yield methane	47
	Found as an intermediate of 3-methoxybenzoic acid degradation in methanogenic cultures	73
	Photometabolized by *Rhodopseudomonas palustris*	
	Degraded by *Pseudomonas* strain PN-1 anaerobically in presence of nitrate	36
	Degraded by *Moraxella* sp. incubated anaerobically nitrate	40
4-Hydroxybenzoic acid	Degraded to yield methane	47, 49
	Methane produced after 12-day lag in medium with 300 mg/ℓ substrate; this culture was simultaneously adapted to ferment phenol and benzoic acid	13
	Found as an intermediate of 4-methoxybenzoic acid degradation in methanogenic cultures	73
	Photometabolized by *Rhodopseudomonas palustris*	30
	Degraded by *Pseudomonas* strain PN-1 anaerobically in presence of nitrate	36
	Degraded by *Moraxella* sp. incubated anaerobically with nitrate	40
2,4-Dihydroxybenzoic acid	Photometabolized by benzoate-grown cells of an unidentified photoheterotroph; however, the bacterium could not grow on this compound	34
3,4-Dihydroxybenzoic acid (protocatechuic acid)	Methane produced after 13-day lag in medium with 300 mg/ℓ substrate; cultures adapted to vanillic acid would also produce methane from protocatechuic acid	13
	Methane produced by benzoate fermenting culture; two degradation pathways detected; the major route had catechol as an intermediate; the minor route had 3-hydroxybenzoic acid and benzoic acid	73

Table 7 (continued)
ANAEROBIC METABOLISM OF CARBOXY- AND METHOXYPHENOLS

Compound	Observations	Ref.
	Degraded by *Pseudomonas* strain PN-1 incubated anaerobically in presence of nitrate	36, 74
	Degraded by *Moraxella* sp. incubated anaerobically with nitrate	40
2,3,4-Trihydroxybenzoic acid	Degraded by a methanogenic consortium enriched on syringic acid; 1,2,3-trihydroxybenzene was a suggested intermediate	71
3,4,5-Trihydroxybenzoic acid (gallic acid)	Intermediate found during the degradation of syringic acid and sinapic acid by methanogenic consortia	71
	In a methanogenic consortium containing 10 mM bromoethanesulfonic acid (to inhibit methane production), 1 mol gallic acid yielded 3 mol acetic acid	53
	Pure cultures of *Pelobacter acidigallici* produced 3 mol acetic acid from 1 mol of gallic acid; growth yield was 10.1 mg/mol substrate utilized; in coculture with *Methanosarcina barkeri*, substrate was converted to methane and CO_2	21
2,4,6-Trihydroxybenzoic acid	Pure cultures of *Palobacter acidigallici* produced 3 mol acetic acid from 1 mol of 2,4,6-trihydroxybenzoic acid; growth yield was 9.6 mg/mol substrate utilized; in coculture with *Methanosarina barkeri*, the substrate was converted to methane and CO_2	21
2-Methoxyphenol (guaiacol)	Completely degraded within 2 weeks in batch cultures; catechol was the initial intermediate; methane was produced	52
	Methanogenic culture enriched on syringic acid converted guaiacol to catechol which accumulated in the medium yielding little or no gas production	71
	Demethylated to catechol by *Pseudomonas* strain PN-1 in the presence of nitrate but would not support growth	74
3-Methoxyphenol	Completely degraded within 1 week in batch cultures; resorcinol was the initial intermediate; methane was produced	52
4-Methoxyphenol	Completely degraded within 1 week in batch cultures; hydroquinone was assumed to be the initial intermediate but it was not detected; methane was produced	52
2,6-Dimethoxyphenol	Enrichment culture adapted to syringic acid degraded 2,6-dimethoxyphenol, likely via 1,2,3-trihydroxybenzene, to methane and CO_2	71

Table 7 (continued)
ANAEROBIC METABOLISM OF CARBOXY- AND METHOXYPHENOLS

Compound	Observations	Ref.
4-Hydroxy-3-methoxybenzyl alcohol (vanillyl alcohol)	Supported growth of *Pseudomonas* strain PN-1 in the presence of nitrate	74
4-Hydroxy-3-methoxyben-zaldehyde (vanillin)	Methane produced after 12-day lag in medium with 300 mg/ℓ; this culture was simultaneously adapted to ferment vanillic acid and syringaldehyde	13
	Methanogenic culture enriched on syringic acid converted vanillin to catechol which accumulated in the medium yielding little or no gas production	71
	Supported growth of *Pseudomonas* strain PN-1 in the presence of nitrate	74
4-Hydroxy-3-methoxyben-zoic acid (vanillic acid)	Methane produced after 9-day lag in medium with 300 mg/ℓ substrate; cultures adapted to vanillic acid would also produce methane from syringaldehyde, syringic acid, vanillin, benzoic acid, catechol, and protocatechuic acid	13
	Readily degraded to methane within 1 week in enrichment cultures containing anaerobic domestic sludge	49
	Methanogenic culture enriched on syringic acid converted vanillic acid to catechol which accumulated in the medium yielding little or no gas production	71
	Supported growth of *Pseudomonas* strain PN-1 in the presence of nitrate	74
3-(4-Hydroxyphenyl)-pro-penoic acid	Degraded by *Moraxella* sp. incubated anaerobically with nitrate	40
3-(3,4-Dihydroxyphenyl)-propenoic acid (caffeic acid)	Degraded by *Moraxella* sp. incubated anaerobically with nitrate	40
3-(4-Hydroxy-3-methoxy-phenyl)-propenoic acid (ferulic acid)	Methane produced after 10-day lag in medium with 300 mg/ℓ substrate; this culture was simultaneously adapted to ferment cinnamic acid and vanillin	13
	Three morphological types of bacteria comprised more than 90% of the organisms in the methane-producing enrichment culture which was simultaneously adapted to cinnamic acid, vanillic acid, and vanillin but not to phenol	14
	Methanogenic culture enriched on syringic acid converted ferulic acid to catechol which accumulated in the medium yielding little or no gas production	71

Table 7 (continued)
ANAEROBIC METABOLISM OF CARBOXY- AND METHOXYPHENOLS

Compound	Observations	Ref.
3-(4-Hydroxy-3-methoxy-phenyl)-propen-1-ol (coni-feryl alcohol)	Degraded in methanogenic cultures en-riched on either coniferyl alcohol or ferulic acid to yield methane and CO_2; the first transformation reaction was a dehydrogenation to give ferulic acid; phenyl propionic acid was also an in-termediate	75
3-(4-Hydroxy-3-methoxy-phenyl)-1-propene (eu-genol)	Degraded to methane when present at 100 but not at 500 mg/ℓ	76
3-(4-Hydroxy-3,5-dimethox-yphenyl)-propenoic acid (sinapic acid)	Degraded to methane and CO_2 by sy-ringic acid enriched consortium; likely intermediates were 3,4,5-trihydroxy-benzoic acid and 1,2,3-trihydroxyben-zene	71
4-Hydroxy-3,5-dimethoxy-benzoic acid (syringic acid)	Methane produced after 2-day lag in medium with 300 mg/ℓ substrate; cul-tures adapted to syringic acid would also produce methane from syringal-dehyde and vanillin	13
	Methanogenic cultures enriched on sy-ringic acid could readily degrade ten other similar phenolic compounds	71
4-Hydroxy-3,5-dimethoxy-benzaldehyde (syringalde-hyde)	Methane produced after 5-day lag in medium with 300 mg/ℓ; this culture was simultaneously adapted to ferment syringic acid	13

Table 8
ANAEROBIC METABOLISM OF CHLOROPHENOLS

Compound	Observations	Ref.
2-Chlorophenol	Not degraded to methane by cultures in-oculated with freshwater lake sediment or with municipal digested sludges	49
	Completely degraded within 3 weeks in batch cultures; dechlorinated to give phenol as initial intermediate; methane was produced	52
	In unacclimated methanogenic cultures 2-chlorophenol was degraded more readily than 3-chlorophenol which was degraded more readily than 4-chloro-phenol; cultures acclimated to 2-chloro-phenol were cross-acclimated to 4-chlorophenol, and 2,4-dichlorophenol but not 3-chlorophenol, 2,3-, 2,5-, or 2,6-dichlorophenol	17
3-Chlorophenol	Not degraded to methane by cultures in-oculated with municipal digested sludges	49
	Completely degraded within 7 weeks in batch cultures; methane was produced	52

Table 8 (continued)
ANAEROBIC METABOLISM OF CHLOROPHENOLS

Compound	Observations	Ref.
	Cultures acclimated to 3-chlorophenol were cross-acclimated to 4-chlorophenol, 3,4-, and 3,5-dichlorophenol but not 2-chlorophenol, 2,3-, or 2,5-dichlorophenol	17
4-Chlorophenol	Not degraded to methane by cultures inoculated with freshwater lake sediment or with municipal digested sludges	49
	Persistent over a period of 8 weeks but completely degraded within 16 weeks; methane was not detected	52
	More resistant to degradation than 2- or 3-chlorophenol; cultures acclimated to 4-chlorophenol were cross-acclimated to other isomers of monochlorophenol and to 2,4- and 3,4-dichlorophenol	17
2,3-Dichlorophenol	Dechlorinated to 3-chlorophenol which was degraded to methane; cultures acclimated to 2- or 3-chlorophenol did not degrade 2,3-dichlorophenol	17
2,4-Dichlorophenol	Dechlorinated to 4-chlorophenol which was degraded to methane; cultures acclimated to 2- or 4-chlorophenol degraded 2,4-dichlorophenol	17
2,5-Dichlorophenol	Dechlorinated to 3-chlorophenol which accumulated in the culture over the 6-week test period; cultures acclimated to 2- or 3-chlorophenol did not degrade 2,5-dichlorophenol	17
2,6-Dichlorophenol	Dechlorinated to 2-chlorophenol which was degraded to methane; cultures acclimated to 2-chlorophenol did not degrade 2,6-dichlorophenol	17
3,4-Dichlorophenol	Persistent over 6-week test period in cultures inoculated with domestic anaerobic sludge; cultures acclimated to 3- or 4-chlorophenol degraded 3,4-dichlorophenol	17
3,5-Dichlorophenol	Persistent over 6-week test period in cultures inoculated with domestic anaerobic sludge; cultures acclimated to 3-chlorophenol degraded 3,5-dichlorophenol	17
Pentachlorophenol	Not degraded to methane during an 8-week test period	49
	^{14}C-labeled substrate added to 100 g soil to give final concentration of 10 ppm; after 24-days incubation in a N_2 atm, 5.3% of the label was recovered as pentachloroanisole and 7% was recovered as tetra- and trichlorophenols; no $^{14}CO_2$ was detected	77

Table 8 (continued)
ANAEROBIC METABOLISM OF CHLOROPHENOLS

Compound	Observations	Ref.
	Semicontinuous 3-ℓ methanogenic culture was acclimated to an influent concentration of 5 mg/ℓ pentachlorophenol; the effluent concentration stabilized at 5 μg/ℓ; the high removal efficiency could not be attributed to sorption nor volatilization; gas chromatography analysis of the effluent showed unidentified chlorinated compounds and the loss of the parent substrate suggesting biological degradation	78
1,2-Dihydroxy-4-chlorobenzene	Not degraded to methane over an 8-week test period in cultures inoculated with municipal digested sludge	49
1,3-Dihydroxy-4-chlorobenzene	Not degraded to methane over an 8-week test period in cultures inoculated with municipal digested sludge	49
3-Chloro-4-hydroxy-benzoic acid	Some methane produced in cultures inoculated with freshwater lake sediment but not in those inoculated with municipal digested sludge	49
5-Chloro-2-hydroxy-benzoic acid	Not degraded to methane over an 8-week test period in cultures inoculated with municipal digested sludge	49

Table 9
ANAEROBIC METABOLISM OF NITROPHENOLS AND AMINOPHENOLS

Compound	Observations	Ref.
2-Nitrophenol	Completely degraded within 1 week in batch cultures; methane was produced from this substrate	52
3-Nitrophenol	Disappeared from batch culture medium within 1 week; inhibited methane production from fermentable organics introduced with inoculum; conversion of 3-nitrophenol to methane and CO_2 could not be ascertained	52
4-Nitrophenol	Disappeared from batch culture medium within 1 week; severely inhibited methane production from fermentable organics introduced with inoculum; cultures given ^{14}C-labeled 4-nitrophenol produced radioactive methane and CO_2 after approximately 5-week incubation	52
2-Aminophenol	Not degraded to methane in cultures inoculated with either municipal digested sludge or freshwater lake sediment	49
4-Aminophenol	Not degraded to methane in cultures inoculated with either municipal digested sludge or freshwater lake sediment	49

undergo some form of degradation under anaerobic conditions. Of 62 tested, 39 were susceptible to degradation to methane and carbon dioxide (mineralization) by methanogenic consortia. Many initial studies showed that particular compounds were not degraded to methane while later investigations demonstrated that they were. Examples are 3-methylphenol (Table 4), 1,2-, 1,3-, and 1,4-dihydroxybenzenes (Table 5), and 2-, 3-, and 4-chlorophenols (Table 8). Thus, it is possible that further studies may find more of the "nonbiodegradable" phenolics susceptible to anaerobic metabolism.

The anaerobic degradation of the naturally occurring aromatic amino acids (tyrosine, tryptophan, and phenylalanine) in methanogenic consortia has been demonstrated.[79] Tyrosine has been shown to yield *p*-cresol and phenol as metabolic end products in pure cultures of anaerobes[80] and as intermediates in the methanogenic consortia.[79] Thus, it is not surprising that inocula from various anaerobic sources are able to metabolize phenol and *p*-cresol.[47,50,51]

As a group, the alkyl phenolics are the least susceptible to anaerobic degradation in methanogenic cultures (Table 5). Of the 16 compounds tested, only 4 have been found to yield gas production. Only the cresols have been tested under nitrate respiration conditions and all three isomers have been found to be degraded. All of the di- and trihydric phenols (Table 6) and carboxy- and methoxyphenols (Table 7) which have been tested under methanogenic conditions have been shown to yield methane. The monochlorophenols and some of the dichlorophenols (Table 8) are also fermented to methane. Dechlorination of pentachlorophenol under anaerobic conditions has been demonstrated (Table 8) but no gaseous end products from these compounds have been detected. The anaerobic degradation of other halophenols has not been reported. However, results from studies using fluoro-, bromo-, and iodo-substituted benzoic acids have been reported.[44,55]

It is unlikely that a phenolic waste water would contain only those phenolics which are degradable. Thus, the ability of the microbial population to remove the degradable compounds from the complex mixture is an important consideration. Figure 8 shows the gas chromatographic analyses of the phenolics in a methanogenic batch culture which received a mixture of phenolics.[64] The mixture contained 250 mg/ℓ phenol and approximately 100 mg/ℓ of each of the following alkylphenols: *o*-cresol, *p*-cresol, 2,6, 2,5-, and 3,5-dimethylphenols. The total phenolic concentration was 731 mg/ℓ. The cultures were inoculated with domestic anaerobic sewage sludge and allowed to acclimate to the phenolics. After 49 days incubation at 37°C, essentially all of the phenol and the *p*-cresol had been selectively removed with concurrent production of methane. The other nondegradable compounds remained. In this case, the fermentable phenolics accounted for 48% of the total phenolics. These results suggest that if a large proportion of the phenolics in a waste water are fermentable, a methanogenic treatment process may result in a high phenolic removal leaving lower demands on subsequent treatment processes.

IV. PHENOLIC CONTENT OF INDUSTRIAL WASTE WATERS

The concentrations of phenolics in domestic waste waters are very low. Phenolics found in municipal waste waters are usually the result of industrial effluents. Ongerth and DeWalle[81] analyzed the influent sewage reaching 11 municipal treatment plants and they found a fairly strong correlation between the concentration of phenol and the proportion of influent flow which originated from industrial sources. A survey of 25 municipal sewage treatment plants identified phenol, dimethylphenol, chlorophenols, and nitrophenols in the raw waste water influents.[82] Baird et al.[83] have used gas chromatographic analyses of phenol and some alkylphenols to "fingerprint" which of two petroleum industries was responsible for discharging phenolic waste waters into municipal sewers.

Numerous industries produce phenolic waste waters.[1,2,84] These include petrochemical industries, petroleum refineries, coal coking operations, coal gasification and liquefaction

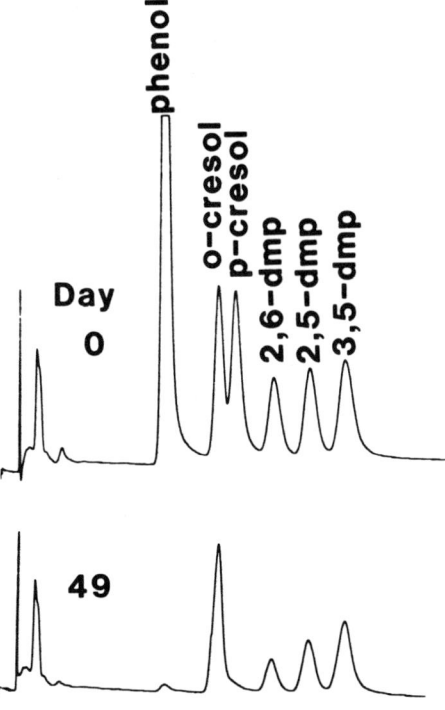

FIGURE 8.   Gas chromatographic analyses of phenolics in a methanogenic batch culture which received a mixture of six phenolics at a total initial concentration of 731 mg/ℓ (Day 0). Phenol and *p*-cresol were biodegraded over the 49-day incubation period. dmp = Dimethylphenol.

processes, resin manufacturing, dye synthesis, wood preserving plants, pulp and paper mills, and aircraft maintenance. Table 10 shows the phenol concentrations in a variety of industrial waste waters. In general, phenolic waste waters contain phenol as the major compound of this group with smaller quantities of phenol derivatives.

The occurrence and approximate overall quantity of phenolic compounds are commonly determined by the nonspecific 4-aminoantipyrine colorimetric method.[85] This method does not respond to para-substituted phenols unless the substituent is a carboxyl, halogen, hydroxyl, methoxyl, or sulfonic acid group.[86] Furthermore, as shown in Tables 4 to 9, only certain phenolics are susceptible to anaerobic degradation. Thus, the identification and quantitation of individual phenolics are required to assess the feasibility of phenolic removal by an anaerobic biological process. Methods such as gas chromatography, high performance liquid chromatography, and gas chromatography-mass spectrometry have been used recently to analyze the phenolics in waste waters.

Of the numerous industries which produce phenolic waste waters, the petroleum and synthetic fuels (bitumens, heavy oils, and coal conversion) and forest products (thermo-mechanical, sulfite, and kraft pulping) industries yield effluents with sufficient anaerobically degradable phenolic content to warrant further consideration. Other industrial waste waters may be amenable to anaerobic treatment but have generally received less attention.

Examples of the phenolic contents in waste streams from petroleum-related industries and coal-conversion processes are summarized in Table 11 and 12, respectively. For each waste water, the proportion of phenolics susceptible to anaerobic degradation has been estimated.

Table 10
INDUSTRIAL SOURCES AND
CONCENTRATIONS OF PHENOL

Industry	Conc (mg/ℓ)
Coking plant	
Weak ammonia liquor without dephenolization	580—10,000
Weak ammonia liquor after dephenolization	4—332
Wash oil-still wastes	30—150
Oil refineries	
Sour water	80—185
General waste water	10—100
API separator effluent	0.3—6.8
Petrochemical	
Benzene refinery	210
Tar distillation	300
Nitrogen works	250
Orlon manufacturing	100—150
Plastics factory	600—2000
Phenolic resin production	1600
Fiberboard factory	150
Fiberglass manufacturing	40—400
Aircraft maintenance	200—400

After Patterson, J. W., *Wastewater Treatment Technology,* Patterson, J. W., Ed., Ann Arbor Science, Ann Arbor, Mich., 1975, chap. 18. With permission.

In the majority of cases, greater than two thirds of the phenolics reported are susceptible to anaerobic degradation.

Analyses provided in the references cited in Tables 11 and 12 considered only the concentrations of monohydric phenolics. It is very likely that dihydric compounds such as catechol, resorcinol, and hydroquinone were also present but not quantitated because different analytical methods were required. Catechols and resorcinols have been shown to account for 1 to 48% of the total phenols in waste waters from coal liquefaction processes.[93] As shown in Table 6, unsubstituted dihydric phenols are very susceptible to anaerobic degradation and they would presumably be removed from waste waters. As well, coal conversion waste waters are often rich in volatile organic acids[12,89,93] which are readily fermented to methane.

In a methanogenic process, the degradable phenolics would be selectively removed as illustrated with a synthetic waste water in Figure 8. Aromatic or aliphatic hydrocarbons present in these waste waters would not be biologically removed in an anaerobic system. These, along with the nondegradable phenolics, would require further biological or physical/chemical treatment for removal. Other components in the waste waters, such as cyanide, sulfide, ammonia, heavy metals, etc., may be present at inhibitory concentrations. The phenolics may occur at concentrations which would be inhibitory to unacclimated methanogenic consortia. Thus, knowledge of the chemical composition of the waste water coupled with a biological treatability study would be required to test the feasibility of using the anaerobic process for phenolic removal.

Another major potential source of phenolic waste waters is the pulp and paper industry. Hydrolysis of lignin, which frees cellulose from wood, can yield a very wide variety of phenolic compounds. Table 13 summarizes the mono- and dinuclear phenols obtained from the hydrolysis of lignin.[94] Many of these compounds are alkyl phenolics which are char-

Table 11
PHENOLIC CONTENT OF SOME WASTE WATER FROM PETROLEUM INDUSTRIES

Activity	Type of sample	Compounds and conc (mg/ℓ unless otherwise stated)	Proportion of phenolics susceptible to anaerobic degradation (%)[a]	Ref.
Petroleum refining	Final effluent	Phenol (0.88); *m*-cresol (0.75)	100	83
Petroleum refining	Discharge to sewer	Phenol (3016); *o*-cresol (5842); *m*- and *p*-cresol (56); 2,3-dmp[b] (8177); 2,4- and 2,5-dmp (963); 2,6-dmp (1547)	16	83
Petroleum refining	Not stated	Phenol (310); *o*-cresol (110); *m*- and *p*-cresol (290)	85	87
Petroleum refining	API separator influent	Phenol (0.78); *p*-cresol (1.2); 2,5-dmp (1.1); *p*-ethylphenol (0.72); 2,3,5-trimethylphenol (0.09); *p*-chloro-*m*-cresol (0.15); 2,3,5,6 tetramethylphenol (0.02)	67	88
Oil shale retorting	By-product water	Phenol (10); *o*-cresol (30); *m*- and *p*-cresol (20)	50	89
Experimental tar sands recovered by *in situ* burning	Water from recovery separator	Phenol (9.0% of alkaline extract); *o*-cresol (2.4%); *m*-cresol (3.1%); *p*-cresol (0.9%); also present were isomers of dmp and C$_3$-phenols	84[c]	90

[a]   Based on entries in Tables 3 and 4, and phenolic concentrations reported.
[b]   dmp = Dimethylphenol.
[c]   Maximum value based on quantitative data available.

Table 12
PHENOLIC CONTENT OF SOME WASTE WATERS FROM COAL CONVERSION INDUSTRIES

Activity	Type of sample	Compounds and conc (mg/ℓ unless otherwise stated)	Proportion of phenolics susceptible to anaerobic degradation (%)[a]	Ref.
Coal gasification	10% Dilution of waste water	Phenol (207); *o*-cresol (57); *m*- and *p*-cresol (139); 2,4-dmp[c] (22); 3,5- and 2,3-dmp (29); 3,4-dmp (9); also detected: C₃ phenols	75[b]	11
H-coal liquefaction	Process waste water	Phenol (4900); *o*-cresol (586); *m*-cresol (1230); *p*-cresol (420); 2,4- and 2,5-dmp (63); 3,5-dmp (213); 3,4-dmp (44)	88	12
Coking	Waste ammonia liquor plus blast furnace blowdown	Phenol (320); *o*-cresol (6); *m*- and *p*-cresol (55)	98	64
Synthane coal gasification	Byproduct water	Phenol (2100); *o*-cresol (670); *m*- and *p*-cresol (1800); 2,6-dmp (40); 2,5-dmp (250); 3,5-dmp (230); 2,3-dmp (30); 3,4-dmp (100); *o*-ethylphenol (30); 1-naphthol (10); 2-naphthol (30)	73	89
Coal gasification	Process waste water	Phenol and *o*-cresol (2209); *m*- and *p*-cresol and 2,4- and 2,5-dmp (1626); 2,6-dmp (24); 2,3-dmp (50); 3,5-dmp and *m*- and *p*-ethyl-phenol (366); 3,4-dmp (119); 3-ethyl-5-methylphenol (66); C₃ phenol (40)	?[d]	91
In situ coal gasification	Product water	Phenol (270); *p*-cresol (10.2); dmp isomers (10); C₃ phenols (1.2)	96	92
In situ coal gasification	Process water	Phenol (50000); *o*-cresol (8000); dmp isomers (4265); C₃ phenols (1395); ethylphenols (855)	78	92
Low-Btu gasification	Aqueous condensate	Phenol (0.28); *o*-cresol (4.8); *m*-cresol (4.5); *p*-cresol (0.04); dmp isomers (7.4); C₃ phenols (0.1); ethylphenols (1.4)	26	92

[a] Based on entries in Tables 3 and 4, and phenolic concentrations reported.
[b] Maximum value based on quantitative data available.
[c] dmp = Dimethylphenol.
[d] Cannot be estimated because the compounds were not adequately resolved by the analytical method.

Table 13
MONO- AND DINUCLEAR PHENOLS OBTAINABLE THROUGH HYDROLYSIS OF LIGNIN

Phenols
Phenol
2-Methylphenol (*o*-cresol)
3-Methylphenol (*m*-cresol)
4-Methylphenol (*p*-cresol)
2-Ethylphenol
3-Ethylphenol
4-Ethylphenol
3-Methyl-4-ethylphenol
4-Hydroxyphenol (hydroquinone)
p-Hydroxybenzoic acid
p-Hydroxybenzaldehyde

Xylenols
2,4-Dimethylphenol (2,4-xylenol)
2,5-Dimethylphenol (2,5-xylenol)
2,6-Dimethylphenol (2,6-xylenol)
3,4-Dimethylphenol (3,4-xylenol)
3,5-Dimethylphenol (3,5-xylenol)

Catechols
2-Hydroxyphenol (catechol)
4-Methyl-2-hydroxyphenol (homocatechol)
4-Ethyl-2-hydroxyphenol
4-Propyl-2-hydroxyphenol

Protocatechuic acid
Protocatechualdehyde
Homoprotocatechualdehyde

Guaiacols
2-Methoxyphenol (guaiacol)
4-Methyl-2-methoxyphenol
4-Ethyl-2-methoxyphenol
4-Propyl-2-methoxyphenol
4-Propenyl-2-methoxyphenol (isoeugenol)
4-Acetyl-2-methoxyphenol (acetylguaiacone)
Vanillin
Vanillic acid
Carboxyvanillic acid
Carboxyvanillin

Syringols
2,6-Dimethoxyphenol
4-Methyl-2,6-dimethoxyphenol
4-Ethyl-2,6-dimethoxyphenol
4-Propyl-2,6-dimethoxyphenol

Syringaldehyde
Syringic acid

Dinuclear phenols
4,4'-Dihydroxy-3,3'-dimethoxystilbene
Dehydrodivanillin
Dehydrodivanillic acid
4,4'-Dihydroxy-3,3'-dimethoxychalcone
4,4'-Dihydroxy-3,3'-dimethoxybenzophenone

From Hanselmann, K. W., *Experientia*, 38, 176, 1982. With permission.

acteristically found in the petroleum and coal conversion waste waters (Tables 11 and 12). In actual waste water analysis, Chernousov et al.[95] found mainly guaiacol (41.8%) with small quantities of phenol, *o*-cresol, 2,5-dimethylphenol, catechol, and 4-methylcatechol in an ether extract of black liquor. The organic matter in sulfate liquor contained approximately 5% phenols of which guaiacol comprised 55 to 60%.[96] McKague[97] also found guaiacol as the predominant phenolic, with vanillin, 4-hydroxy-3-methoxy-acetophenone, and others present in unbleached white water. Kringstad and Lindstrom[98] extensively reviewed the chemical composition of spent liquors from pulp bleach plants. They described the processes leading to formation of chlorinated phenolics through the reaction of chlorine with phenolic constituents of lignin. Much of the resulting chlorinated organic matter exhibits a high molecular weight with only 30% of the originally bound chlorine in spent chlorination liquor occurring below a molecular weight of 1000.

Voss et al.[99] reported the occurrence of chlorinated phenols, guaiacols, vanillins, and catechols in spent bleach liquors from softwood kraft pulps. In addition to those compounds, chlorinated syringols and syringaldehydes were reported for hardwood (birch) kraft pulps. Analysis of bleach kraft effluents from three Finnish kraft pulp mills found total chlorinated phenolics (occurring as chlorophenols, chlorocatechols, and chloroguaiacols) ranging from 100 to 258 g/ton of pulp bleached.[100] Other studies[101,102] have detected similar compounds, generally at waste water concentrations less than 500 μg/ℓ. The total content of chlorinated phenolic compounds in spent chlorination liquor has been shown to be very sensitive to process parameters. Specifically, the quantity of chlorine applied per tonne of unbleached kraft pulp and the final pH of the bleaching process are important.[103]

V. TREATMENT OF PHENOLIC WASTE WATERS

Anaerobic biological treatment technology has historically been used mainly for municipal sewage sludge digestion and high-strength food processing wastes. Rapid advances in process design concepts have encouraged consideration of anaerobic technology for a wider range of industrial waste waters.[104] Process design engineers have traditionally regarded anaerobic processes as too sensitive and unstable to consider the treatment of complex organic waste waters. Consequently, the majority of research on the anaerobic biological treatment of phenolic waste waters has emerged recently. To date, there have only been reports on laboratory investigations. So far, no reports of implementation of full-scale anaerobic treatment technology on high phenolic strength industrial discharges have been located.

A. Treatment of Synthetic Waste Waters

Khan et al.[66] studied the removal of phenol from a synthetic nutrient-supplemented waste water using a three-stage, anaerobic-activated carbon filter with intermediate effluent recycle. They performed experiments with empty bed contact times of 9.3 and 18.6 hr. Phenol loading rates from 0.26 to 2.58 kg phenol per cubic meter per day (feed concentrations from 200 to 1000 mg/ℓ of phenol) were studied. Removal efficiencies ranging from 92.5 to essentially 100% were observed with no apparent dependence of removal efficiency on loading rate over this range. These experiments also included glucose-phenol mixtures providing various ratios of phenol to glucose chemical oxygen demand (COD), resulting in COD loadings varying from 0.66 to 9.33 kg COD per cubic meter per day. Overall, COD removal efficiencies ranged from 80.3 to 97.3%. A COD mass balance (COD applied vs. COD in effluent and methane produced) was attempted to estimate COD losses which may represent substrate diverted to biomass generation. While feeding 200 mg/ℓ phenol, the steady-state-estimated biomass yield (on a COD basis) was 24% while at 400 mg/ℓ phenol feed this dropped to 11%.

Khan et al.[105] compared the performance of granular activated carbon with anthracite as a medium for two single-stage, completely mixed anaerobic column reactors. The synthetic nutrient-supplemented waste water was qualitatively similar to that reported above with glucose and phenol as major carbon sources.[66] They found that the activated carbon filter provided much better performance than the anthracite filter. The latter medium had negligible adsorption properties. In particular, a dynamic loading experiment was conducted wherein the phenol concentration was increased in steps of 500 mg/ℓ/day from the steady-state level of 500 up to 2000 mg/ℓ and back down again. For both filters, the effluent phenol concentration mirrored the influent feed change but the maximum phenol concentration in the anthracite filter reached 1140 mg/ℓ, whereas the activated carbon reactor only reached 70 mg/ℓ. However, both the anthracite and activated carbon reactors returned to steady-state effluent concentrations comparable to the preloading levels (4 and <1 mg/ℓ, respectively).

Suidan et al.[69,70,106] studied phenol, *o*-cresol, and catechol treatment in multistage, anaerobic activated carbon columns with an intermediate effluent recycle. The synthetic waste medium was qualitatively similar to that reported by Khan et al.[66,105] As would be expected from batch studies, phenol and catechol were degraded with methane production, whereas *o*-cresol was removed but no evidence of degradation to methane was found. In the particular case of *o*-cresol, the adsorption capacity of the activated carbon was able to remove this compound from the effluent. Because relatively long contact times were provided (9.3 to 11.6 hr, empty bed contact time per column), essentially all of the phenolic substrate was removed in the first activated carbon column for phenol and catechol at feed concentrations up to 1000 mg/ℓ. In the case of *o*-cresol, the lack of anaerobic degradation led to the breakthrough of this compound as the carbon became saturated. An adsorption capacity over the first two columns of 0.27 g *o*-cresol per gram of carbon was observed.

Suidan et al.[107] evaluated a two-stage anaerobic filter on a synthetic mixture intended to simulate a coal conversion waste water. The waste water composition including nutrient supplementation is summarized in Table 14. The treatment system consisted of a first-stage contactor packed with polypropylene berl saddles providing a microbial attachment surface area of 2.2 m^2 in the reactor column (volume of 0.011 m^3). The second-stage contactor was charged with granular activated carbon providing an external surface area of 36 m^2. Both reactors were provided with an effluent recycle to maintain complete mix conditions in the first stage and to provide a 24% bed expansion in the carbon column. The latter provided an empty bed contact time of 23.6 hr. No significant treatment was provided by the first-stage contactor with steady-state dissolved organic carbon (DOC) and COD removals of 9.2 and 11.1% respectively. The activated carbon contactor achieved DOC and COD removals of 91 and 88%, respectively. Likewise, only 7.6% of the phenol was removed in the first stage compared with 99.8% in the second stage. The cresols were also monitored in this study. They were not removed as effectively (*o*-cresol, 85% removal and *m/p*-cresol, 87% removal) suggesting some breakthrough of the more resistant compounds. On the other hand, 2,4-dimethylphenol did not appear to break through the activated carbon contactor. However, this compound has a stronger affinity for the activated carbon than do the cresols. At steady state, a sample of the carbon was withdrawn and extracted for analysis by gas chromatography-mass spectrometry. This analysis indicated the presence of dimers of the three cresols, compounds which were not present in the feed waste water. An attempt to cause dimer formation on a sterile activated carbon column failed, suggesting that this mechanism may be microbially mediated.

B. Treatment of Authentic Industrial Waste Waters
1. General
The pioneering work on treatability of phenolic waste waters was conducted on Chmielowski and Kusznik[9] using a draw and feed procedure with 1 ℓ of anaerobic inoculum in a sealed 3-ℓ container. Three waste waters were evaluated. One waste water was from a

Table 14
COMPOSITION OF SYNTHETIC COAL
GASIFICATION WASTE WATER

Group	Compound	Conc (mg/ℓ)
Volatile acids	Acetic acid	50
	Propionic acid	10
	Butanoic acid	5
	Pentanoic acid	5
Monohydric phenols	Phenol	200
	o-Cresol	60
	m-Cresol	40
	p-Cresol	70
	2,4-Dimethylphenol	4
Dihydric phenols	Catechol	80
	4-Methylcatechol	25
	Resorcinol	80
Polycyclic hydroxy compounds	α-Naphthol	1
	β-Naphthol	10
	5-Indanol	8.5
	o-o'-Biphenol	5
Monocyclic N-aromatic compounds	Pyridine	10
	α-Picoline	10
	Aniline	2
Polycyclic N-aromatic compounds	Quinoline	5
	Quinaldine	5
	Indole	5
Inorganics	Thiocyanate	10
	Cyanide	1.5
	Sodium carbonate	1000—1300
Nutrient salts	$FeCl_3$	9.48
	$MnCl_2 \cdot 4H_2O$	2.31
	$ZnCl_2$	1.59
	$CuCl_2 \cdot 2H_2O$	1.00
	$CoCl_2 \cdot 6H_2O$	1.39
	$Na_2B_4O_7 \cdot 10H_2O$	0.56
	$Na_3C_6H_5O_7$	86.08
	$(NH_4)_6Mo_7O_{24} \cdot 4H_2O$	1.01
	KH_2PO_4	201.15
	$NaH_2PO_4 \cdot H_2O$	122.38
	$(NH_4)_2SO_4$	78.04
	NH_4Cl	727.18
	$CaCl_2$	65.62
	$MgCl_2 \cdot 6H_2O$	120.16
Vitamins	Biotin	0.027
	Folic acid	0.027
	Pyridoxine hydrochloride	0.136
	Riboflavin	0.068
	Thiamin	0.068
	Nicotinic acid	0.068
	Pantothenic acid	0.068
	Cyanocobalamin-B12	0.001
	p-Aminobenzoic acid	0.068
	Thioctic acid	0.068

After Suidan, M. T., Siekerka, G. L., Kao, S. W., and Pfeffer, J. T., *Biotechnol. Bioeng.*, 25, 1581, 1983. With permission.

phenol-synthesizing plant; it contained mainly phenol. It was used both in the "raw" state with 17,000 mg/ℓ phenol and after solvent extraction which left 920 mg/ℓ phenol. The second waste water came from the conversion of coal to heating gas and contained 5645 to 7750 mg/ℓ phenolics. The third waste water was from a coking process and it contained phenolics in the range of 5320 to 7270 mg/ℓ.

Cultures which had been acclimated to pure phenol were each fed one of the waste water types. For the phenol-synthesis waste water reactor, phenol concentrations were stepped up to 1000 mg/ℓ and gas generation was achieved in 3 to 4 days resulting in effluent phenol concentrations below 100 mg/ℓ. A similar experiment with the gasification waste yielded a lower average rate of gas generation and effluent phenol concentrations were found to be several hundred milligrams per liter. Likewise, the coking liquor produced a lower methane generation and much higher residual phenol concentrations. Attempts to acclimate the cultures directly in the waste waters rather than on pure phenol failed. Consequently, the authors[9] speculated on the possibility of the waste waters containing nonmetabolized inhibitory substances which could build up to inhibitory concentrations during the acclimation process.

Fedorak[64] studied the batch anaerobic treatability of a steel plant waste water consisting of 30% blast furnace blowdown and 70% ammonia-stripped waste ammonia liquor from coking pyrolysis gas-quenching drums. The waste water contained 320 mg/ℓ of phenol, 6 mg/ℓ of o-cresol, 55 mg/ℓ of m/p-cresol, and a COD of 5500 mg/ℓ. Batch dilutions ranging from 10 to 50% were studied and inhibition of methanogenesis occurred at all concentrations above 10%. Complete phenol removal was achieved at 30% or less but no phenol removal occurred at 40 or 50% by volume. Ether-extracted coke waste water was added to solutions containing 200 mg/ℓ of pure phenol to determine if the inhibitory substances were extractable. Essentially the same results were obtained suggesting the possibility of inorganic inhibition.

Waste water from an oil shale retorting process was treated in a laboratory-scale anaerobic digester by Ossio and Fox.[108] This waste water would be expected to contain significant phenols but no measured values were reported. COD concentrations ranged from 9440 to 29,000 mg/ℓ. Digested sludge from a treatment plant receiving both municipal and industrial wastes was acclimated by stepwise increases in the proportion of retort water fed to the laboratory digester. Ammonia removal and neutralization of the waste water were found to be essential. Carbon dioxide was used to reduce the pH after ammonia stripping at pH 11. Prior use of H_2SO_4 during the early stages of the study was thought to produce sulfide toxicity through the activity of sulfate-reducing bacteria and was therefore discontinued. The addition of the nutrients calcium, magnesium, and phosphorus was required to achieve a 65 to 70% COD removal and 90% biochemical oxygen demand (BOD) removal.

Liquid wastes from the pyrolysis of municipal refuse were subjected to anaerobic degradation in a variety of experiments reported by Dague.[109] The major organic components of the waste water were phenols, alcohols, and amines. Phenols comprised 100 to 600 mg/ℓ in the waste exhibiting a COD of 42,000 mg/ℓ. The waste also contained polynuclear aromatics and heavy metals including zinc, lead, chromium, and cadmium. Both suspended growth and attached growth reactors were used. In all cases, the pyrolysis waste water had to be diluted with a more readily degradable waste such as domestic primary effluent. Also, acclimation to these wastes was achieved only by gradually increasing the proportion of pyrolysis waste water in the influent mixture. In the anaerobic contact mode (solids retention time of 15 days and a hydraulic retention time of 5 days), the suspended growth method would handle a feed which contained 30% of COD from the pyrolysis waste water and would remove up to 70% of the COD from that source. The two 1.22- × 0.14-m diameter anaerobic filters were operated in series and handled the same strength influent with the same COD removal efficiency as the suspended growth reactor.

Khan et al.[105] fed diluted and nutrient-supplemented aircraft paint-stripping wastes to reactors which had been acclimated to phenol fermentation. The influent waste water contained 500 mg/ℓ phenol and 1460 mg/ℓ dissolved COD providing a loading of 9.9 kg COD per cubic meter per day with an empty bed contact time of 3.6 hr. Stable conditions were attained after 22 days of operation with the new waste water. Over the next 26 days of operation, the anthracite coal-packed column reduced the phenol and dissolved COD concentrations by 54 and 81%, respectively, while the activated carbon-packed column reduced these parameters by 82 and 90%, respectively. The rates of methane production were 84 and 160 mℓ/g COD applied in the anthracite and activated carbon columns, respectively.

2. Coal Conversion Waste Waters

Umfleet[110] assessed the treatability of a coal gasification waste water using the Celrobic®* anaerobic fixed film process. The raw waste water contained phenols at 4900 mg/ℓ which accounted for 45% of the total COD. Based on a 7-month investigation, Umfleet concluded that waste water was not treatable by this process even at dilutions of 5 to 7% by volume. No significant phenolic biodegradation was achieved at these loading conditions. Organic loading based on theoretical oxygen demand (TOD) was generally <3.2 kg/m³/day.

Fedorak and Hrudey[12] studied the batch anaerobic treatability of a phenolic waste water from an H-coal conversion demonstration plant (Catlettsburg, Ky.). The waste water had a COD of 21,100 mg/ℓ and phenolics at phenol 4900 mg/ℓ; o-cresol, 586 mg/ℓ; m-cresol, 1230 mg/ℓ; p-cresol, 420 mg/ℓ; 2,4/2,5-dimethylphenol, 63 mg/ℓ; 3,5-dimethylphenol, 213 mg/ℓ; and 3,4-dimethylphenol, 44 mg/ℓ. Batch methanogenesis of the waste water was obtained at up to 6% v/v dilution. Dilutions of 8 and 10% (v/v) were inhibitory to methane generation. Waste water subjected to ether extraction followed by replacement of identified phenolic compounds to original concentration was tested. Batch methanogenesis was successful at 9% (v/v) and only minor inhibition relative to control methanogenesis at 16% (v/v). At the latter concentration, the phenolic substrates were present at concentrations known to be inhibitory to methanogenesis in batch culture.[51] These results indicated that the substance(s) responsible for inhibition of batch methanogenesis in batch culture were ether extractable (i.e., low polarity organics).

Cross et al.[10] studied an anaerobic reactor system consisting of two 183- × 10-cm (internal diameter) columns in series providing empty bed contact times of 24 hr in each column. The first was packed with Raschig rings and the second contained activated carbon. Settled digested sludge was used as an inoculum for the first column. Waste water from a coal gasification plant (Grand Forks Energy Technology Center) containing 5600 mg/ℓ phenol with a COD of 26,900 mg/ℓ was diluted to 10% (v/v) with phosphate solution and fed to the system. At a loading rate of 2.5 kg COD per cubic meter per day, COD removals were near 80% while phenol and cresol removals were in excess of 90 and 99%, respectively. Only 10 to 20% of the phenol removal and approximately 58% of the cresol removal occurred in the first column. The majority of the methane produced (86%) came from the second column. The total quantity of methane produced was near that expected from the degradation of the influent phenol.

Suidan et al.[11] operated laboratory-scale, two-stage anaerobic filters followed by an activated sludge nitrification unit to treat coke oven waste water and coal gasification waste water (Morgantown Energy Technology Center). The anaerobic filter apparatus was similar to that described for Suidan et al.[107] in Section V.A. During their initial studies, a berl saddle-packed anaerobic filter was fed 1000 mg/ℓ glucose and 5% coke oven waste water supplemented with nutrients. Vigorous methane production was observed during this 61-day period. However, when the influent was switched to a nutrient-supplemented 10% dilution of the coal gasification waste water, and the glucose was eliminated from the feed, methane production decreased markedly within 27 days. An expanded bed, granular activated

---

* Celrobic® is a service mark for the anaerobic waste water treatment system of the Celanese Corporation.

carbon anaerobic filter was then placed between the berl saddle-packed unit and the nitrification unit. This system was operated for 118 days on nutrient-supplemented 10% coal gasification waste water. During this time, the berl saddle-packed unit removed little or no total organic carbon (TOC) or DOC while the activated carbon filter removed most of the TOC and DOC. Over the first 63 days when both anaerobic systems were in use, much of the organic removal was by adsorption to the activated carbon. However, between days 63 and 91, vigorous methane production was observed along with a marked improvement in the effluent quality. After 118 days of operation, the berl saddle-packed unit was removed leaving the activated carbon unit as the sole anaerobic reactor. This was maintained for a further 184 days. Methane production remained stable while the concentrations of organic material in the effluent slowly increased. This material was thought to be nonbiodegradable compounds which were escaping because of the gradual loss of the adsorptive capacity of the activated carbon.

Analysis of the effluent from the anaerobic activated carbon filter showed that phenol was reduced from 207 to 0.08 mg/ℓ.[11] Other phenolic compound reductions included *o*-cresol, from 57 to 14.2 mg/ℓ; *m/p*-cresol, from 139 to 18.2 mg/ℓ; 2,4-dimethylphenol, from 22 to 0.46 mg/ℓ; 3,5/2,3-dimethylphenol, from 29 to 0.14 mg/ℓ; and 3,4-dimethylphenol, from 9 to 0.06 mg/ℓ. While the removals of phenol could be clearly attributed to anaerobic decomposition, the more-resistant cresols showed some indications of breakthrough and the resistant dimethylphenols were likely removed strictly because of the strong affinity for the activated carbon. These observations were supported by an analysis of the ether extract of the activated carbon after 302 days of operation with coal gasification waste water feed. The analysis indicated 100 to 300 times higher mass retention of the cresols compared with phenol on the carbon vs. the waste water feed ratios (cresol/phenol) of only 0.28 to 0.67. Likewise, dimethylphenols retained on the carbon ranged up to 60 times the mass of retained phenol despite a corresponding waste water feed ratio of only 0.14 (dimethylphenol/phenol). These findings demonstrate the key role played by activated carbon in adsorbing and retaining nondegradable substrates. This role would be particularly important where ether-extractable inhibitory compounds, such as reported by Fedorak and Hrudey,[12] are encountered since such compounds would be expected to readily adsorb to activated carbon.

The dynamics of anaerobic degradation of phenolic substrates of varying degradability were explicitly demonstrated by Fedorak and Hrudey.[111] Using semicontinuous cultures fed 2 and 4% (v/v) dilutions of an H-coal conversion waste water, the effluent concentrations of phenol, *m/p*-cresol, and *o*-cresol were monitored.

As expected, *o*-cresol was not degradable and appeared in the effluent at feed concentrations. Under a variety of loading conditions stable operating periods of 17 to 53 days were obtained during which phenol and *m/p*-cresol was essentially completely removed and converted to methane. However, in all cases, the semicontinuous reactors ultimately stopped degrading *m*-cresol. In the reactors receiving higher loadings, *p*-cresol and ultimately phenol degradation ceased as well. Considering the findings of Suidan et al.[11] and Cross et al.,[10] the presence of activated carbon in reactors will apparently delay and possibly prevent this type of inhibitory failure.

3. Denitrification Treatment of Coking Waste Waters

As noted in Table 2, phenol can be degraded in the absence of oxygen with nitrate as the electron acceptor. Denitrification processes using anoxic reactors seek to convert nitrate to nitrogen gas and thereby reduce waste water nitrogen content. Early studies of this process involved a synthetic carbon source, usually methanol, as the electron donor. Beccari et al.[112] explicitly studied the use of phenol as the carbon source in a denitrification reactor. In batch tests, they obtained maximum denitrification rates which were only 40% of those reported with methanol. The average phenol-carbon consumed to oxidized nitrogen consumed ratio (phenol-C/NO$_T$-N) in batch trials was 1.2 mg C per milligram N with a coefficient of variation of 23%. During continuous feed trials, the phenol-C/NO$_T$-N ratio ranged from 1.1 to 2.4. The anoxic reactor of the two-stage, anoxic-aerobic system achieved phenol removals of 88 to 100% for a phenol feed of 200 mg/ℓ. Studies[113] on 16% (v/v) dilutions of a coke plant waste water providing a diluted feed of 400 mg/ℓ total phenols showed 93% phenol removal at a phenol-C/NO$_T$-N ratio of 1.3.

Several other studies[114-117] have confirmed that the organic carbon present in coking waste waters can act as the carbon source for biological denitrification. Bridle et al.[114] studied a single sludge predenitrification-nitrification process for treating ammonia-still effluent from a coking process. Predenitrification in an anoxic reactor was achieved without addition of a supplemental carbon source. Phenolic carbon represented approximately 30% of raw waste water-filtered organic carbon corresponding to feed-phenolic concentrations of around 200 mg/ℓ. No data was presented on phenolic removal in the anoxic reactor but, as expected, the overall removal of phenolics, including the aerobic reactor, exceeded 99.9%.

Nutt et al.[115,116] applied a fluidized bed fixed-film process to perform predenitrification-nitrification on a coke plant waste water. Phenolic carbon represented 42 to 54% of raw waste water filtered organic carbon (FOC) corresponding to a feed-phenolic concentration of around 400 mg/ℓ. They found a requirement for approximately 3 g FOC per gram of oxidized nitrogen removed (NO$_T$-N). Although removal of phenolics was not reported for the anoxic reactor, overall process removal of phenolics exceeded 99.9%.

Melcer et al.[117] studied the same fluidized bed system for the treatment of a combination of blast furnace blowdown water with coke plant waste water at ratios of 2:1 and 3:1. These dilutions reduced total phenol concentrations to an average of 120 mg/ℓ with phenol carbon comprising around 40% of combined waste water FOC. They found a requirement for 0.84 g FOC per gram NO$_T$-N when relying on waste water organics and 1.26 g FOC per gram NO$_T$-N where supplemental carbon as methanol was provided. These values are much lower than those reported for coke plant waste water alone but are closer to the experimental values reported by Beccari et al.[112] As in the related studies, phenol removal across the anoxic reactor was not reported but overall phenolic removal exceeded 99.9%.

Considering the finding of various phenolic substrates which are nondegradable in anaerobic fermentations and the common occurrence of nitrogen in many phenolic waste waters, degradation of phenolics by nitrate respiration seems to offer an attractive option in the overall treatment of phenolic waste waters. Provision of an intermediate anoxic stage to denitrify recycled effluent from a final polishing aerobic nitrification stage should warrant consideration when treatment of all waste water pollutants (anaerobically degradable and otherwise) is considered.

4. Pulp and Paper Industry Waste Waters

As noted in Section IV, waste waters from various pulp and paper industry processes often contain phenolics as minor or trace components. Consequently, studies in the anaerobic treatability of pulp and paper wastes have generally not focused explicitly on removal of phenolic compounds. Norrman[118] treated a kraft mill, black liquor evaporator condensate in fixed bed, expanded bed, and fluidized bed anaerobic film reactors. More than 80% COD reduction was achieved at COD loadings of 2 kg COD per cubic meter per day (fixed bed) up to 13 kg COD per cubic meter per day (fluidized bed). An earlier study[119] quoted an expectation of guaiacol concentrations of about 10 mg/ℓ which was approximately 0.1 to 1% of the total organic content of this waste type.

Benjamin et al.[76,120,121] evaluated the anaerobic treatability of evaporator condensate from the sulfite pulping process. Guaiacol was generally present at <25 mg/ℓ in the waste water with a soluble COD of 4000 to 7000 mg/ℓ comprised mainly of short-chain acids and alcohols. Acetic acid and methanol contributed 70 to 80% of the condensate COD. Treatment efficiencies of 79 to 90% COD reduction at COD loading of 16 kg COD per cubic meter per day were achieved. A yield coefficient of only 0.09 g biomass COD removed was estimated with the balance of COD removed being stoichiometrically converted to methane.

Hakulinen and Salkinoja-Salonen[122-125] have described the application of a two-stage, anaerobic-aerobic pilot process to bleach kraft mill effluents. As noted in Section IV, those waste waters contain small concentrations of a wide variety of chlorinated phenolic compounds. In their studies, the raw waste water contained less than 5 mg/ℓ of chlorinated phenolic compounds while TOC ranged from 500 to 1000 mg/ℓ. Chlorinated phenols were removed by 99 and 81% at hydraulic loadings of 3 and 9 m^3/m^3-day, respectively in the anaerobic reactor.[122] Corresponding TOC removals were poor at 34 and 27%. Woods et al.[126] studied the anaerobic removal of chlorinated phenols from caustic extract from kraft bleaching waste water. They found that highly chlorinated phenols such as pentachlorophenol, tetrachloroguaiacol, and trichlorophenols were removed to very low levels (<10 μg/ℓ). In turn, o-chlorophenol, 2,4-dichlorophenol, and 4,5-dichloroguaiacol were produced in the anaerobic process.

VI. SUMMARY

Full-scale anaerobic processes for the microbiological removal of phenolics from high-strength waste waters may soon evolve. Of the four biochemical mechanisms for anaerobic degradation of phenolics, only two are likely candidates for industrial application. These are nitrate respiration and the methanogenic fermentation.

To date, the studies dealing with the degradation of phenolics via nitrate respiration have attempted to demonstrate this phenomenon in microbial cultures. These investigations have usually included a limited survey of phenolics which are metabolized by this mechanism. More information is required in this area since many common compounds found in waste waters (i.e., dimethylphenols and di- and trihydric phenols) have not been tested. More recent studies have explicitly demonstrated that phenol can serve as an electron donor for the denitrification. Other investigations have shown that the organics present in phenolic-containing waste waters can serve as a carbon source for denitrification, thus indirectly suggesting that phenolics can be removed by this mechanism. In situations where denitrification is being used to treat phenolic waste waters, the process is typically designed to maximize nitrogen removal with the extent of organic carbon removal in the anaerobic reactor being of secondary concern. However, the utilization of phenolics in this reactor reduces the need for supplementation with another suitable carbon source and thereby reduces the operating costs. Thus, the ability of the denitrification process to remove phenolics warrants further investigation.

Methanogenic treatment of diluted phenolic waste waters from numerous coal conversion processes has been demonstrated in several laboratory-scale systems. However, there is still relatively little known about the microbiology of the degradation of phenolic compounds. The methanogenic process likely requires a consortium of anaerobic bacteria. One or more types would be responsible for the endergonic metabolism of the phenolics yielding hydrogen and acetate.[53] These end products would then serve as substrates for hydrogenotrophic and acetotrophic methanogens, respectively, which must be active to ensure that the phenolic degradation continued. This process would be similar to the degradation of propionate in methanogenic ecosystems.[44]

A better understanding of the interactions within the microbial consortium will indicate which segment of the population is limiting the overall rate of phenolic degradation. This will help establish maximum loading rates. The sensitivity of the process to inhibitory compounds which are often found in complex waste waters must also be evaluated and overcome. For example, the use of activated carbon to adsorb inhibitory organics and the complexing of cyanide are methods which have been demonstrated to reduce toxicity.

The methanogenic fermentation is not capable of degrading all of the wide range of phenolics found in industrial waste waters. However, the process may be developed as a means of removing the majority of the organic carbon which is present as fermentable phenolics. Subsequent treatment would be required to remove the remaining phenolics. The successful use of activated carbon in conjunction with the methanogenic process has been demonstrated in many studies. Conventional aerobic processes could also be used to polish the relatively low-strength effluents from the anaerobic process.

ACKNOWLEDGMENTS

This work was supported by the Alberta/Canada Energy Resources Research Fund through the Hydrocarbon Research Center and the Natural Sciences and Engineering Research Council of Canada.

REFERENCES

1. **Babich, H. and Davis, D. L.**, Phenol: a review of environmental and health risks, *Regul. Toxic Pharmacol.*, 1, 90, 1981.
2. **Buikema, A. L., Jr., McGinniss, M. J., and Cairns, J., Jr.**, Phenolics in aquatic ecosystems: a selected review of recent literature, *Marine Environ. Res.*, 2, 87, 1979.
3. **Harborne, J. B.**, Plant phenolics, in *Secondary Plant Products*, Bell, E. A. and Charlwood, B. V., Eds., Springer-Verlag, New York, 1980, chap. 6.
4. **Ou, L. T. and Sikka, H. C.**, Extensive degradation of silvex by synergistic action of aquatic microorganisms, *J. Agric. Food Chem.*, 25, 1336, 1977.
5. **Hopper, D. J.**, Microbial degradation of aromatic hydrocarbons, in *Developments in Biodegradation of Hydrocarbons*, Vol. 1, Watkinson, R. J., Ed., Applied Science, London, 1978, chap. 3.
6. **Evans, W. C.**, Biochemistry of bacterial catabolism of aromatic compounds in anaerobic environments, *Nature*, 270, 17, 1977.
7. **Ehrlich, G. G., Goerlitz, D. F., Godsy, E. M., and Hult, M. F.**, Degradation of phenolic contaminants in ground water by anaerobic bacteria: St. Louis Park, Minnesota, *Ground Water*, 20, 703, 1982.
8. **Godsy, E. M., Goerlitz, D. F., and Ehrlich, G. G.**, Methanogenesis of phenolic compounds by a bacterial consortium from a contaminated aquifer in St. Louis Park, Minnesota, *Bull. Environ. Contam. Toxicol.*, 30, 261, 1983.
9. **Chmielowski, J. and Kusznik, W.**, Preliminary trials on the methane fermentation of some phenolic wastewaters, *Zesz. Nauk. Politech. Slask., Inz. Sanit.*, 9, 123, 1966.
10. **Cross, W. H., Chian, E. S. K., Pohland, F. G., Harper, S., Kharkar, S., and Lu, F.**, Anaerobic biological treatment of coal gasifier effluent, in *Biotech. Bioeng. Symp. No. 12*, John Wiley & Sons, New York, 1982, 349.
11. **Suidan, M. T., Strubler, C. E., Kao, S. W., and Pfeffer, J. T.**, Treatment of coal gasification wastewater with anaerobic filter technology, *J. Water Pollut. Control Fed.*, 55, 1263, 1983.
12. **Fedorak, P. M. and Hrudey, S. E.**, Batch anaerobic methanogenesis of phenolic coal conversion wastewater, *Water Sci. Technol.*, 17, 143, 1985.
13. **Healy, J. B., Jr. and Young, L. Y.**, Anaerobic biodegradation of eleven aromatic compounds to methane, *Appl. Environ. Microbiol.*, 38, 84, 1979.
14. **Healy, J. B., Jr., Young, L. Y., and Reinhard, M.**, Methanogenic decomposition of ferulic acid, a model lignin derivative, *Appl. Environ. Microbiol.*, 39, 436, 1980.
15. **Colberg, P. J. and Young, L. Y.**, Biodegradation of lignin-derived molecules under anaerobic conditions, *Can. J. Microbiol.*, 28, 886, 1982.
16. **Balba, M. T. and Evans, W. C.**, The methanogenic biodegradation of catechol by a microbial consortium: evidence for the production of phenol through *cis*-benzenediol, *Biochem. Soc. Trans.*, 8, 452, 1980.
17. **Boyd, S. A. and Shelton, D. R.**, Anaerobic biodegradation of chlorophenols in fresh and acclimated sludge, *Appl. Environ. Microbiol.*, 47, 272, 1984.
18. **Bakker, G.**, Anaerobic degradation of aromatic compounds in the presence of nitrate, *FEMS Lett.*, 1, 103, 1977.

19. **Tsai, C. G. and Jones, G. A.,** Isolation and identification of rumen bacteria capable of anaerobic phloroglucinol degradation, *Can. J. Microbiol.,* 21, 794, 1975.

20. **Tsai, C. G., Gates, D. M., Ingledew, W. M., and Jones, G. A.,** Products of anaerobic phloroglucinol degradation by *Coprococcus* sp. Pe₁5, *Can. J. Microbiol.,* 22, 159, 1976.

21. **Schink, B., and Pfennig, N.,** Fermentation of trihydroxybenzenes by *Pelobacter acidigallici* gen. nov. sp. nov., a new strictly anaerobic non-sporeforming bacterium, *Arch. Microbiol.,* 133, 195, 1982.

22. **Patel, T. R., Jure, K. G., and Jones, G. A.,** Catabolism of phloroglucinol by the rumen anaerobe *Coprococcus, Appl. Environ. Microbiol.,* 42, 1010, 1981.

23. **Fina, L. R. and Fiskin, A. M.,** The anaerobic decomposition of benzoic acid during methane fermentation. II. Fate of carbons one and seven, *Arch. Biochem. Biophys.,* 91, 163, 1960.

24. **Ferry, J. G. and Wolfe, R. S.,** Anaerobic degradation of benzoate to methane by a microbial consortium, *Arch. Microbiol.,* 107, 33, 1976.

25. **Keith, C. L., Bridges, R. L., Fina, L. R., Iverson, K. L., and Cloran, J. A.,** The anaerobic decomposition of benzoic acid during methane fermentation. IV. Dearomatization of the ring and volatile fatty acids formed on ring rupture, *Arch. Microbiol.,* 118, 173, 1978.

26. **Shlomi, E. R., Lankhorst, A., and Prins, R. A.,** Methanogenic fermentation of benzoate in an enrichment culture, *Microb. Ecol.,* 4, 249, 1978.

27. **Mountfort, D. O. and Bryant, M. P.,** Isolation and characterization of an anaerobic syntrophic benzoate-degrading bacterium from sewage sludge, *Arch. Microbiol.,* 133, 249, 1982.

28. **Dagley, S.,** The microbial metabolism of phenolics, in *Soil Biochemistry,* Vol. 1, McLaren, A. D. and Peterson, G. H., Eds., Marcel Dekker, New York, 1967, chap. 12.

29. **Truper, H. G. and Pfennig, N.,** Characterization and identification of the anoxygenic phototrophic bacteria, in *The Prokaryotes: A Handbook on Habitats, Isolation and Identification of Bacteria,* Vol. 1, Starr, M. P., Stolp, H., Truper, H. G., Balows, A., and Schlegel, H. G., Eds., Springer-Verlag, New York, 1981, chap. 18.

30. **Dutton, P. L. and Evans, W. C.,** The metabolism of aromatic compounds by *Rhodopseudomonas palustris, Biochem. J.,* 113, 525, 1969.

31. **Guyer, M. and Hegeman, G.,** Evidence for a reductive pathway for the anaerobic metabolism of benzoate, *J. Bacteriol.,* 99, 906, 1969.

32. **Whittle, P. J., Lunt, D. O., and Evans, W. C.,** Anaerobic photometabolism of aromatic compounds by *Rhodopseudomonas* sp., *Biochem. Soc. Trans.,* 4, 490, 1976.

33. **Pfennig, N.,** *Rhodocyclus purpureus* gen. nov. sp. nov., a ring-shaped, vitamin B₁₂-requiring member of the family *Rhodospirillaceae, Int. J. Syst. Bacteriol.,* 28, 283, 1978.

34. **Tanaka, H., Maeda, H., Suzuki, H., Kamibayashi, A., and Tonomura, K.,** Metabolism of thiophene-2-carboxylate by a photobacterium, *Agric. Biol. Chem.,* 46, 1429, 1982.

35. **Oshima, T.,** On the anaerobic metabolism of aromatic compounds in the presence of nitrate by soil microorganisms, *Z. Allg. Mikrobiol.,* 5, 386, 1965.

36. **Taylor, B. F., Campbell, W. L., and Chinoy, I.,** Anaerobic degradation of the benzene nucleus by a facultatively anaerobic microorganism, *J. Bacteriol.,* 102, 430, 1970.

37. **Taylor, B. F. and Heeb, M. J.,** The anaerobic degradation of aromatic compounds by a denitrifying bacterium, *Arch. Mikrobiol.,* 83, 165, 1972.

38. **Taylor, B. F.,** Aerobic and anaerobic catabolism of vanillic acid and some other methoxy-aromatic compounds by *Pseudomonas* sp. strain PN-1, *Appl. Environ. Microbiol.,* 46, 1286, 1983.

39. **Bache, R. and Pfennig, N.,** Selective isolation of *Acetobacterium woodii* on methylated aromatic acids and determination of growth yields, *Arch. Microbiol.,* 130, 255, 1981.

40. **Williams, R. J. and Evans, W. C.,** The metabolism of benzoate by *Moraxella* species through anaerobic nitrate respiration. Evidence for a reductive pathway, *Biochem. J.,* 148, 1, 1975.

41. **Widdel, F., Kohring, G.-W., and Mayer, F.,** Studies on dissimilatory sulfate-reducing bacteria that decompose fatty acids. III. Characterization of the filamentous gliding *Desulfonema limicola* gen. nov. sp. nov., and *Desulfonema magnum* sp. nov., *Arch. Microbiol.,* 134, 286, 1983.

42. **Balba, M. T. and Evans, W. C.,** The anaerobic dissimilation of benzoate by *Pseudomonas aeruginosa* coupled with *Desulfovibrio vulgaris*, with sulfate as a terminal electron acceptor, *Biochem. Soc. Trans.,* 8, 624, 1980.

43. **Pfenning, N., Widdel, F., and Truper, H. G.,** The dissimilatory sulfate-reducing bacteria, in *The Prokaryotes: A Handbook on Habitats, Isolation and Identification of Bacteria,* Starr, M. P., Stolp, H., Truper, H. G., Balows, A., and Schlegel, H. G., Eds., Springer-Verlag, New York, 1981, chap. 74.

44. **Boone, D. R. and Bryant, M. P.,** Propionate-degrading bacterium, *Syntrophobacter wolinii* sp. nov. gen. nov. from methanogenic ecosystems, *Appl. Environ. Microbiol.,* 40, 626, 1980.

45. **McInerney, M. J., Bryant, M. P., Hespell, R. B., and Costerton, J. W.,** *Syntrophomonas wolfei* gen. nov. sp. nov., an anaerobic syntrophic fatty acid-oxidizing bacterium, *Appl. Environ. Microbiol.,* 41, 1029, 1981.

46. **Wolin, M. J. and Miller, T. L.,** Interspecies hydrogen transfer: 15 years later, *Am. Soc. Microbiol. News,* 48, 561, 1982.

47. **Chmielowski, J., Grossman, A., and Labuzek, S.,** Biochemical degradation of some phenols during methane fermentations, *Zesz. Nauk. Politech. Slask., Inz. Sanit.,* 8, 97, 1965.

48. **Tarvin, D. and Buswell, A. M.,** The methane fermentation of organic acids and carbohydrates, *J. Am. Chem. Soc.,* 56, 1751, 1934.

49. **Horowitz, A., Shelton, D. R., Cornell, C. P., and Tiedje, J. M.,** Anaerobic degradation of aromatic compounds in sediments and digested sludge, *Dev. Ind. Microbiol.,* 23, 435, 1982.

50. **Healy, J. B., Jr. and Young, L. Y.,** Catechol and phenol degradation by a methanogenic population of bacteria, *Appl. Environ. Microbiol.,* 35, 216, 1978.

51. **Fedorak, P. M. and Hrudey, S. E.,** The effects of phenol and some alkyl phenolics on batch anaerobic methanogenesis, *Water Res.,* 18, 361, 1984.

52. **Boyd, S. A., Shelton, D. R., Berry, D., and Tiedje, J. M.,** Anaerobic biodegradation of phenolic compounds in digested sludge, *Appl. Environ. Microbiol.,* 46, 50, 1983.

53. **Kaiser, J.-P. and Hanselmann, K. W.,** Fermentative metabolism of substituted monoaromatic compounds by a bacterial community from anaerobic sediments, *Arch. Microbiol.,* 133, 185, 1982.

54. **Suflita, J. M., Horowitz, A., Shelton, D. R., and Tiedje, J. M.,** Dehalogenation: a novel pathway for the anaerobic biodegradation of haloaromatic compounds, *Science,* 218, 1115, 1982.

55. **Horowitz, A., Suflita, J. M., and Tiedje, J. M.,** Reductive dehalogenations of halobenzoates by anaerobic lake sediment microorganisms, *Appl. Environ. Microbiol.,* 45, 1459, 1983.

56. **Miller, T. L. and Wolin, M. J.,** A serum bottle modification of the Hungate technique for cultivating obligate anaerobes, *Appl. Microbiol.,* 27, 985, 1974.

57. **Owen, W. F., Stuckey, D. C., Healy, J. B., Jr., Young, L. Y., and McCarty, P. L.,** Bioassay for monitoring biochemical methane potential and anaerobic toxicity, *Water Res.,* 13, 485, 1979.

58. **Fedorak, P. M. and Hrudey, S. E.,** A simple apparatus for measuring gas production by methanogenic cultures in serum bottles, *Environ. Technol. Lett.,* 4, 425, 1983.

59. **Shelton, D. R. and Tiedje, J. M.,** General method for determining anaerobic biodegradation potential, *Appl. Environ. Microbiol.,* 47, 850, 1984.

60. **Pearson, F., Shiun-Chung, C., and Gautier, M.,** Toxic inhibition of anaerobic biodegradation, *J. Water Pollut. Control Fed.,* 52, 472, 1980.

61. **Johnson, L. D.,** Inhibition of Anaerobic Digestion by Priority Pollutants, Ph.D. thesis, Iowa State University, Ames, 1981.

62. **Johnson, L. D. and Young, J. C.,** Inhibition of anaerobic digestion by organic priority pollutants, *J. Water Pollut. Control Fed.,* 55, 1441, 1983.

63. **Chmielowski, J. and Wasilewski, W.,** A study of the dynamics of anaerobic decomposition of some phenols in methane fermentation, *Zesz. Nauk. Politech. Slask., Inz. Sanit.,* 9, 95, 1966.

64. **Fedorak, P. M.,** Anaerobic Biological Treatment of Phenolic Wastewaters, Ph.D. thesis, University of Alberta, Edmonton, 1984.

65. **Neufeld, R. D., Mack, J. D., and Strakey, J. P.,** Anaerobic phenol biokinetics, *J. Water Pollut. Control Fed.,* 52, 2367, 1980.

66. **Khan, K. A., Suidan, M. T., and Cross, W. H.,** Anaerobic activated carbon filter for the treatment of phenol-bearing wastewater, *J. Water Pollut. Control Fed.,* 53, 1519, 1981.

67. **Balba, M. T. M., Senior, E., and Nedwell, D. B.,** Anaerobic metabolism of aromatic compounds by microbial associations isolated from salt marsh sediment, *Biochem. Soc. Trans.,* 9, 230, 1981.

68. **Blum, D. J. W., Hergenroeder, R., Parkin, G. F., and Speece, R. E.,** Anaerobic treatment of coal conversion wastewater constituents: biodegradability and toxicity, in Proc. 57th Annu. Conf. Water Pollution Control Federation, New Orleans, 1984.

69. **Suidan, M. T., Cross, W. H., and Fong, M.,** Continuous bioregeneration of granular activated carbon during the anaerobic degradation of catechol, *Prog. Water Technol.,* 12, 203, 1980.

70. **Suidan, M. T., Cross, W. H., Fong, M., and Calvert, J. W.,** Anaerobic carbon filter for degradation of phenols, *Environ. Eng. Div., Am. Soc. Civ. Eng.,* 107, 563, 1981.

71. **Kaiser, J.-P. and Hanselmann, K. W.,** Aromatic chemicals through anaerobic conversion of lignin monomers, *Experientia,* 38, 167, 1982.

72. **Chou, W. L., Speece, R. E., and Siddiqi, R. H.,** Acclimation and degradation of petrochemical wastewater components by methane fermentation, in *Biotech. Bioeng. Symp. No. 8,* John Wiley & Sons, New York, 1978, 391.

73. **Balba, M. T., Clarke, N. A., and Evans, W. C.,** The methanogenic fermentation of plant phenolics, *Biochem. Soc. Trans.,* 7, 1115, 1979.

74. **Taylor, B. F.,** Aerobic and anaerobic catabolism of vanillic acid and some other methoxy-aromatic compounds by *Pseudomonas* sp. strain PN-1, *Appl. Environ. Microbiol.,* 46, 1286, 1983.

75. **Grbić-Galić, D.**, Anaerobic degradation of coniferyl alcohol by methanogenic consortia, *Appl. Environ. Microbiol.*, 46, 1442, 1983.

76. **Benjamin, M. M., Woods, S. L., and Ferguson, J. F.**, Anaerobic toxicity and biodegradability of pulp mill waste constituents, *Water Res.*, 18, 601, 1984.

77. **Murthy, N. B. K., Kaufman, D. D., and Fries, G. F.**, Degradation of pentachlorophenol (PCP) in aerobic and anaerobic soil, *J. Environ. Sci. Health*, B14, 1, 1979.

78. **Guthrie, M. A., Kirsch, E. J., Wukasch, R. F., and Grady, C. P. L., Jr.**, Pentachlorophenol biodegradation. II. Anaerobic, *Water Res.*, 18, 451, 1984.

79. **Balba, M. T. and Evans, W. C.**, Methanogenic fermentation of the naturally occurring aromatic amino acids by a microbial consortium, *Biochem. Soc. Trans.*, 8, 625, 1980.

80. **Barker, H. A.**, Amino acid degradation by anaerobic bacteria, *Annu. Rev. Biochem.*, 50, 23, 1981.

81. **Ongerth, J. E. and DeWalle, F. B.**, Pretreatment of industrial discharges to publicly owned treatment works, *J. Water Pollut. Control Fed.*, 52, 2246, 1980.

82. **DeWalle, F. B., Kalman, D. A., Dills, R., Norman, D., Chian, E. S. K., Giabbai, M., and Ghosal, M.**, Presence of phenolic compounds in sewage, effluent, and sludge from municipal sewage treatment plants, *Water Sci. Technol.*, 14, 143, 1982.

83. **Baird, R. B., Carmona, L. G., and Jenkins, R. L.**, The direct-injection GLC analysis of xylenols in industrial wastewater, *Bull. Environ. Contam. Toxicol.*, 17, 764, 1977.

84. **Patterson, J. W.**, Treatment technology for phenols, in *Wastewater Treatment Technology*, Patterson, J. W., Ed., Ann Arbor Science, Ann Arbor, Mich., 1975, chap. 18.

85. American Public Health Association, *Standard Methods for the Examination of Water and Wastewater*, 15th ed., American Public Health Association Washington, D.C., 1980, 508.

86. American Society for Testing and Materials, Standard test methods for phenolic compounds in water, in *Annual Book of ASTM Standards 11.02*, American Society for Testing and Materials, Philadelphia, 1984, 220.

87. **Baird, R. B., Kuo, C. L., Shapiro, J. S., and Yanko, W. A.**, The fate of phenolics in wastewater-determination by direct-injection GLC and Warburg respirometry, *Arch. Environ. Contam.*, 2, 165, 1974.

88. **Robertson, J. L. and Tierney, D. R.**, Fate of selected trace organics in Canadian petroleum refinery wastewater treatment process, in Proc. Industrial Wastes Symposia, 56th Annu. Conf. of Water Pollution Control Federation, Atlanta, 1983.

89. **Ho, C. H., Clark, B. R., and Guerin, M. R.**, Direct analysis of organic compounds in aqueous by-products from fossil fuel conversion processes: oil shale retorting, synthane coal gasification and COED coal liquefaction, *J. Environ. Sci. Health*, A11(7), 481, 1976.

90. **Barbour, F. A. and Guffey, F. D.**, Organic and inorganic analysis of constituents in water produced during *in situ* combustion experiments for the recovery of tar sands, in *Analysis of Waters Associated with Alternative Fuel Production*, ASTM STP 720, Jackson, L. P. and Wright, C. C., Eds., American Society for Testing and Materials, Philadelphia, 1981, 38.

91. **Neufeld, R. D. and Spinola, A. A.**, Ozonation of coal gasification plant wastewater, *Environ. Sci. Technol.*, 12, 470, 1978.

92. **Pellizzari, E. D., Castillo, N. P., Willis, S., Smith, D., and Bursey, J. T.**, Identification of organic components in aqueous effluents from energy-related processes, in *Measurement of Organic Pollutants in Water and Wastewater*, ASTM STP 686, Van Hall, C. E., Ed., American Society for Testing and Materials, Philadelphia, 1979, 256.

93. Dearborn Environmental Consulting Services, Environmental Implications of Coal Gasification/Liquefaction Technologies in Canada, unpublished report prepared for the Water Pollution Control Directorate, Environmental Protection Service, Hull, Canada, 1983.

94. **Hanselmann, K. W.**, Lignochemicals, *Experientia*, 38, 176, 1982.

95. **Chernousov, Y. I., Murskaya, M. L., Stanyukovich, I. Y., Piyalkin, V. N., and Nikitin, V. M.**, Ether-soluble part of the black liquor obtained during cooking of sulfate cord pulp with prehydrolysis in the vapor phase, *Chem. Abstr.*, 80, 52118r, 1972.

96. **Chernousov, Y. I., Ivanov, N. A., and Piyalkin, V. N.**, Organic substances in wastewaters from sulfate pulp production. III. Phenols, *Chem. Abstr.*, 84, 155281a, 1975.

97. **McKague, A. B.**, Phenolic constituents in pulp mill process streams, *J. Chromatogr.*, 208, 287, 1981.

98. **Kringstad, K. P. and Lindstrom, K.**, Spent liquors from pulp bleaching, *Environ. Sci. Technol.*, 18, 236A, 1984.

99. **Voss, R. H., Wearing, J. T., and Wong, A.**, A novel gas chromatographic method for the analysis of chlorinated phenolics in pulp mill effluents, in *Advances in the Identification and Analysis of Organic Pollutants in Water*, Keith, L. H., Ed., Ann Arbor Science, Ann Arbor, Mich., 1981, 1059.

100. **Salkinoja-Salonen, M., Saxen, M.-L., Pere, J., Jaakkola, T., Saarikoski, J., Hakulinen, R., and Koistinen, O.**, Analysis of toxicity and biodegradability of organochlorine compounds released into the environment in bleaching effluents of kraft pulping, in *Advances in the Identification and Analysis of Organic Pollutants in Water*, Keith, L. H., Ed., Ann Arbor Science, Ann Arbor, Mich., 1981, 1131.

101. **Lindstrom, K. and Nordin, J.,** Gas chromatography-mass spectrometry of chlorophenols in spent bleach liquors, *J. Chromatogr.,* 128, 13, 1976.

102. **Kovacs, T. G., Voss, R. H., and Wong, A.,** Chlorinated phenolics of bleached kraft mill origin, *Water Res.,* 18, 911, 1984.

103. **Voss, R. H., Wearing, J. T., and Wing, A.,** Effect of softwood chlorination conditions on the formation of toxic chlorinated compounds, *Pulp Pap. Can.,* 82, 97, 1981.

104. **Speece, R. E.,** Anaerobic biotechnology for industrial wastewater treatment, *Environ. Sci. Technol.,* 17, 416A, 1983.

105. **Khan, K. A., Suidan, M. T., and Cross, W. H.,** Role of surface active media in anaerobic filters, *J. Environ. Eng. Div. Am. Soc. Civ. Eng.,* 108, 269, 1982.

106. **Suidan, M. T., Cross, W. H., Khan, K. A., and Fong, M.,** Treatment of phenol and substituted phenols with an anaerobic activated carbon filter, in *Chemistry in Water Reuse,* Vol. 2, Cooper, W. J., Ed., Ann Arbor Science, Ann Arbor, Mich., 1981, chap. 23.

107. **Suidan, M. T., Siekerka, G. L., Kao, S. W., and Pfeffer, J. T.,** Anaerobic filters for the treatment of coal gasification wastewater, *Biotechnol. Biceng.,* 25, 1581, 1983.

108. **Ossio, E. and Fox, P.,** Anaerobic biological treatment of *in situ* retort water, Report No. LBL-10481, Lawrence Berkeley Laboratory, University of California, Berkeley, 1980.

109. **Dague, R.,** Anaerobic biological treatment of liquid wastes from pyrolysis processes, in Proc. 54th Annu. Conf. of the Water Pollution Control Federation, Detroit, 1981.

110. **Umfleet, R. A.,** *Evaluation of the Treatability of Coal Gasification Waste by the Celrobic Process,* Report DOE/PL/30189-1, Pittsburgh Energy Technology Center, NTIS Publ. DE82 010099, National Technical Information Service, Springfield, Va., 1981.

111. **Fedorak, P. M. and Hrudey, S. E.,** Anaerobic treatment of phenolic coal conversion wastewater in semicontinuous cultures, *Water Res.,* 20, 113, 1986.

112. **Beccari, M., Passino, R., Ramadori, R., and Tandoi, V.,** Denitrification kinetics in the treatment of phenol wastes in a single sludge system, *Environ. Technol. Lett.,* 4, 163, 1983.

113. **Beccari, M., DiPinto, A. C., Passino, R., Ramadori, R., and Tandoi, V.,** Single sludge anoxic-aerobic systems for biological treatment of coke plant wastewaters, *Water Sci. Technol.,* 17, 421, 1985.

114. **Bridle, T. R., Bedford, W. K., and Jank, B. E.,** Biological nitrogen control of coke plant wastewaters, *Prog. Water Technol.,* 12, 667, 1980.

115. **Nutt, S. G., Melcer, H., Marvan, I. J., and Sutton, P. M.,** Treatment of coke plant wastewater in the coupled predenitrification-nitrification fluidized bed process, in *Proc. Purdue Industrial Waste Conference,* Vol. 37, Ann Arbor Science, Ann Arbor, Mich., 1982, 527.

116. **Nutt, S. G., Melcer, H., and Pries, J. H.,** Two-stage biological fluidized bed treatment of coke plant wastewater for nitrogen control, *J. Water Pollut. Control Fed.,* 56, 851, 1984.

117. **Melcer, H., Nutt, S., Marvan, I., and Sutton, P.,** Combined treatment of coke plant wastewater and blast furnace blowdown water in a coupled biological fluidized bed system, *J. Water Pollut. Control Fed.,* 56, 192, 1984.

118. **Norrman, J.,** Anaerobic treatment of a black liquor evaporator condensate from a kraft mill in three types of fixed film reactors, *Water Sci. Technol.,* 15, 247, 1983.

119. **Norrman, J. and Hakansson, H.,** Anaerobic-aerobic treatment of specific pulp and paper wastewaters, in Proc. 9th IAWPR Int. Conf., Post Conference Seminar PS4, Water Pollution Control in Pulp and Paper Industries, Stockholm, 1978.

120. **Benjamin, M. M., Ferguson, J. F., and Buggins, M. E.,** Treatment of sulfite evaporator condensate with an anaerobic reactor, *Tappi,* 65(8), 96, 1982.

121. **Benjamin, M. M., Ferguson, J. F., and Buggins, M. E.,** Treatment of sulfite evaporator condensate with an anaerobic reactor, in *Proc. TAPPI Environmental Conference,* TAPPI Press, Atlanta, 1981, 307.

122. **Hakulinen, R. and Salkinoja-Salonen, M.,** An anaerobic fluidized bed reactor for the treatment of industrial wastewater containing chlorophenols, in *Proc. Biological Fluidized Bed Treatment of Water and Wastewater Conference,* Cooper, P. F. and Atkinson, B., Eds., Ellis Horwood, Chichester, England, 1981, 374.

123. **Hakulinen, R., Salkinoja-Salonen, M. S., and Saxelin, M. L.,** Purification of kraft bleach effluent by an anaerobic fluidized bed reactor and aerobic trickling filter at semi-technical scale (ENSO-FENOX), in *Proc. TAPPI Environmental Conference,* TAPPI Press, Atlanta, 1981, 197.

124. **Hakulinen, R. and Salkinoja-Salonen, M. S.,** Treatment of pulp and paper industry wastewaters in an anaerobic fluidized bed reactor, *Process Biochem.,* 17(2), 18, 1982.

125. **Salkinoja-Salonen, M. S., Hakulinen, R., Valo, R., and Apajalahti, J.,** Biodegradation of recalcitrant organochlorine compounds in fixed film reactors, *Water Sci. Technol.,* 15, 309, 1983.

126. **Woods, S. L., Ferguson, J. F., and Benjamin, M. M.,** The fate of chlorinated phenols in anaerobic wastewater treatment, in Abstr. 56th Annu. Conf. of the Water Pollution Control Federation, Atlanta, Session 29, 1983.

PUBLICATIONS WHICH HAVE APPEARED SINCE THIS REVIEW WAS COMPLETED

127. **Andreoni, V., Nali, M., Rindone, B., Tollari, S., Treccani, V., and Villa, M.,** Anaerobic degradation of caffeic acid by mixed bacteria cultures, *Ann. Microbiol. Enzymol.*, 34, 53, 1984.

128. **Barik, S., Brulla, W. J., and Bryant, M. P.,** PA-1, a versatile anaerobe obtained in pure culture, catabolizes benzenoids and other compounds in syntrophy with hydrogenotrophs, and P-2 plus *Wolinella* sp. degrades benzenoids, *Appl. Environ. Microbiol.*, 50, 304, 1985.

129. **Blum, D. J. W., Hergenroeder, R., Parkin, G. F., and Speece, R. E.,** Anaerobic treatment of coal conversion wastewater constituents: biodegradability and toxicity, *J. Water Pollut. Control Fed.*, 58, 122, 1986.

130. **Bossert, I. D., Rivers, M. D., and Young, L. Y.,** p-Cresol biodegradation under denitrifying conditions: isolation of a bacterial coculture, *FEMS Microbiol. Ecol.*, 38, 313, 1986.

131. **Corbo, P. and Ahlert, R. C.,** Anaerobic treatment of concentrated industrial wastewater, *Environ. Prog.*, 4, 22, 1985.

132. **Donovan, E., Jr.,** *Pilot Anaerobic Biological Treatment of Pulp Mill Evaporator Foul Condensate*, EPA Project Summary, EPA/600/S2-86/002/, National Technical Information Service, Springfield, Va., 1986.

133. **Drummond, C. J., Noceti, R. P., Miller, R. D., Feeley, T. J., III, and Cook, J. A.,** Fate of contaminants during treatment of H-coal process wastewaters, *Environ. Prog.*, 4, 26, 1985.

134. **Dwyer, D. F., Krumme, M. K., Boyd, S. A., and Tiedje, J. M.,** Kinetics of phenol biodegradation by an immobilized methanogenic consortium, *Appl. Environ. Microbiol.*, 52, 345, 1986.

135. **Edeline, F., Lambert, G., and Fatticcioni, H.,** Anaerobic treatment of coke plant wastewater, *Process Biochem.*, 21(2), 58, 1986.

136. **Fedorak, P. M. and Hrudey, S. E.,** Nutrient requirements for the methanogenic degradation of phenol and p-cresol in anaerobic draw and feed cultured, *Water Res.*, 20, 929, 1986.

137. **Fedorak, P. M., Roberts, D. J., and Hrudey, S. E.,** The effects of cyanide on the methanogenic degradation of phenolic compounds, *Water Res.*, 20, 1315, 1986.

137a. **Fedorak, P. M. and Hrudey, S. E.,** Inhibition of anaerobic degradation of phenolics and methanogenesis by coal coking wastewater, *Water Sci. Technol.*, 19, 219, 1987.

138. **Frazer, A. C. and Young, L. Y.,** Anaerobic C_1 metabolism of the O-methyl-^{14}C-labeled substituent of vanillate, *Appl. Environ. Microbiol.*, 51, 84, 1986.

139. **Giabbai, M. F., Cross, W. H., Chian, E. S. K., and Dewalle, F. B.,** Characterization of major and minor organic pollutants in waste waters from coal gasification processes, *Intern. J. Environ. Anal. Chem.*, 20, 113, 1985.

140. **Gibson, S. A. and Suflita, J. M.,** Extrapolation of biodegradation results to groundwater aquifers: reductive dehalogenation of aromatic compounds, *Appl. Environ. Microbiol.*, 52, 681, 1986.

141. **Godsy, E. M., Goerlitz, D. F., and Ehrlich, G. G.,** Effects of pentachlorophenol on methanogenic fermentation of phenol, *Bull. Environ. Contam. Toxicol.*, 36, 271, 1986.

142. **Grbić-Galić, D. and Young, L. Y.,** Methane fermentation of ferulate and benzoate: anaerobic degradation pathways, *Appl. Environ. Microbiol.*, 50, 292, 1985.

143. **Grbić-Galić, D.,** Anaerobic production and transformation of aromatic hydrocarbons and substituted phenols by ferulic acid-degrading BESA-inhibited methanogenic consortia, *FEMS Microbiol. Ecol.*, 38, 161, 1986.

144. **Harper, S. R. and Pohland, F. G.,** Enhancement of anaerobic treatment efficiency through process modification, in Proc. 58th Annu. Conf. Water Pollution Control Federation, Kansas City, Mo., 1985.

145. **Kim, B. R., Chian, E. S. K., Cross, W. H., and Cheng, S.-S.,** Adsorption, desorption, and biodegradation in an anaerobic, granular activated carbon reactor for the removal of phenol, *J. Water Pollut. Control Fed.*, 58, 35, 1986.

146. **Kitunen, V., Valo, R., and Salkinoja-Salonen, M.,** Analysis of chlorinated phenols, phenoxyphenols and dibenzofurans around wood preserving facilities, *Intern. J. Environ. Anal. Chem.*, 20, 13, 1985.

147. **Krumholz, L. R. and Bryant, M. R.,** *Eubacterium oxidoreducens* sp. nov. requiring H_2 or formate to degrade gallate, pyrogallol, phloroglucinol and quercetin, *Arch. Microbiol.*, 144, 8, 1986.

148. **Mikesell, M. D. and Boyd, S. A.,** Complete reductive dechlorination and mineralization of pentachlorophenol by anaerobic microorganisms, *Appl. Environ. Microbiol.*, 52, 861, 1986.

149. **Mountfort, D. O. and Asher, R. A.,** Isolation from a methanogenic ferulate degrading consortium of an anaerobe that converts methoxyl groups of aromatic acids to volatile fatty acids, *Arch. Microbiol.*, 144, 55, 1986.

150. **Pfeffer, J. T. and Suidan, M. T.,** Anaerobic-aerobic process for treating coal gasification water, in Proc. 58th Annu. Conf. Water Pollution Control Federation, Kansas City, Mo., 1985.

151. **Sahm, H., Brunner, M., and Schoberth, S. M.,** Anaerobic degradation of halogenated aromatic compounds, *Microb. Ecol.*, 12, 147, 1986.

152. **Sleat, R. and Robinson, J. P.,** The bacteriology of anaerobic degradation of aromatic compounds, *J. Appl. Bacteriol.,* 57, 381, 1984.
153. **Suidan, M. T., Fox, P., and Pfeffer, J. T.,** Anaerobic treatment of coal gasification wastewater, *Water Sci. Technol.,* 19, 229, 1987.
154. **Szewzyk, U., Szewzyk, R., and Schink, B.,** Methanogenic degradation of hydroquinone and catechol via reductive dehydroxylation to phenol, *FEMS Microbiol. Ecol.,* 31, 79, 1985.
155. **Tschech, A. and Schink, B.,** Fermentative degradation of resorcinol and resorcylic acids, *Arch. Microbiol.,* 143, 52, 1985.
156. **Wang, Y. T., Suidan, M. T., and Rittman, B. E.,** Anaerobic treatment of phenol by an expanded-bed reactor, *J. Water Pollut. Control Fed.,* 58, 227, 1986.
157. **Young, L. Y.,** Anaerobic degradation of aromatic compounds, in *Microbial Degradation of Organic Compounds,* Gibson, D. T., Ed., Marcel Dekker, New York, 1984, chap. 16.
158. **Young, L. Y. and Rivera, M. D.,** Methanogenic degradation of four phenolic compounds, *Water Res.,* 19, 1325, 1985.

Chapter 5

BIOLOGICAL TREATMENT OF TOXIC INDUSTRIAL WASTES

P. Kumaran and N. Shivaraman

TABLE OF CONTENTS

I.   Introduction..228
      A.   Historical Background..228
            1.   Environmental Deterioration During the Industrial
                  Revolution..229
            2.   The Stockholm Conference 230
            3.   Developments in Controlling Pollution Due to Industrial
                  Discharges...230

II.   Biological Waste Water Treatment Systems 233
      A.   Biological Filters 234
      B.   Activated Sludge Processes.....................................235
            1.   Aeration in Activated Sludge Units 236
      C.   Parameters Employed for the Assessment of Biological Waste
            Treatment Processes..237

III. Waste Water Characteristics 238
      A.   Nontoxic and Toxic Wastes 238
            1.   Waste Water Characteristics from Coal Conversion
                  Processes 241
            2.   Waste Water Characteristics from Refineries 243
            3.   Waste Water Characteristics from the Petrochemical
                  Industry...244
            4.   Waste Water Characteristics from the Pesticide Industry....246
            5.   Waste Water Characteristics from the Plating Industry......247

IV.   Toxicity of Pollutants 247
      A.   Toxicity of Phenolic Compounds.................................248
      B.   Toxicity of Ammonia 248
      C.   Toxicity of Cyanide 249
      D.   Toxicity of Thiocyanates 249
      E.   Toxicity of Hydrocarbons 249
      F.   Toxicity of Pesticides 250
      G.   Toxicity of Heavy Metals 250

V.   Microbial Degradation of Toxic Pollutants............................251
      A.   Biochemical Pathways for Catabolism of Toxic Aromatic
            Hydrocarbons...252
            1.   Biochemical and Genetic Regulation.......................255
      B.   Biodegradation of Surfactants 255
      C.   Biochemical Aspects of Nitrogen Metabolism.....................256
            1.   Nitrification..256
                  a.   Biochemical Pathway of Autotrophic Nitrification....256
                  b.   Biochemical Pathway of Heterotrophic Nitrification....257

2.   Denitrification 257
        a.   Biochemical Pathway of Denitrification 257
    D.   Biological Transformation of Sulfur-Containing Pollutants 258
    E.   Biochemical Aspects of Cyanide Metabolism 258

VI.   Biological Treatment of Toxic Industrial Wastes............................ 259
    A.   Biodegradability of Toxic Pollutants 262
        1.   Microbiological Aspects of Toxic Waste Treatment.............. 264
            a.   Microbial Interactions 265
            b.   Biochemical Augmentation 268

VII.   Full-Scale Biological Waste Treatment Plants for Toxic Wastes 272

VIII.   Conclusions and Future Perspectives...................................... 274

Acknowledgments 276

References.. 276

I. INTRODUCTION

A. Historical Background

Many books, treatises, and reviews have been written on the biological treatment of industrial wastes at various times. Research and development in theoretical science advancements in applied science have resulted in updating the status of current knowledge on the subject from time to time. A review of the subject "biological treatment of industrial waste" is appropriate, especially when everyone concerned is aiming at the "zero pollution" level. Our efforts in this regard are to write a state-of-the-art review on the systems in general and discuss in detail the status of biotransformation of toxic organic pollutants, discharged through industrial wastes into waste treatment plants. It is best to trace the causes that lead to environmental pollution and why it is necessary to control pollution.

From the time terrestrial life began, approximately 395 million years ago, the face of the earth had undergone immense changes. Pollution in the most primitive form existing then as excreta could have undergone putrification. The clock counted down 350 million years before the ancestors of upright man, "the Apes", evolved from reptiles which bred for the first time in the terrestrial habitat. A tailless primitive ape, possibly related to the ancestors of man, appeared at this time. It took another 15 million years for him to transform the four-legged stature to the upright, two-legged posture. He developed enough intelligence to make tools from stones for killing animals for food. Domestication of animals took place only during the last ten million years. From the time man had domesticated animals such as cattle for milk and meat, horses and camels for transport, and elephants for waging wars, the journey for colonization has started.

As a result of this process, civilization originated mostly in river valleys, thus creating environmental deterioration in the most primitive form. Archaeological findings from various river basins of the world prove that man progressed intelligently through the ages while other animals perished. The findings from the Yellow River valleys in China have shown that the sage kings were the first to develop medicine and agriculture, between 10,000 to 5000 years B.C. Excavation in the Indus valley (India-Pakistan) revealed that a great civilization existed there, parallel to the Chinese civilization. Buildings constructed of bricks, paved streets, well-laid sewers for collection of wastes from private and public latrines, and treatment of waste water in cesspools were the remnants found at Mohenjo-Daro and Harappa. Agriculture was the main profession of the people who lived there, as in other regions of the world. Even though environmental pollution was in existence in those periods, the effect was never felt because of the availability of vast areas of land and watersheds. Transformation of man from the Bronze Age to the Iron Age could be the nonreversible starting point of environmental deterioration. From here on, the use of bronze for making agricultural implements and weapons was replaced by the use of iron.

About 2500 years ago, the Celts developed improved farming and working techniques. Science and philosophy were considered to have been developed during the golden age of the Romans which later spread to the British Isles. Science gave the materialistic comforts while philosophy, the spiritualistic. Scientific thinking led to experimentation while spiritualism developed into religion.

Religious persecution, followed by the urge to find new land, resulted in travel by sea and land to different parts of the world. Ship building and astronomical knowledge were essential for traveling by sea. As centuries passed, man's activity also progressed to a considerable extent. Between the 15th and 17th century several exploratory voyages were made from the European continent to the east and west. Those who traveled to the east established colonies and empires while those who traveled to the west migrated and settled in the new-found lands.

Modern American history originated at this time. Mass migration from Europe to the Americas started in the 15th century — the French settled in Quebec, the Spanish dominated the migration to South America, the Portuguese and the Italians settled in Brazil and Argentina, and the British traveled west to establish an empire in Canada and east to many lands including the Indo-Burma peninsula. It was the migration of the European community that caused the industrial revolution and the resultant environmental deterioration in the European continent. The needs for the industrial revolution were many; two of the most important were wars and transportation of materials by sea routes. The industrial revolution saw great changes in the countries where it took place. In the U.S., environmental deterioration originated in the middle of the 18th century. Natural resources were exploited. Large factories were built and transcontinental railroads were constructed for transportation.

Looking back on the epoch-making events of world history from the present to the past, one could find that there were two World Wars, the industrial revolution, the Reformation, the great voyages for the discovery of new lands, the Renaissance, and the golden age of the Greeks. If one looked back into time and space, evidences could be obtained for over a period of 6000 years. In the future, the field of vision expands to take mankind over all the habitable and transversable surface of this planet and the whole stellar system.

1. Environmental Deterioration During the Industrial Revolution

In the early stages of industrialization, the work force migrated from the countryside to the manufacturing towns. Historical evidence from England[1] narrated that wastes were dumped in the courtyards (if one was available) or on streets where it decomposed. Davis[2] wrote that from the earliest papers read to the Royal Society it was evident that, what one had called technology or the application of scientific knowledge to industry, in turn had contributed to the problem of environmental pollution. Primitive engineering, building of factories, and extensive use of coal for energy and steel production were the real causes of environmental deterioration. In Europe the Rhine River was the dumpyard for the iron and steel industry at the time of the industrial revolution. The early period of the Civil War (1860s) in the U.S. saw the industrial revolution. Large factories and steel mills, transcontinental railroads, flourishing cities, and vast agricultural holdings marked the landscapes. The accompanying evils of pollution came as a result of the industrialization. Capitalistic corporations owned by individuals controlled the financial institutions who in turn dictated terms to the county councils and civil administrations. The process of environmental deterioration was enhanced by the two World Wars and large-scale agricultural outputs after the second World War. Awareness of pollution came late in the industrialized nations. It started only when the signs of pollution started showing its effects on human health. Environmental protection legislation in those countries was merely related to public health aspects. From the history of environmental protection legislation (Table 1) it could be seen that pollution control from industrial discharges was recognized only after 1960. Earlier, environmental legislation for preventing industrial discharge existed in many countries. It was only after the Stockholm Conference that large-scale industrial waste pollution and deterioration of human environment attained global importance.

2. The Stockholm Conference

International awareness was aroused as a result of the 1972 historic conference on Human Environment at Stockholm. The recommendations of this conference included identification and control of pollutants of broad international significance and informational, social, and cultural aspects of environmental issues.[3] As a result of the conference, several governments constituted action plans to restrict environmental deterioration. Separate departments have been constituted in Britain, India, and Canada. The U.S., the federal government entrusted the task to the Environmental Protection Agency (EPA). In the U.S.S.R., the duties were coordinated by several ministries and departments. Legal measures were taken in many countries by enacting environmental protection laws and making amendments to the existing laws from time to time (Table 1). Enforcement of these legislations was the responsibility of the River Authority in Britain, State Department in the U.S., and Pollution Control Board in India.

Technologically advanced countries realized the danger and gravity of pollution very late. Industrial production was given priority and pollution control had the lowest importance, especially in the case of industrial effluents. As a result, toxic industrial waste discharges caused environmental hazards. The "Minamata Bay" incident in Japan, and similar mercury poisonings in Pakistan and Iraq, the "Kepone" discharge into the James River in the U.S., and the polychlorinated biphenyls into Hudson River are some of the recorded incidents of environmental disasters relating to industrial waste discharges. Even though these countries have learned from the experience that environmental hazards could occur, the enforcement of laws was most primitive in these nations. The report of the U.N. Conference at Stockholm mentioned that among the most important sources of pollution in developing countries were unsatisfactory methods of disposal and treatment of domestic sewage, household and industrial wastes, excessive use of pesticides, herbicides, and chemical fertilizers, and excessive land use causing soil erosion.[3]

3. Developments in Controlling Pollution Due to Industrial Discharges

In the early stages of industrialization, wastes were collected and drained into open cesspools, where it became putrified by natural decomposition.[1] When industrial growth surpassed area development, domestic and industrial wastes were collected and discharged into rivers, lakes, and oceans. Natural purification in lakes was a slow process whereas in running streams and rivers the process was enhanced by atmospheric reaeration. Simultaneously, agricultural utilization of domestic wastes at the outskirts of thickly populated areas have been reported in Britain. The right to discharge trade wastes into domestic sewers of local authorities are given to industries by the Public Health Act of 1875. The River Pollution Act of 1876 prevented the local authorities from permitting the discharge of the waste without treatment. As a result, various methods have been experimented by the local authorities, rather than by the industries. The same was the case in the U.S. where either metropolitan or county councils were given the responsibility to deal with waste water treatment. The

Table 1

HISTORY OF ENVIRONMENTAL PROTECTION LEGISLATION

U.S.	Britain	India	Japan	U.S.S.R.	European Economic Community
Refuse Act of 1899	Public Health Act of 1818	Northern India Canal and Drainage Act of 1873	Laws for the Preservation of Water Quality of 1958	Decree on Environmental Sanitation by Lenin, 1922	Council Directives of 1976 on pollution caused by certain dangerous substances discharged into aquatic environment of the community
Oil Pollution Act of 1924	Public Health Act of 1875			Protection of Environment in U.S.S.R., 1960	
Water Pollution Control Act of 1948	River Pollution Prevention Act of 1876	The Obstruction of Fairways Act of 1881	Maritime Pollution Act of 1970		
Water Pollution Control Amendment Act of 1956	Public Health Act of 1936	Indian Fisheries Act of 1897	Public Cleaning Law of 1976		
Federal Water Quality Act of 1965	Public Health (Drainage of Trade Premises Act of 1937	Damodar Valley Corporation (Prevention of Pollution Act of 1948)			
Water and Environmental Quality Improvement Act of 1970	Public Health Act of 1961	The Water (Prevention and Control of Pollution) Act of 1974			
Federal Water Pollution Control Amendment Act (PL-92-500) of 1972	Water Act of 1973	Water (Prevention and Control of Pollution Cess) Act of 1977			
	Control of Pollution Act of 1974	Environmental Protection Act of 1986			

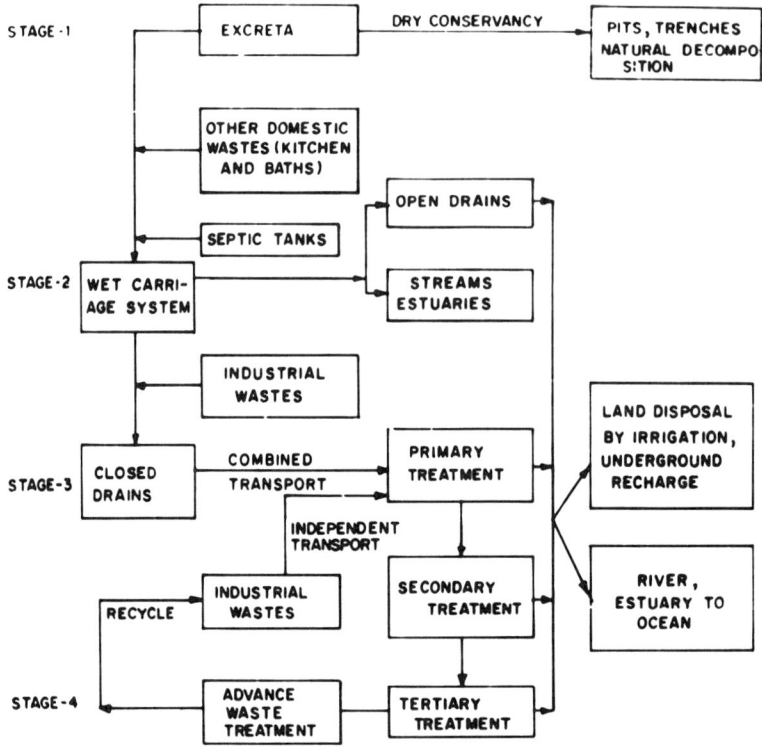

FIGURE 1.   Developments in waste water collection, transport and disposal. Stage 1: collection of night soil (excreta) alone, transport to the site of disposal by head carriage, and disposal by dumping in pits and trenches. Stage 2: introduction of wet carriage, treatment in septic tanks, discharge into open drains, and land disposal by agriculture. Stage 3: domestic sewage and industrial wastes mixed in closed drains, transported to disposal sites, sedimented, and discharged into land, rivers, or oceans. Stage 4: industrial wastes transported independently, treated in different units and disposed into rivers, lakes, or oceans or is recycled.

Public Health Act of 1936 of Britain stipulated that any industrialist discharging waste into inland waterways would be liable for prosecution under civil or criminal laws.[4] The discharge into municipal sewers was controlled by the local authorities under the Public Health (drainage of trade premises) Act of 1937 and Public Health Act of 1961. Consequent upon the formation of water authorities in 1973, waste water treatment became the responsibility of the water authorities.[5] For the treatment of the trade waste access is charged, based on the strength and volume of the waste. The 1972 amendment of the Clean Water Act switched the U.S. National water quality management from the ambient basis to dominantly effluent limitation orientation.[6] The goals of this national program were to employ the best available technology economically achievable (BATEA) for industrial discharges of toxic materials, irrespective of the waste characteristics, by 1984. The amended law could be enforced on those industries which were established after the amendments while others continued to discharge the toxic and nontoxic trade waste into the publicly owned treatment works (POTW) in the U.S. and the waste water treatment plants of the water authorities in Britain. Developments in waste water collection, transport, treatment, and disposal can be summarized as shown in Figure 1.

The migration of the work force to the site of the industrial nucleus formed the first stage of industrialization of an area. The dry-conservancy system was adopted in British practice for transport of excreta followed by burying in pits and trenches. When the habitat became

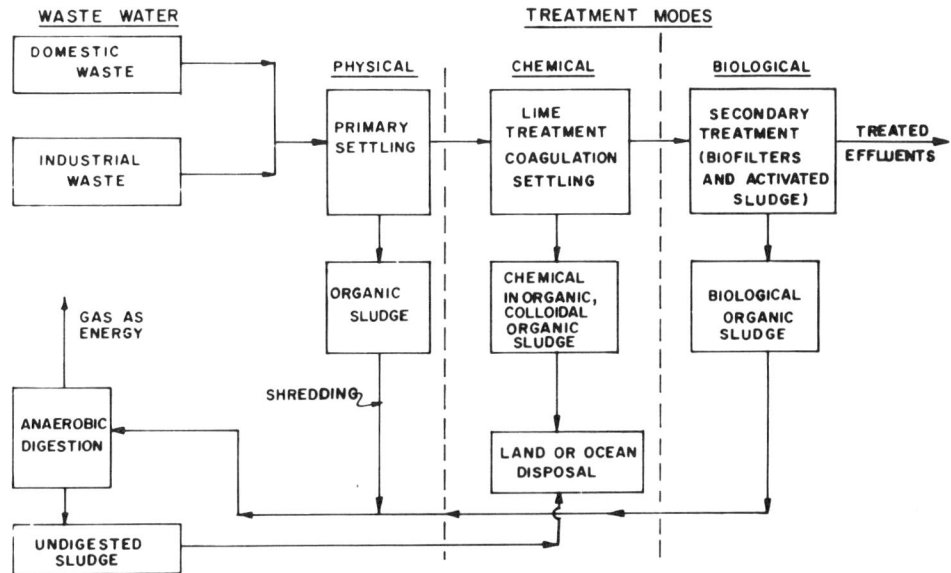

FIGURE 2.   Treatment flow sheet for combined treatment of domestic and industrial wastes.

congested the wet-carriage system was introduced. The privies discharged into the open drains which carried the waste into lakes and rivers. Industrial effluents and domestic wastes were then carried in closed sewers to a common facility. This facility is called the POTW in the U.S.

II. BIOLOGICAL WASTE WATER TREATMENT SYSTEMS

Biological waste water treatment fits into the secondary treatment block of the flow diagram. There are several modes related to biological treatment processes. These modes often form part of the physical, chemical, and biological functions of the total diagram. Waste water from domestic as well as industrial areas are carried together to a common treatment facility with the primary consideration that domestic waste can contribute to some of the nutritional deficiencies of the industrial waste.[7] Cost-sharing benefits between industries, consumers, and the municipalities have been taken as another consideration. Combined waste water treatment systems were preferred more often because it was cheaper to combine several projects than to build small units for each project separately.[8] The primary considerations were economics of the process and convenience, rather than the waste characteristics and influence on the waste treatment processes.[9] As a result, most of the metropolitan sanitation districts treated both wastes in the POTWs. The waste treatment plants in Britain constructed before 1950 were mostly combined treatment plants. Figure 2 explains the various steps of the combined waste treatment plant.

The treatment modes can be divided into three major steps:
1.   Primary settling is a physical process in which coarse particulates are removed by plain settling.
2.   Coagulation of fine particulates may be necessary at times. When the pH of the waste is high, addition of lime along with a coagulation process is considered necessary.
3.   Biological treatment consisting of several steps before the water is fit for final disposal. In the biological mode the units can be broadly classified into two categories: biofilters and activated sludge processes. These are oxidative processes in which oxygen acts

as the stabilizing agent. During the anaerobic process stabilization of waste materials is carried out in the absence of oxygen.

A. Biological Filters

The terms biological filters, percolating filters, sprinkling filters, sparkling filters, etc. refer to the process in which liquid waste passes through a solid medium. The liquid may be pumped to take downward or upward flow. The medium acts as a support in which the microbial flora is fixed. Civil engineers call these processes as mentioned above while chemical engineers call them "fixed film" reactors. They not only act as filters but also, at the same time, remove the soluble organics and, to a certain extent, the inorganic materials. Biological filters originated as a result of attempts made to reduce the area required for treating sewage by irrigation. The earliest references available were on the work carried out by Frankland and Silvester in 1868 to 1870. (See Reference 1). They used soil as the filter medium. Simultaneously, studies were also conducted at the Lawrence Experimental Research Station at Massachusetts during 1887 and 1890. In 1882, Warrington reported that "it would be possible to construct filter beds having oxidizing power rather than it would be possible with oridinary soil." By saying this he would have meant the use of coarser material as the filter media. Experimental evidences to show that oxidation in the filters was a biological function, was given by Stoddart.[10] He said that "the thinnest layer of liquid travelled over an extended surface, charged with appropriate micro-organisms and exposed to air" oxidized ammonium sulfate to nitrate. Several types of support media have been used in experimental studies. They included coarse sand to gravel, calcined to ordinary bricks, and plastic medium. The widely used support medium in biological filters, however, is gravel and pebbles, a material which is available at sites where the plants are constructed. Selection of the support medium is very important as it has a direct impact on the life of the material. It is influenced by the waste characteristics. For most of the corrosive wastes plastic medium-filled biofilters have been found to withstand disintegration. "Flocor", is a versatile system that has been claimed to have high performance efficiency in treating a variety of industrial wastes. The drawing of this system is shown in Figure 3.

The support medium in this system consists of alternate flat and vacuum-formed PVC sheets bonded together to form a light, strong, rigid structure. The liquid is distributed from the top surface. It, along with air, flows downwards, but has no free fall. The microbial film is supported by the plastic medium. It provides a large surface area in comparison with the gravel-filled filters. In addition to the modular sheet packings, randomly packed plastic media have also been developed. Plastic-packed filters have been used in the treatment of a variety of industrial wastes.[11,12] The treatment of high-strength industrial wastes requires packing of a large specific surface area as well as high voidage. Also it should withstand the chemical attack of the waste water. Naturally occurring filter media such as gravel and pebbles do not create high voids and also weather fast on loading with corrosive wastes. In the case of plastic-medium filters, extensive civil engineering cost is avoided. These advantages have proven the application of high-rate biological filters for treatment of industrial effluents. An alternative method of recent development is the emergence of rotating biological contactors (RBC). In principle, they are horizontal biofilters, wherein the microbial film-supporting medium also moves, so that the film is exposed alternatively to air and soluble food in the liquid. Development of RBC is the accumulation of many separate but related ideas.[13] Borchart enumerated the advantages of this unit as a secondary treatment process. It has excellent capability to withstand shock loading at a low F/M ratio, minimum retention time, low head loss, low power consumption, less short circuiting, and low sludge growth. Biofilters are preceded by primary sedimentation tanks or other treatment processes in which the coarser suspended particulates are separated and removed. The filter units receive wastes with less-suspended solids. Depending upon the effective treatment desired, the operational

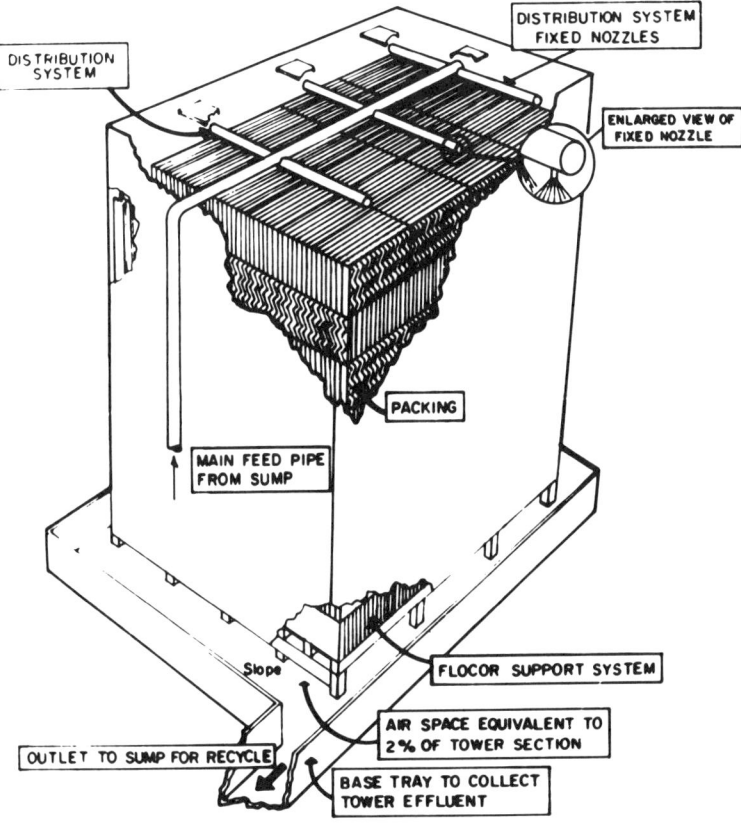

FIGURE 3. Flocor biofilter for effluent treatment.

parameters vary. When the treated effluent does not achieve the desired quality, it is recirculated back into the unit, making it a closed-loop operation.

B. Activated Sludge Processes

The introduction of biological filters proved that sewage purification could be augmented by the oxidative process. In an endeavor to further accelerate this oxidative processes Angus Smith carried out experiments on aeration of sewage. He was not successful. (See Reference 14.) However, in 1912, Fowler succeeded in his attempts. Under his direction, Arden and Lockett[15] were the first to present their results. The authors commented that "for reference purposes and failing a better term, the deposited solids resulting from the complete oxidation of sewage have been designated as activated sludge." The activated sludge process (ASP) is an oxygen-induced biological system. In chemical engineering terminology it is known as suspended-growth reactors, fluidized-bed biological reactors, etc. Based on the flow pattern and the mixing characteristics, ASPs are known by the names plug or tapered flow and completely mixed activated sludge (CMAS) process. On the basis of organic loading rates, they are called high-rate, conventional, and extended-aeration processes.

In the initial stages of the development of ASP, it was thought that the requirements were an aeration tank and compressed air. Settling tanks, constructed for primary sedimentation of domestic waste were converted into ASPs. Compressed air was passed through difusers made of foundry slag and cement. The first ASP in Britain was commissioned in 1914 by Arden and Lockett.[15] The developments in the designs of aeration tanks, types of aeration systems, and clarification of the aerated waste water to remove the suspended solids have undergone considerable changes by now. However, the basic units remain the same. An

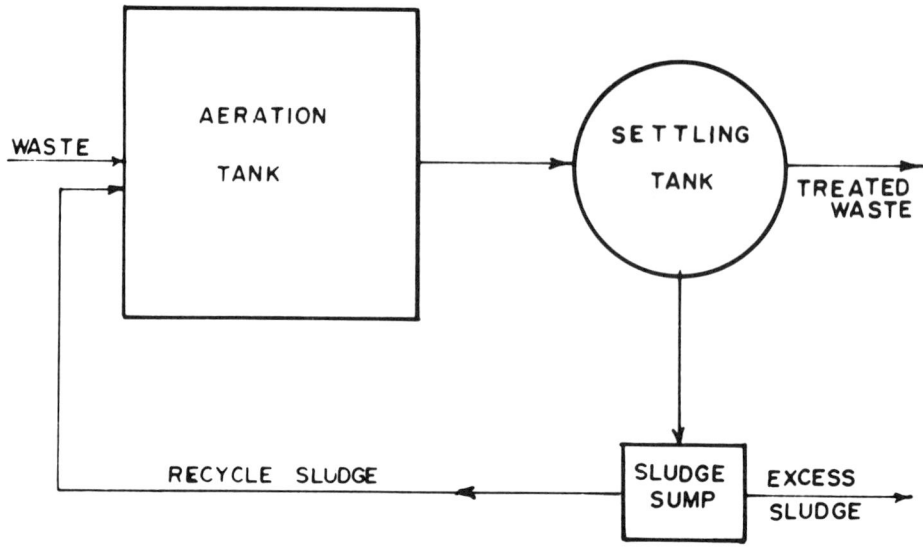

FIGURE 4. Flow diagram of ASP.

ASP consists basically of an aeration basin and a suspended-solid clarifier, whether it is a laboratory model or a full-scale waste treatment plant. The basic diagram of the process is shown in Figure 4.

The number of aeration tanks and clarifiers may vary according to the volume and strength of waste to be treated. Raw waste water which has a variable flow and characteristics needs proper balancing before pumping into the aeration basins. The microbial mass (activated sludge) is returned to the aeration basin from the clarifier to (1) keep the food to microorganisms ratio constant and (2) digest the sludge for stability. Biomass synthesis in high-rate activated sludge is generally high. It exhibits a dispersive growth. In conventional activated sludge, biomass synthesis and oxidation to stable end products are on the order of 50% while in the extended aeration the stability of the synthesized sludge is high with less sludge synthesis.

1. Aeration in Activated Sludge Units

Artificial aeration was introduced for effective and enhanced purification. Different types of aeration were experimented. One of the earliest and most widely employed systems is compressed air diffusion through liquid. Plate aerators were used in the early years.[16] The plates were later replaced by domes, mounted on air mains, and placed on the longitudinal furrows. The domes made out of allundum porous material have been used in most of the plants. These diffusers can provide fine air bubbles which are more effective in transferring oxygen into the microorganism through the liquid film. Ridge and furrow arrangements having a liquid depth of 4 m have been most commonly used. According to the Lewis and Whiteman two-film theory, the rate of oxygen absorption by water is governed by the rate of diffusion through thin films of gas and liquid at the interface. There can be no concentration buildup in either film and transfer across the interface must be steady. However, transfer of oxygen from the gas phase is influenced by several factors, such as the rheology of the liquid and dissolved-solids concentration.

Mechanical surface aerators have been used in ASPs in recent times. A comparative evaluation of fine-bubble aeration and mechanical aeration shows that the transfer efficiency of bubble aeration is on the order of 1.16 kg oxygen per kilowatthour, while that of vertical shaft mechanical aerators is about 2.27 kg oxygen per kilowatthour.[17] Aeration in deep tanks, up to a depth of 8 m has, however, been found to transfer about 5 kg oxygen per

**Table 2
AERATION CAPACITY OF A FEW
SELECTED MECHANICAL AERATORS**

System number	Type of rotors	Oxygenation rate (kg O$_2$ per kWh)	Ref.
1	Symplex	1.5—2.5	23
2	BSK turbine	2.0—2.8	23
3	TNO, Pasveer	1.7—1.85	22
4	Mammoth	1.6—2.8	24
5	Carrousel	1.6—2.6	24

kilowatthour, in clean water experiments.[18] The choice and selection of aeration systems depend upon the geometry of the aeration tank, nature of the waste water, availability of installation space, and topography of the site conditions. The deep-shaft compressed air systems occupy the minimum space and seem to be the reactor of the future in congested areas. This process is being used for the treatment of domestic as well as industrial wastes.[19,20]

Experiments on mechanical aeration was started by Kessenger in 1925.[21] Since then, several types of vertical-shaft rotors have been developed. The brush aerator developed by Pasveer[22] is being used in shallow-channel aeration basins. Oxygenation capacities of different types of mechanical aeration systems are compiled and presented in Table 2.

One of the major defects in the evaluation of the oxygenation capacity of any aeration system is that the experiments are carried out in batch processes, using clean water and waste water. The classic sulfite oxidation test is a chemical test which cannot represent the microbial respiration rate. The kinetics of the biochemical reaction under continuous culturing is quite different from the batch process. The concept is well established. Very few attempts have been made to field test the laboratory observations or batch-scale field studies. It is not clear whether the aerobic biological process in which a mechanical aerator is installed is able to keep the dissolved-oxygen level for which it is installed.

Aeration in ASPs serve three purposes: (1) to provide sufficient oxygen to the microorganisms for the oxidation of organic materials, (2) to keep the suspended solids in suspension in the liquid medium, and (3) to provide intimate contact between the soluble food and the microorganisms. The technological innovations, starting from the compressed-air diffusion to pure oxygen-induced aerations, essentially serve the same purpose, except that the latter can provide a higher transformation rate at extra cost. When sufficient agitation is made, the reactor content becomes completely mixed. This system is called CMAS. In an incompletely mixed basin the solids as well as the soluble substrate and the oxygen profile show a concentration gradient. These reactors are known as plug-flow systems.

C. Parameters Employed for the Assessment of Biological Waste Treatment Processes

The classic, 20°C, 5-day biochemical oxygen demand (BOD$_5$) has been used as a basic parameter in designing and assessing biological waste treatment plants the world over. The authenticity of BOD$_5$ has never been questioned. BOD$_5$ represents the demand of oxygen required for the oxidation of carbonaceous organic compounds present in a waste, provided that the microorganisms present in the test condition are not subjected to any inhibition from the biochemical activity and that food and other nutrients are available. When an English professor was asked, "Why is the BOD test conducted at 20°C for 5 days?" he answered, "The mean temperature in River Thames at the time of conducting the test was 20°C and it took 5 days for the flow to reach the ocean from the point of its origin." This only testifies the scientific principles under which the test was designed. The objectives of the waste water treatment plants are to remove soluble organic and inorganic materials from the feed streams.

In the biological processes soluble carbon components are transformed into insoluble cellular mass, certain metabolic intermediates, and gaseous carbon dioxide. Nitrogenous constituents are incorporated into cellular proteins and nucleic acids, oxidized to inorganic nitrates, and then to gaseous nitrogen. Sulfur reaches cellular proteins and the rest is oxidized into inorganic sulfate while phosphorus gets fixed in nucleic acids and energy-rich ATP bonds. The stable end product of phosphorus transformation is PO_4. The BOD_5 only represents the stabilization of carbonaceous materials and hence authenticity in assessing total stabilization of waste material has become a controversy. Chemical oxygen demand (COD) has been used as a parameter for assessing the waste strength. These tests have become "surrogate indicators".[25]

III. WASTE WATER CHARACTERISTICS

A. Nontoxic and Toxic Wastes

Incomplete characterization of industrial effluents has resulted in a poor understanding of the biodegradability of significant pollutants in the waste water treatment plants. The failure to remove the pollutants has resulted in accumulation and persistance in the environment. Russel et al.[26] reported that 12 organic priority pollutants have been detected in POTW effluents even though the concentrations were less than the threshold values recommended by the U.S. EPA. However, some of the heavy metals such as cadmium, copper, lead, nickel, and zinc had much higher concentrations than threshold values. Industrial waste water characterization has gained significant priority because of the biomagnification of the persistant pollutants.

Waste water treatment plants were constructed several years ago by the sanitation districts and metropolitan councils to treat domestic and industrial wastes in the combined system. The U.S. Federal Water Pollution Control Amendment Act of 1972 is a sequel of the realization that these facilities failed to meet the desired goals for which they were installed. Pretreatment programs at the "end-of-the-pipe-line" at factory sites had to be made obligatory on the part of industries, especially those which discharge toxic industrial waste. Simultaneously, developments in analytical techniques have enabled one to understand the chemical nature of the waste waters. Based on this understanding, the waste waters can now be classified as nontoxic and toxic wastes. Table 3 illustrates the different types of wastes, based on the influence it has on microorganisms present in biological treatment processes.

Liquid industrial waste discharged from various manufacturers contains both organic and inorganic constituents. The composition of the constituents depends upon the raw materials used, the intermediates formed during the process, and the finished products obtained. These constituents are objectionable in the sense that the readily biodegradable materials under anaerobic conditions release a foul smell while slowly biodegradables and nonbiodegradables persist and accumulate in the environment, resulting in bioaccumulation and biomagnification. The slow process of microbial transformation results in the disappearance of many unnatural organic molecules derived from industrial processes as well as from natural deteriorative phenomenon. Many of the synthetic polymeric molecules, xenobiotics, and fossil fuels derived compounds which are transformed very slowly by the natural microbial population. The easily assimilable constituents such as the organic matter present in domestic wastes, food and dairy products, and meat- and sugar-processing wastes are easily oxidized into stable end products by the biological population present in the municipal waste treatment plants. A variety of synthetic, organic molecules are finding ways either through industrial effluents or through intentional application into agriculture or forestry. On the basis of chemistry, these molecules can be classified as aliphatic alicyclic, aromatic, or heterocyclic, with or without substitutions. The natural constituents present in the waste are biodegradable, while most of the xenobiotics and the products derived from processing naturally occurring

Table 3
CLASSIFICATION OF INDUSTRIAL WASTES ON THE BASIS OF TOXIC AND NONTOXIC NATURE

Toxic	Nontoxic
Coal processing	Domestic sewage
Steel mills, gasification, liquification, and domestic coking	Food processing
	Meat processing
Petroleum refining	Sugar, brewery, and distillery (nuisance)
Oil	Textile and synthetic fibers (nuisance)
Petrochemicals	
Surfactants	Pulp and paper mills (nuisance)
	Laundry (nuisance)
Synthetic organic polymers	
Pesticides, herbicides, and insecticides	
Pharmaceuticals	
Fertilizers	
Plating and metal finishings	
Institutional (military and other) explosives	
Basic organic chemicals	

materials are either resistant to biodegradation or nonbiodegradable. Organic molecules, especially the aromatics and heterocyclics, discharged through industrial effluents from fossil fuel processing plants, synthetic chemical manufacturing plants, pharmaceuticals, plastics and polymers, and metal finishings have been reported to be toxic to the microbial population present in domestic waste treatment plants.

Waste water discharged from fossil fuel processing industries has great significance as these fuels form the primary sources of energy. Fuel is most essential in everyday life. Energy balance is dependent upon the population shift. In developed countries, it is stabilized because of the population stabilization while in developing countries the demand is ever increasing because of two major factors: rapid industrialization and failure to curb the population explosion. The least-developed countries are yet to realize the need for their energy bill. The world energy consumption increased from 3.3 gigatonne oil equivalent (GTOE) in 1960 to 6.8 GTOE in 1978.[27] It could be doubled or even tripled between 1978 to 2020. Less-developed countries are expected to demand over 60% of the total world need during the coming years. The advanced nations will have to be satisfied with the rest of the available oil or tap other sources of energy. The world energy demand between 1960 and 2020 is compiled from Frisch's paper and given in Table 4.

Energy sources cover mostly coal, oil, nuclear, and water powers. Natural gas and renewable energy sources are also the possibilities of the future. A review of the data presented in Table 4 shows that the depressive general climate, characterizing the world economy and the energy sector since the second oil shock, is not favorable for any long-term outlook especially when oil prices are gyrating. Any prediction under such circumstances may invalidate the results of any forecasting efforts. However, it is clear that energy and economic growth have a direct relationship.

Extraction processes from the oil resources are mainly dependent on the minimum support price for crude oil. Production costs enforced by the major oil producing and exporting countries (OPEC) are some of the measures taken for a stable price structure. Statistical data

Table 4
WORLD ENERGY DEMAND (1960—2020)

Parameters	1960	1978	2000	2020	Increase/decrease (%) 1978	Increase/decrease (%) 2020
Demographic/economic situation (billions of people)		4.3	6	7.7		
Gross national production per capita (U.S. dollars)		2000		3200	5	2.9
Energy consumption (demand)/energy supply						
Coal (GTOE)	3.3	6.8	10.1—12.0	14.0—18	{ 36 → { 25	{ 32 ← { 28
Oil (GTOE)		2.7	2.8	2.4	{ 40 → { 30	{ 30 → { 20
Natural gas (GTOE)		1.2	1.8	3.2	17	20
Nuclear power (GTOE)		0.8	1.0	2.3	2	12
Water power (GTOE)		0.4	0.6	1.3	6	8
Unconventional (MTOE)	560	735		1100	11	5

provided in Table 4 clearly indicate that oil consumption had reduced from 40 to 30% by 1978.

In the event of dwindling production of oil, coal is predicted to play a major role in the world energy market. Experts from more than 75 countries, during the Second World Energy Conference had predicted that coal would become the leading energy source of the world by the fall of the present century. More than 10,000 billion tonnes of coal are spread underground in over 80 coutries. The U.S.S.R., U.S., China, and Australia are believed to share nearly 90% of the total estimate. The U.S. national energy plan estimated an annual increase from 660 million tonnes (1977) to 1.2 billion tonnes by 1985[28] The Indian target is expected to increase from 135 to 400 million tonnes by the turn of the century.

1. Waste Water Characteristics from Coal Conversion Processes

The importance of coal as an energy source is second to none. Thermal powerhouses consume the major share of coal produced. The second largest consumer is the steel industry. An emerging area of coal utilization in recent times are the coal gasification and liquifaction processes. Low temperature carbonization of coal for producing smokeless coke for use in urban areas and in villages have been developed in India.[29] Coal gasification and liquifaction processes may rank as the most significant coal-refining operations in the U.S. in the coming years.[30]

Coal conversion processes yield highly contaminated waste water. These wastes contain high concentrations of organic and inorganic constituents. Waste water characteristics from various coal conversion processes, compiled from different sources are shown in Table 5.

Coke-oven effluents undergo a by-product recovery scheme. Low-temperature carbonization (LTC) waste has high concentrations of phenolics and ammonia, as by-product recovery is economically infeasible in this process.[29] Coal gasification entails the reaction between coal, steam, and air or oxygen to produce low- or high-Btu gas. High-Btu gasification processes operate at temperature ranges of 815 to 1040°C. The production of organic contaminants during coal gasification is related to the gasifier physical configuration and operating conditions.[30] The data presented on producer gas have been from a plant in which coal is burned at a temperature of 400°C. The gas composition in this case is nitrogen: 51 to 53%, carbon monoxide. 23 to 27% hydrogen: 9%, carbon dioxide: 5 to 7%, and methane: 1 to 2%.[31] Combined waste water recovered from coke-oven by-products contains phenolics, ammonia, and cyanide as the major pollutants.[32] Liquid waste from coal carbonization processes contains aromatic compounds as the dominant organic pollutant. Aldehydes and unsaturated fatty acids are also present in traces. Mono-, di-, and trihydric phenols and methyl-substituted phenols form the core of the phenolics. Coke-oven effluents contain about 50% monohydric phenol out of the total phenols. Catechols and resorcinols are in equal proportions, while xylenols and quinols are in traces.[33] Monohydric phenols and catechols constitute to more than 70% of LTC waste water.[34] A detailed spectral analysis of coke-oven effluents has shown that there are over 130 organic compounds.[35] In addition to acidic phenolic groups, the waste has neutral and basic aromatic fractions. The basic fraction contains pyridines, picolines, and other heterocyclic fractions. LTC waste water has 15 mg/ℓ pyridine. Picolines and lutidines are on the order of 3 to 5 and 2 to 3 mg/ℓ, respectively. Coke-plant, ammonia-still effluent contains up to 2000 µg/ℓ polycyclic aromatic hydrocarbons.[36] Ammonium salts comprised of carbonates, carbamates, sulfide, hydrosulfide polysulfide, and cyanides are present in the waste. They liberate free ammonia on boiling. Fixed ammonium salts represent chlorides, thiocyanates, ferrocyanides, sulfates, and thiosulfates.[37] The polluting nature of the liquid effluents from the coal processing industry is due to the substances which exert toxicity to the flora and fauna of the streams or domestic waste water treatment plants. The waste water from the producer gas plant and LTC plant behave in a peculiar manner. Fresh waste water from both these sources were found to exhibit a straw

Table 5

WASTE WATER CHARACTERISTICS FROM COAL CONVERSION PROCESSES

Parameters	Low temperature carbonization[29]	Hygas[30]	Synthane[30]	METC[30]	GFETC[30]	Hi-coal[30]	Producer gas[31]	Coke oven byproduct recovery effluent[32]
pH	8.98—9.97	7.8—8.4	8.6—9.3	8.5—8.8	8.2—8.6	9.5	8.2—8.9	8.6—9.0
Color	13,000—23,000	NA	NA	NA	NA	NA	3,000—10,000	NA
Dissolved solids (mg/ℓ)	NA	NA	NA	NA	NA	NA	NA	125—800
Suspended solids (mg/ℓ)	200—800	NA	NA	NA	NA	NA	320—680	50—500
COD (mg/ℓ)	40,000—43,000	7,000—14,000	15,000—43,000	16,900—87,000	21,000—30,000	88,000	4,200—16,440	800—1,800
BOD (mg/ℓ)	15,200—21,750	NA	NA	NA	NA	53,000	1,400—4,800	200—1,100
Phenolics (mg/ℓ)	9,250—17,500	1,300—2,000	1,700—6,600	250—300	3,500—6,500	6,800	870—2,020	100—1,000
NH_3N (mg/ℓ)	7,000—7,500	11,000—13,000	7,200—11,000	3,400—7,740	4,000—7,500	14,000	230	100—1,500
CN^-(total)(mg/ℓ)	11.4—22	3—85	0.1—0.6	16—110	~50	NA	0.5	10—60
SCN^- (mg/ℓ)	117—350	16—150	22—200	8—200	18—200	NA	1.0	NA
$S^=$ (mg/ℓ)	67—74	60—560	NA	560—750	60—300	29,000	NA	NA
SO_4^- (mg/ℓ)	25—75	70—370	NA	NA	90—230	NA	NA	NA
Akalinity ($CACO_3$) (mg/ℓ)	15,000—39,000	39,000—49,000	10,000—20,000	800—25,000	14,000—24,000	NA	NA	NA
Hexane extractables (mg/ℓ)	172—350	40—90	NA	NA	200—430	600	NA	NA

Note: NA, results not available.

yellow color. The intensity of color was never above 500 units (470 nm) in both cases. When these wastes were stored in closed tubes, devoid of oxygen and exposed to sunlight, the color intensity remained unchanged and, when stored in open tubes, the surface layer started acquiring a deep brownish color. The intensity increased upon aeration by several-fold. The producer gas-plant effluent had a color concentration up to 10,000 units.[31] In the case of LTC waste it was on the order of 23,000 units.[29] The chemistry of this transformation is less understood. Ashmore et al.[38] reported that pure catechol solution, made alkaline with ammonium hydroxide, upon aeration developed a brown color. A repetition of this experiment could not produce the same color intensity that was observed with both the wastes. However, a solution of hydroquinone having a very low concentration on exposure to atmospheric oxygen in a test tube could produce the color intensity as that observed in carbonization waste water. Chemical properties of quinonic molecules exist in three forms. The alkaline solution darkens upon the exposure to air. Oxidation occurs in which the first stage is the formation of quinone and H_2O_2. Hydroxy-*para*-benzoquinone is a further stage in the oxidation which eventually results in the formation of darker products of less-defined structure. If ammonia and sulfite are present, aminoquinosulfonates are formed.[39] The presence of quinonic compounds even at the smallest concentration in coal carbonization waste water can cause high-color intensities. Complexation of iron with thiocyanates, cyanides, and catechol can also impart color to the waste water.

2. Waste Water Characteristics from Refineries

The world oil supply in 1978 was estimated as 2.78 GTOE. It is approximately 40% of the total energy supply from various sources as mentioned in Table 4. The consumption of oil may reach 10 to 11 GTOE by the turn of the 20th century and touch the figure 18 GTOE by 2020 as predicted by the experts. However, the growth will be reduced to 30 and 20% by 2000 and 2020, respectively.[27] Oil is used (for the greater part) for bunkering and nonenergy consumption. It may have a second preference in the energy market in the developed countries by the end of this century. Developing countries and the Third World will share more than 50% of the total oil production by that time.

Oil-refining processing in Europe jumped from 231 million tonnes/year in 1960 to 951 million tonnes/year by 1974. There are about 135 refineries located in different countries of the OECD region,[40] of which 34% are located inland. In India there are ten refineries having a processing capacity of 16 million tonnes/year (1982). Five are located inland discharging effluents into rivers and lakes. Major pollutants as identified by analyses of refinery effluent are oil and grease, phenols, BOD, and COD. Ammonia and sulfides are also sometimes included in the analyses chart.

Oil refineries generally have a manufacturing schedule to produce gas, gasoline, aviation turbine fuel, kerosene, gas oil, and fuel oil. More complex refineries produce lubricating oil, wax, and asphaltic bitumen. Naphtha is another by-product of petroleum cracking. Crude oil is a mixture of complex hydrocarbons. It is processed in crude distillation units, liquid extraction units, and crystallization and blendings units. Waste water from the refinery operation has dissolved, emulsified, or suspended oils, dissolved organic salts, light hydrocarbons such as phenols and thiopenes, and other oxygenated products such as organic acids.[40] It is also reported to have pyridines, naphthalenes, sulfides, and certain carcinogenic compounds.[41] Water is used for cooling, boiler feeding, crude processing, fire protection, and general housekeeping. Polluted waste water is discharged from the crude processing and housekeeping areas. The waste water characteristics reported in literature have been compiled and given in Table 6.

Oil is the most obvious pollutant in refinery waste. Most components of oil are insoluble in water. It may form dispersions or films that cannot be separated easily. When water is contacted with the multicomponent oil phase, the water-soluble constituents of oil dissolve

Table 6
WASTE WATER CHARACTERISTICS OF PETROLEUM REFINING[41]

Characteristics	Processing operations				
	Topping	Cracking	Petrochemical	Lube	Integrated
COD	50—150	150—400	300—600	400—700	300—600
BOD	10—50	30—600	52—800	100—700	100—800
TOC	10—50	50—500	100—250	100—400	50—500
TSS	10—40	10—100	50—200	80—300	20—200
Phenol	0—200	0—100	0.5—50	0.1—25	0—50
NH$_3$N	0.05—20	0.5—200	4—300	1—120	1—2.50
Sulfide S$^=$	0—5	0—400	0—200	0—40	0—50
Oil and grease	10—50	15—300	20—250	40—400	20—500

From Doctor, T. R., *Proc. Natl. Workshop on Microbial Degradation of Industrial Wastes*, Sundaresan, B. B., Kumaran, P., Joshi, S. R., and Shivaraman, N., Eds., NEERI, Nagpur, India, 1981, 78. With permission.

in it. The solubility of a 10% oil-water mixture equilibration is shown in Table 7. The results presented provide an example of a water sample which had been in contact with gas oil for several days.[43]

Toluenes, benzenes, naphthalenes, and substituted derivatives can be present in oil-polluted water bodies in submilligram concentrations, as is clear from the data reported by the authors. More than 50 refinery-derived organic pollutants could be detected in oil-polluted water.

3. Waste Water Characteristics from the Petrochemical Industry

Petrochemical industries are ancillary units which receive raw materials from the refineries. The naphtha from crude oil cracking is further thermally cracked to produce a range of low molecular weight aromatics and olefins. The primary products are liquid petroleum (fuel), hexane, cumene, and toluene. Mixed xylenes are derived from the aromatic plant.

The products from olefin plants are fuel gas, ethylene, propylene, butadiene, C$_4$ alcohols, ketones, and petrol. These products are the feed stock for a variety of industries such as surfactant, pharmaceutical, synthetic fibers, plastics and polymers, antioxidants, and synthetic rubber. The post-World War II period of reconstruction the world over witnessed an unprecedented growth of the petrochemical industry. Synthetic chemicals which could be used as starting materials for life-saving drugs and pharmaceuticals to the smallest sticker pin of the most complicated artificial heart are being manufactured from petrochemical products.

Waste water from the petrochemical industry can contain certain traces of all these constituents that are manufactured provided that they are soluble in an aqueous medium. Organic acids, aldehydes, long- and short-chain aromatics, and chlorinated residues of long- and short-chain aromatics are soluble in water. Free and emulsified oils are present as suspensions as they are insoluble in water. Mineral acids and alkali salts are also present in the waste water. Nonionic detergents can also be present if they are manufactured in the ancilliary industries situated in the same premises. Even though the chemical nature of the waste products are known, petrochemical wastes are characterized in terms of BOD and COD estimation,[44] or as PV and TOC estimations.[45] An Indian petrochemical plant has a product range of ethylene glycol, surfactants, PVC resins, polystyrenes, SBR rubber, abrasives, agrochemicals, phenolic resins, phenols, methyl methacrylate, and plasticizers. Process waste water in terms of chemical characteristics is shown in Table 8. The constituents that have concentrations between 10 to 100 mg/ℓ can exert high BOD values. Davey and Harkness[46] described that a synthetic resin manufacturing waste contains a high concentration of phenol.

Table 7
DISSOLVED HYDROCARBONS IN WATER EQUILIBRATED WITH 10% VOLUME GAS OIL (AS DETERMINED BY HIGH RESOLUTION GAS CHROMATOGRAPHY).[43]

Component	Conc (mg/ℓ)	Component	Conc (mg/ℓ)
Benzene	0.36	1,4-Dimethyl-2-ethyl-benzene	0.026
Toluene	0.61		
n-Octane	0.016	1-3-Dimethyl-4-ethylbenzene	0.026
Ethylbenzene	0.13		
1,4-Dimethylbenzene ⎫	0.44	1,2-Dimethyl-4-ethylbenzene	0.026
1,3-Dimethylbenzene ⎬			
1,2-Dimethylbenzene ⎭	0.24		0.034
n-Nonane	0.004	1-3-Dimethyl-2-ethylbenzene	0.004
Isopropylbenzene	0.011		
n-Propylbenzene	0.022	1-2-Dimethyl-3-ethylbenzene	0.006
1-Methyl-3-ethylbenzene	0.079		
1-Methyl-4-ethylbenzene	0.029	1,2,4,5-Tetramethylbenzene	0.02
1,3,5-Trimethylbenzene	0.054		
1-Methyl-2-ethylbenzene	0.056	1,2,3,5-Tetramethylbenzene	0.032
tert-Butylbenzene ⎫		5-Methylindan	0.029
1,2,4-Trimethylbenzene ⎭	0.19	1,2,3,4-Tetramethylbenzene	0.12
Isobutylbenzene ⎫	0.003	4-Methylindan	0.032
n-Decane ⎭			
sec-Butylbenzene	0.007	Tetralin	0.046
1-Methyl-3-isopropylbenzene ⎫	0.012	Napthalene	0.39
1-Methyl-isopropylbenzene ⎭			
		2-Methylnaphthalene	0.027
1,2,3-Trimethylbenzene	0.11	1-Methylnaphthalene	0.22
Indan	0.02	Biphenyl	0.026
1-Methyl-2-isopropylbenzene	0.002	2-Ethylnaphthalene	0.039
1,3-Diethylbenzene	0.005	1-Ethylnaphthalene	0.027
1-Methyl-3-*n*-propylbenzene	0.013	1-4-Dimethylnaphthalene	0.024
n-Butylbenzene ⎫	0.007		
1-Methyl-4-propylbenzene ⎭			
1-2-Diethylbenzene	0.004	Balance	0.42
1,3,Dimethyl-5-ethylbenzene ⎫	0.02		
1,4-Diethylbenzene ⎭			
1-Methyl-2-propylbenzene	0.012	Total	4.2

From Devos, L. L., Bridie, A. L., and Herberg, S., *H₂O*, 10, 277, 1977. With permission.

Table 8
COMBINED PROCESS WASTE WATER CHARACTERISTICS FROM A PETROCHEMICAL COMPLEX[a]

Constituents	Conc range (mg/ℓ)
Ethylene glycol, acetic acid butanol, formic acid, naphtha, and fatty acids	10—100
Acrilonitrile and acetonitrile	1—10
Acetaldehyde, ethyl methylbenzoate, heptanes, and phenol acetamide	>1.0

[a]   IPCL at Baroda, India.

4. Waste Water Characteristics from the Pesticide Industry

Intensive farming with an abundant use of chemical fertilizers and an insecticide umbrella in agriculture has resulted in grains, a "plentiful", in the U.S., Philippines, and India. Preservation of postharvest grains also necessitated a high demand for these chemicals. Insecticide manufacture gained importance because of the acceptance of DDT as a chemical agent to eradicate mosquitoes during World War II. Later, the green revolution warranted postharvest treatment of grains. However, it is a fact to be recognized that a large-scale use of pesticides has given rise to serious environmental problems. These arose as a result of the persistance of the toxic pesticides in nature. Pollution from these chemicals constitutes one of the most serious challenges for public health and environmental aspects in developed and developing countries, the manifestations having different dimensions. Public concern arose because of invalidity of the previous views that soil and water are the perfect sinks for waste disposal.

The organochlorine pesticide, DDT, was used probably for the first time on a mass scale in the Indo-Burma peninsula during World War II to eliminate the mosquito menace for the advancing British troops. Widespread use of pesticides began after the 1948 Bengal famine.[47] Today, India is the largest manufacturer of pesticidal chemicals in South Asia, having a total licensed capacity of 78,000 tonnes/year.[48] The most common pesticides, manufactured in India are DDT , BHC, 2,4-D, the carbamates organophosphorus compounds such as quinolphos, malathion, and parathion. Though DDT and BHC are manufactured in America for restricted use and in limited quantities, the case is entirely reversed in India. Two factories produce 14,000 tonnes of DDT and 41,000 tonnes of BHC per year in India.[49] Synthetic organic pesticide production in the U.S. rose from 0.0211 billion kg/year in 1951 to 0.67 billion kg/year in 1980.[50]

Environmental hazards from the pesticide industry are due not only to wide application for crop protection and disease control but also accidental spillage and leakage from the manufacturing plants; disposal into rivers and estuaries of industrial effluents has become a great concern. Malathion poisoning killed at least five persons in Pakistan in 1976.[51] In the same year a burst in the manufacturing plant at Seveso, Italy caused the death of several cattle.[52] One of the most recent industrial accidents occurred at Bhopal, India on December 2, 1984. This worst disaster of the world claimed an estimate of 2500 human lives, impaired vision of 7000 people, and killed several thousand cattle. The factory was manufacturing the carbamate pesticide Sevin using methylisocyanate as a raw material.

Pesticides can be classified broadly on the basis of the target action and chemical nature as shown in Table 9.

Organohalogenated pesticides are practically insoluble in water.[53] Organophosphates have comparatively high solubilities in water. Malathion has a solubility as high as 145 mg/ℓ.[54] Waste water discharged by the pesticide industry where water-soluble pesticides are manufactured can contribute to the pollution load in the waste. One of the major reasons for switching over to the synthesis and manufacture of water-soluble pesticides is that the water-insoluble pesticides have slow microbial degradability. Second, they are more toxic to higher forms of aquatic life than other groups of chemicals used for pest control. Water-soluble pesticide-containing waste can exhibit high BOD values, provided that a suitable-acclimated sludge is used for the test. The waste water from these industries may have a wide variety of equally active intermediates of the manufacturing process. The effluent discharged from plants manufacturing 2,4-D, 2,4,5-T, and MCPA also contain phenol and the solvents used in the industry. DDT waste will have other aromatics and chlorobenzenes. It is highly acidic in nature. Waste characterization from these industries is restricted to the estimation of COD, BOD, dissolved solids, pH, etc. Efforts are very seldom made to identify the individual pollutants.

Table 9
CLASSIFICATION OF PESTICIDES

```
                                    Pesticide
                    ┌──────────────────┴──────────────────┐
            Target oriented                        Chemical composition
                    │                        ┌───────────┴───────────┐
            Algicides                     Organic                 Inorganic
            Insecticides            ┌────────┴────────┐               │
            Herbicides           Natural          Synthetic    Mercuric chloride
            Rhodenticides           │                          Selenium compounds
            Fungicides          Rhotenone                      Lime sulfur (CAX)
            Nematocides         Pyrethrin I                    Paris green
            Ascaricides         Pyrethrin II                   Lead arsenate
                                                               Calcium arsenate

        ┌──────────────┬──────────────┴──────────────┬──────────────┐
  Organohalogenated   Organophosphorus          Carbamate         Others
        │                   │                       │                │
      DDT               Malathion               Carbaryl         Quinones
      BHC               Parathion               Sevin            Quinolphos
      Methoxychlor      Diazinon                Arsenater        Phenols
      Aldrin            Systox                  Coptain
      Dieldrin          Chlorthion              IPC
      Endrin            Disystox                CIPC
      Heptachlor        Metasystox
      Lindane
      Toxaphene
      Chlordiane
      2, 4-D
      2, 4, 5-T
```

5. Waste Water Characteristics from the Plating Industry

The electroplating industry accounts for the major contribution of heavy metals such as cadmium, nickel, zinc, and copper. Cadmium coatings are used primarily as protection against corrosion for steel, brass, and alloys. A detailed review of the waste water characteristics from the plating industry could not be done because the literature available on this subject pertains to the toxicity of these heavy metals to microorganisms in an aquatic environment or accumulation of these metals in sludge of the ASP. The reason for the lack of information could be that plating industries are small units that have been discharging the liquid waste into domestic sewers where it was thought to achieve enough dilution and would not cause any deleterious effect on the receiving waters. However, studies on the effects of these metals started with a view to conduct monitoring in the environment and, later, laboratory bioassay studies to assess the toxicity to fish. Toxicity of organic and inorganic pollutants, mentioned in the previous pages, to microorganisms that are normally present in the sewage-acclimated sludge in biological treatment facilities has been a subject of detailed study during the past few decades.

IV. TOXICITY OF POLLUTANTS

The majority of the pollutants described in the previous sections has been reported to inhibit biochemical functions in aquatic and terrestrial life. In the case of many toxic wastes, the most important question about the treatment process is the efficiency with which it protects the receiving waters by removing the toxic material from the waste.[55] The sensitivity of biological treatment processes to the effect of poison is reduced by several factors. The microorganisms that are not directly exposed to the toxic compound in the waste water treatment plant may be protected to some extent from transitory shocks and the absence of the population that is responsible for the oxidation of the organic toxicants.[56] It is believed that the bacterial population has a remarkable capability to acclimate to the point of utilizing the toxic compounds by evolving, adaptive enzymes. This concept many times has mislead the designers and operators resulting in the discharge of partially treated effluents. Some of the materials that are let into the receiving waters, waste treatment plants, and soil are actually toxic to the native microbial population.

A. Toxicity of Phenolic Compounds

Phenol has been reported to inhibit thiocyanate oxidation.[57] Jones and Carrington[58] studied the behavior of three bacterial strains isolated from activated sludge. Pure culture studies on thiocyanate degradation have shown that phenol up to 100 mg/ℓ did not inhibit thiocyanate oxidation. When phenol- and thiocyanate-degrading organisms were cultured together, thiocyanate reduction was found to be less than that with pure culture studies. They attributed it to the overgrowth of phenol-utilizing bacteria. Phenol at 5.6 mg/ℓ and other related aromatics having a concentration between 12.8 and 16.5 mg/ℓ were reported to inhibit oxidation of ammonia.[59] The effect of phenols and heterocyclic bases on nitrification was studied by Stafford.[60] Activated sludge, receiving ammonium thiocyanate (500 mg/ℓ), was able to nitrify. The rate of ammonia oxidation was decreased when >3 mg/ℓ of phenol of cresol was added. At 10 mg/ℓ, the biochemical process was completely stopped. However, a concentration up to 100 mg/ℓ of phenol was not inhibitory to nitrite oxidation. 2- and 4-Methylpyridine also inhibited ammonia oxidation. Transformation of cyanide to ammonia was found to be inhibited by a concentration of phenol about 25 mg/ℓ.[61] All these observations tend to make a statement that the presence of phenolic compounds in the waste water treatment process even in low concentrations can inhibit or eliminate bacteria that are responsible for the transformation of other equally toxic pollutants in biological waste water treatment processes. Even though phenol acts as a substrate to certain bacteria, yeasts, and fungi, it exhibits substrate inhibition also.[62-66]

Toxicity of phenol to aquatic life was studied individually and in combination with ammonia, zinc, and copper[67,68] with a view to assess the influence on stream survey bioassay. Brown and Dalton[68] concluded that the combined toxicities of the mixtures could be adequately predicted by summation of fractional toxicities of the particular poison. The lethal concentration $(LC)_{50}$ (6 to 96 hr) value for phenol to fish has been reported to be 9 to 25 mg/ℓ.[69] Bioaccumulation and derivatives of phenol in fish can produce an unpleasant taste.[70] The phenolics are absorbed through gills, skin, and food-chain organisms. Human illness, characterized by mouth sores, diarrhea, dark urine, and burning mouth have been reported when individuals consumed phenol up to 240 mg/ℓ.[71] Even though PAH is less soluble in the aqueous medium, the carcinogenic nature of these constituents has been recognized by the World Health Organization, the U.S. EPA, and other countries.[36]

B. Toxicity of Ammonia

Toxicity and oxidative products of ammonia and nitrite and nitrate have been recognized in the recent past, though studies of fish toxicity have been reported earlier. Ammonia toxicity to fish bears a direct relationship to the pH and temperature of the aqueous medium. Only undissociated NH_3N is toxic.[72] Depending upon the species of fish used, the LC of NH_3N varies between 0.2 and 2.0 mg/ℓ.[73] The concentration of NH_3N increases with an increase in pH. However, at pH 7.0 the fish could tolerate ammonia concentration of more than ten times to that at pH 8.0. Prakasam et al.[74] reported that in biological waste water treatment free ammonia, rather than total ammonia and nitrite, has inhibitory effects on nitrification. These investigators observed that free nitrous acid between 0.22 and 2.8 mg/ℓ was inhibitory to nitrification. Ford et al.[75] also reported similar observations while working on nitrification of chemical processing waste water.

C. Toxicity of Cyanide

Cyanide is an extremely toxic substance. In the mammalian cells, hydrocyanic acid prevents the normal process of oxidation in tissues and paralyzes the respiratory centers of the brain cells. The lethal effect of the cyanide ion is due to the inability of the tissue cells to utilize oxygen. Respiration consists of the terminal oxidation of various organic intermediates by means of molecular oxygen. Cyanide forms a highly stable complex with cytochrome oxidase. Hydrogen cyanide above 0.3 mg/ℓ upon inhalation and hydrocyanic acid upon injection or ingestion of 1.0 mg/ kg body weight has been reported to be toxic.[76] However, the lethal dose to animals has been reported to be higher, (9 mg/kg body weight).[77] Aquatic life has been reported to be susceptible to cyanide. Hydrocyanic acid concentrations of 0.02 to 0.04 mg/ℓ were found to be lethal to fish life.[76]

Biological waste treatment plants, receiving cyanide-bearing waste are affected by cyanide shock. Downing et al.[78] reported that cyanide showed inhibitory effects on nitrification in ASPs. Cyanide in the presence of sulfide had shown an inhibitory effect on thiocyanate oxidation.[35] The presence of cyanide above 10 mg/ℓ had shown inhibition on phenol metabolism by *Candida tropicalis*. Toxicity of cyanide to the activated sludge mass is due primarily to the undissociated HCN and, to a lesser extent, to the cyanide ion. The dissociation of HCN to CN^- and H^+ depends upon the pH value. The lower the pH, the higher the toxicity level because of decreased dissociation. Temperature also enhances dissociation of HCN.

D. Toxicity of Thiocyanates

Thiocyanate is relatively less toxic than cyanide. A concentration above 15 mg/100 mℓ in the bloodstream is critically dangerous. Chronic absorption of thiocyanate can cause dizziness, skin erruption, running nose, vomiting, and nausea. Chlorination of water containing thiocyanate can split off sulfur to release cyanide. Inorganic thiocyanate can complex with iodine of thyroid glands.[80] Fish life has a higher tolerance rate to thiocyanate. Orange-spotted sunfish could survive for 1 hr to a concentration of 300 mg/ℓ thiocyanate, while goldfish could survive in water containing 1600 mg/ℓ of ammonium thiocyanate.[81] Newfeld and Valiknac[82] reported that thiocyanate inhibited phenol biodegradation, while Shivaraman et al.[79] observed that thiocyanate even at 1400 mg/ℓ had no inhibitory influence while experimenting on phenol metabolism by respirometric studies. Hydrogen sulfide like hydrocyanic acid is a respiratory poison. It causes the coupling of cytochrome oxidase. Acute toxicity and salts of H_2S have been documented in literature.[83] The minimum LC of H_2S to fish life has been reported to be 1 to 3 mg/ℓ as S^-.[84]

E. Toxicity of Hydrocarbons

Oils and other hydrocarbons derived from coal carbonization and petroleum refineries form emulsions. These are reported to be toxic to vertebrates of aquatic environment and benthic invertebrates.[85] Hydrocarbons, specifically the light-oil fractions, have been found to be toxic to fish, even at lower concentrations.[86] Fishery damage upon introduction of oil-containing waste waters arises as a result of resinification, weathering, and flocculation of oils, by sinking to the bottom or by wetting of the growth areas and sediments. The toxicity of refinery waste water has been studied as a whole system rather than with the individual constituents present in the waste. For that matter, at present, technical literature does not include any publication dealing with the identification of individual components in the overall hydrocarbon fraction of the refinery waste water. The effect of hydrocarbons on bacteria that are incapable of utilizing the hydrocarbons has been studied by Patrick and Whipple.[87] Aqueous extracts containing 10 to 12 mg/ℓ hydrocarbon content definitely impaired the uptake and utilization of glucose in a bacterial population present in fresh water. Exposure of *Vaucheria* to light gas oil proved fatal at a concentration range of 20 to 50 mg hydrocarbon per gram of dry weight. Winters et al.[88] identified a polycyclic aromatic compound phenalen-1-one occurring in small amounts in light gas oil which inhibited the growth of green algae *Dunaleilla* and *Chlorella* at concentrations as low as 0.25 mg/ℓ while diatoms *Ankistrodesmus* could tolerate up to 5 mg/ℓ.

The toxicological behavior of hydrocarbons in water could be determined to quite an appreciable extent by the physical state of the hydrocarbon phase; fine distribution, emulsification, or solution increases the harmful effect considerably. In some cases the acute toxicity could increase by several-fold and it could be due to the truly dissolved hydrocarbon oils and the aromatics. Toxicological behavior of refinery waste water showed that untreated waste waters possess acute toxicity to fish.

F. Toxicity of Pesticides

Studies on the toxic effect of pesticides have been directed mainly on the persistence in nature, bioaccumulation, and biomagnification in a nontarget group of organisms. The public awareness of the possible adverse effects of pesticides to these groups dates back to early 1960s when DDT was traced in mothers milk, cereals, and other food products. This signifies the potential capability of producing injurious and deleterious physiological effects. Exposure of the nontarget organisms to pesticides is due mainly to application sites rather than the waste water from the manufacturing plants. However, the toxic effects of pesticide pollution is greatly exemplified by a number of incidents especially the fish kill in the Mississippi River, northern Alabama streams (U.S.), and the Hubli River (India).[89] Toxicity tests of the pesticides are carried out mostly using fish bioassay techniques. In the laboratory, Henderson et al.[90] studied the effects of ten of the most commonly used pesticides (chlorinated hydrocarbons and organophosphorus compounds). Organochlorine pesticides have been found to be more toxic than others. These pesticides have high solubility in lipids which results in the bioaccumulation in fat deposits. It is difficult to metabolize these compounds. Organophosphorus compounds, on the contrary, are readily soluble in water and are not retained in the tissues. Toxicity of pesticides may be chronic when the nontarget organisms are exposed to the sublethal concentrations of the toxic compounds. Pathological and behavioral changes occur in the organs, tissues, or cells. Pathological manifestations such as swelling in the organs and rupturing of cells and tissues have been reported.[91] The very words "pesticides, herbicides, or insecticides" indicate that they are toxic to these target organisms. Most of them act as metabolic analogues in the DNA replication or cause hormonal imbalance in the organisms which are exposed to the compound. It should be mentioned that these chemical entities cannot distinguish between the target organisms and nontarget organisms such as the analogue antibiotics.

G. Toxicity of Heavy Metals

Toxicity of heavy metals to biochemical reactions in living cells is due to complexation with proteins. Cd(II) forms disulfides with SH bonds of sulfur amino acids containing proteins in which sulfur amino acids are present. They act as competitive inhibitors in metalloenzyme-catalyzed biochemical reactions. Sherrard et al.[92] stated that heavy metals are known to be very toxic to life at low concentrations. The effect of heavy metals on biological treatment plants can reduce the treatment efficiencies and even cause complete failure. Further, they have shown that cadmium-shock loading in the batch-culture activated sludge could tolerate up to 100 mg/ℓ.[93] It is due to the high concentration of sludge solid that such a high concentration did not prove toxic to the sludge. Webber and Sherrard[94] reported that COD removal efficiency was independent of the sludge age between Cd^{2+} concentrations of 5.15 and 9.98 mg/ℓ. The rate of nitrification for a given sludge age decreased as the influent metal concentration increased. Nitrifying organisms were reported to be more sensitive to nickel than the heterotrophic population.[95] Cadmium and nickel, when present together, had more inhibitory effects on nitrifying organisms than the heterotrophic population.[96]

V. MICROBIAL DEGRADATION OF TOXIC POLLUTANTS

It is strange to observe in nature that many of the natural, anthropogenic, or xenobiotic organic pollutants are toxic to some population (microbial or macrobial) at the same time; these compounds are being biologically transformed or utilized by a certain class of microorganisms to which some of these compounds form the sole source of carbon and energy. Many organic compounds that are degraded slowly by plants and animals can be oxidized to carbon dioxide and water by microorganisms. A great number of these compounds are being synthesized, transformed, and decomposed continuously. During this process of transformation, as life evolved, many organic compounds made an appearance in nature. Most of these were secondary metabolites (end products of secondary metabolism of biosynthetic reactions). Such compounds came to serve as energy sources for microbes to be oxidized completely during the course of evolution. These compounds were xenobiotics to other living beings. For the elimination of these organic compounds, the microbes had to acquire newer properties by undergoing genetic evolution resulting in the synthesis of specific enzymes for the catabolism of the newer compounds. If the xenobiotics were toxic to the microbial cells, they developed resistance to the toxic nature of the alien substance as a first step for survival. Bacterial antibiotic (drug) and DDT resistance by mosquitoes can be cited as two of the most common examples to this situation. From the resistance status, the organisms acquired the genetic trait of dependence. In this case, the cells had to synthesize proteins (enzymes) that could transform the new molecules into food. In the course of evolution, the organisms could have acquired new capabilities by natural events at the genetic material. Environmental microbiologists called this process "adaptation" while environmental engineers termed it "acclimation". In principle, it is a process in which the newer class of organisms survived while the others were killed by the toxic effect of the organic pollutants. Bacteria, yeast, other fungi, and a few algae could acquire the specialized properties during the process of adaptation to the newer environment. Isolation of the cultures from the natural or adapted environment was the first step in establishing the facts that toxic chemicals could be catabolized by microorganisms. It was followed by the elucidation of biochemical pathways during the metabolism of the molecules. The regulation of the pathways at the genetic level and the engineering of the genes were further steps toward the advancement of our understanding and manipulative capabilities in this direction. This had both academic as well as applied interest. In the first instance it could be established that the toxic pollutants could be biologically oxidized to stable molecules. This has led to the treatment of toxic organic molecules by biological methods, along with domestic and other industrial wastes in the combined waste water treatment systems after achieving enough dilution. Later on, these wastes were treated independently in separate treatment works at the industrial site.

Waste water characteristics in all the industries referred to earlier shows that the major pollutants have aromatic rings. The benzenoid nucleus is one of the most widely distributed structural units in nature. Woody tissues of plants contain 18 to 35% lignin. It is a polymer of three aromatic alcohols — *p*-coumaryl, coniferyl, and sinapyl — which seems to have been enzymatically polymerized during the formation of wood. Fossil fuels, the products of plant tissues, are coal, oil, and natural gas. Processing of these fuels releases important products such as benzene, toluene, naphthalene, and phenols as well as other carbonaceous materials. Apart from this, many of the man-made synthetic organic pesticides and herbicides have mostly aromatic molecules. They reach the environment every minute. For this reason, and for the academic interest involved, studies on microbial degradation of these molecules received greater attention during the past 2 decades. However, studies on isolation of microorganisms and biodegradation of the aromatic molecules date back to 1908. Bacteriologists directed their efforts toward understanding the biochemical aspects of the metabolism of the organic molecules including the genetic regulation. Environmental scientists and engineers

Table 10
MICROORGANISMS KNOWN TO DEGRADE
TOXIC ORGANIC POLLUTANTS

Organic pollutants	Organism
Phenolic compounds	*Achromobacter, Alcalegenes, Acenitobacter, Arthrobacter, Azotobacter, Bacillus cereus, Flavobacterium, Pseudomonas putida, P. aeruginosa* and *Nocardia*
	Candida tropicalis, Debaromyces subglobosus, and *Trichosporon cutaneoum*
	Aspergillus, Penicillium, and *Neurospora*
Benzoates and related compounds	*Arthrobacter, Bacillus* sp., *Micrococcus, Moraxella, Mycobacterium, P. putida,* and *P. fluorescence*
Hydrocarbons	*Escherichia coli, P. putida, P. aeruginosa,* and *Candida*
Surfactants	*Alcaligenes, Achromobacter, Aerobacter aeruginosa, Bacillus, Citrobacter, Clostridium resinae, Corynebacterium, Flavobacterium, Nocardia, Pseudomonas, Candida,* and *Cladosporium*
Pesticides	
DDT	*P. aeruginosa,* 640X
Linurin	*B. sphaericus*
2,4-D	*Arthrobacter* and *P. cepacia*
2,4,5-T	*P. cepacia*
Parathion	*Pseudomonas* sp. and *E. coli; P. stutzeri* and *P. aeruginosa*

were interested in pursuing the disappearance of these pollutants from the aqueous phase of the systems. Several microbial species have been isolated from different habitats by microbiologists and used in pure culture studies. Most commonly referred organisms are listed in Table 10.

Dagley[97] gave three major reasons for giving attention to the studies on the enzymatic degradation of the benzene nucleus. First, the plant kingdom synthesizes considerable amounts of natural products that are biochemically inert but are degraded by microbial enzymes. The benzene nucleus is an example of the chemically stable, inert molecule. If this continuously synthesized material would not have been reopened by the microbial enzymes, vast quantities of carbon would have been blocked in the stable ring of the 6-carbon molecule. The second line of thought which justifies a continued interest in aromatic degradation has been the disturbance of the natural cycle due to the activities of man. The increased introduction of man-made and anthropogenic molcules such as detergents, pesticides, and fossil fuel derivatives could have accumulated in the environment as recalcitrants if they would not have been attacked by the microbial enzymes. The third reason for the interest was that they provided a convenient system for studying the conditions that determine the derepression of functionally related enzymes.

Biochemical pathways, through which the multi- to mononucleoid benzene ring fission mechanism have been elucidated, employing several bacterial strains and other microorganisms. The involvement of molecular oxygen in the aerobic process and the mechanism of gene expression and regulation in a few cases have also been elucidated.

A. Biochemical Pathways for the Catabolism of Toxic Aromatic Hydrocarbons

Figure 5 is a compilation of the biochemical pathways. Each arrow in the figure represents a separate enzyme reaction step. Microorganisms enzymatically attack the aromatic com-

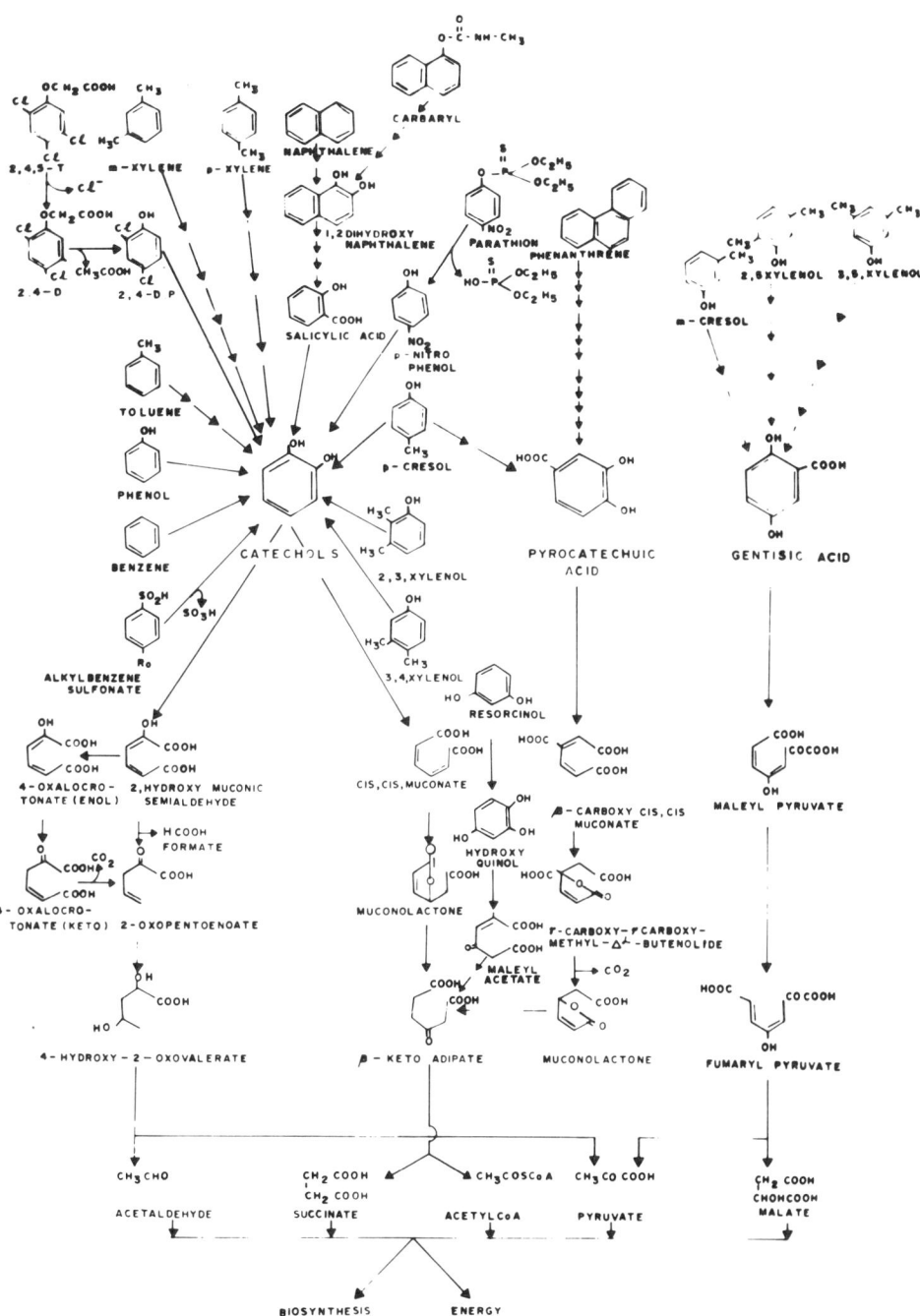

FIGURE 5.   Biochemical pathways for the metabolism of aromatic hydrocarbons and pesticides.

pounds during the degradative processes. The elucidation of the biochemical pathway using individual benzenoid molecules has been a contribution made by many groups of workers. This compilation illustrates what types of organisms are required to bring about biochemical stabilization of the pollutants in a complex system of waste water containing either one or more molecules. It also illustrates the fate of these molecules during the catabolism. Aromatic molecules are rich in carbon content and, once they are cleaved by the aromatic degradative enzymes, the product (organic acid) can enter into the energy cycle, provided that correct conditions are available to the biosynthetic cycle. The mechanisms of detoxifications of the toxicant can also be known by referring to Figure 5. As an example, the pesticidal property of Parathion is eliminated by hydrolysis of the compound to form *p*-nitrophenol and dithiophorate.[98] *p*-Nitrophenol amented growth medium to which bacterial cultures were inoculated by Barik et al.[99] They observed that nitrite and phenol were formed during the growth of the culture. *Pseudomonas* sp., tetra-, tri-, or dicyclic aromatic hydrocarbons were hydroxylated to form a dihydrodiol. This is followed by the cleavage of the ring. Aerobic biochemical pathways for the ring fission have been elucidated and the fundamental studies have generated the largest volume of literature. The degradative pathways involve the occurrence of oxygen-involved reactions to form a dihydroxy aromatic compound as a preparation to the ring cleavage. Phenols are directly hydroxylated to catechol.[100] Biochemical pathways for naphthalenes, carbaryl (naphthalene-based pesticide), phenanthrenes, salicylic acid, and other substituted aromatics converge to one of the three dihydroxy aromatic compounds, namely catechol and substituted derivatives of catechol, protochatechuic acid, or homogentisic acid. The hydroxylases have been purified from a variety of microbial systems and the properties were studied. The function of these enzymes is to incorporate one of the two atoms of oxygen into the aromatic ring, while the second atom is reduced to water by an appropriate hydrogen donor. Reduced pyridine nucleotides or tetrahydropterine act as the hydrogen donors.[101] Following hydroxylation, the aromatic ring fission takes place between one and two or two and three positions by the mediation of oxygenases. The properties of these enzymes have become a textbook commodity. The only significance of these enzymes is that it directly incorporates molecular oxygen into the aromatic ring before the cleavage. These are iron- or copper-containing metalloenzymes.

The biochemical pathways diverge through two fission routes of the aromatic ring — *ortho* or *meta* pathways — resulting in the formation of the organic acids which in turn enter into the energy or biosynthetic cycle of the metabolism. Fewson[102] described the five phases in the metabolism of aromatics as (1) energy into the cell through specific transport mechanism, (2) manipulation of side chains and formation of substrate for ring cleavage, (3) ring cleavage, (4) conversion of the cleavage product into amphibolic intermediates, and (5) utilization of the amphibolic intermediates.

Biochemical pathways related to pesticide biodegradation have been a subject of great interest in the recent past. At least three reviews have appeared in the literature.[103-105] A series of papers were presented at the 12th FEMs Symposium at Zurich. Cook and Hütter[106] observed that *Pseudomonas*, strains A and D, and *Klebsiella pneumoniae*, strain 99 could utilize nitrogen from *S*-triazine(ametryne). These authors proposed a degradative pathway in which ammonia and carbon dioxide formed the end products. However, from the biochemical pathway, it is evident that a compound (cyanuric acid) still remains to be metabolically transformed.

Chloridazon biodegradation was studied by Eberspacher and Lingens[107] in soil bacteria. These authors traced a metabolite which they called heterocyclic chloridazon in the medium and in soil. The soil bacteria had the dioxygenase system for the cleavage of the benzenoid ring and the utilization of the metabolic intermediates of this pathway. Munnecke[108] used cell-free enzyme preparations derived from microorganisms to hydrolyze organophosphate, phenyl-carbamate, phenylurea, phenoxy acetate, and anilide class of pesticides. He claimed that by hydrolytic cleavage the pesticide could be detoxified. The product of the cleavage was the respective organic acid containing the dithiophorate and phenol. Both are still toxic. The DDT degradation pathway in *P. aeruginosa* was studied by Golovleva and Skrabin.[109] A degradation pathway has been proposed by these authors; accordingly, it leads to homogentisic acid via benzhydrol. The first six reactions involve no carbon or energy. Anoxic conditions and nitrate were the requirements for dechlorination of DDT along with other carbon sources. The strain was observed to grow heavily on benzoic acid and *p*-hydroxybenzoic acid. The final end product of DDT degradation by *P. aeruginosa* 640X was phenylacetic acid which then entered into the homogentisic acid pathway for the catabolism.

Molecular oxygen is obligatory for these aerobic processes. It serves a dual purpose; the first being incorporation into the aromatic molecule and, second, acting as an electron acceptor. The literature on this subject has been extensively reviewed.[110]

1. Biochemical and Genetic Regulation

Different strains of *P. putida* have been used to study the biochemical pathway as well as the gene order and regulation. Genetic analyses in these strains have shown that the regulatory and structural genes for genetic expression for the catabolism of several aromatic compounds are plasmid borne.[111-114] Biochemical expression of these genes, at least in the case of phenol, is induced by phenol for the entire pathway of mRNA. Catechol induced the enzymes of the *ortho-* or *meta-*fission pathway only and not phenol hydroxylase.[115] The plasmid *Xyl-* and *Tol-*encoded enzymes have been shown to degrade at least three of the hydrocarbons: toluene, *m-*, and *p-*xylene.[116] Christopher et al.[113] reported that the hyper-degradative mutants structured in the laboratory have multiple copies of the plasmids.

Recently, it was thought that microbes had billions of years to acquire the ability to degrade organic natural products. This view has to be changed in light of the new findings that, in soil and an aqueous environment, microbes had acquired newer capabilities within a span of a few decades. Organic chemists have been contributing novel substances for only over a century. The biosphere received the so-called xenobiotics during the last quarter of the century. They are no longer alien to microbes as evidenced by the versatile nature of microbes. Williams[116] foresaw that it will not be difficult to construct a gene bank in bacteria for the genetic expression to catabolize a wide variety of structually related organic molecules.

Anaerobic catabolism of the aromatic compounds has received attention in recent times. Theoretical consideration would suggest that in the primitive biosphere, lacking molecular oxygen, other methods of obtaining energy for the survival of bacteria must have existed and they might have used biochemical fermentative reactions. In the anoxygenic atmosphere bacteria would have followed the hydration or hydrogenation pathway followed by the nonoxidative (reductive) pathway for the aromatic ring fission.[117] The benzene nucleus is first reduced and then cleaved to aliphatic acids by facultative Gram-negative bacteria. Several examples of anaerobic degradation of benzoate through nitrate respiration have been cited by Fewson.[102] In *Moraxella* sp. benzoate was shown to be degraded by an intradiol catechol pathway under aerobic conditions. Anaerobically, both *Moraxella* sp. and *Pseudomonas* sp. used the denitrification pathway to support the reductive conversion of benzoate.

B. Biodegradation of Surfactants

The biodegradation of surfactants received attention, not because it is present in large quantities in industrial effluents, but because of large-scale use in domestic washings of clothes and laundries which caused foaming in waste treatment facilities. A bloom in petrochemical by-products, especially the synthetic detergents, has completely replaced the classic washing soaps. These surfactants are chemically linear alkylbenzene sulfonates (LAS) and aryl benzene sulfonates (ABS). The alkylbenzene sulfonates have three likely sites for metabolic attack: the alkyl chain, the sulfonate group, or the aromatic ring. Cain[118] suggested that microbial oxidation of surfactants followed a combination of ω and β oxidation and, in some cases, α oxidation for the degradation of the alkane chain. The aromatic ring cleavage followed the oxygenase-mediated reactions of the aromatic pathway.

The purpose of explaining the biochemical pathways and the genetic mechanism for the transformation of toxic pollutants is to advise the designer of the waste water treatment facilities and the operator of such plants how intricate the mechanisms of the transformation is. It also serves as a guideline to those who wish to know at a glance whether a particular

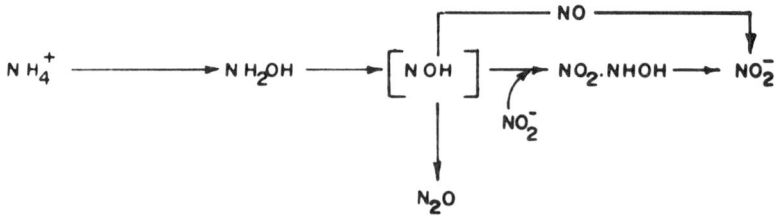

FIGURE 6.   Ammonia oxidation to nitrite.

toxicant is biotransformable and what end products that one can expect in the aqueous medium.

C. Biochemical Aspects of Nitrogen Metabolism

Nitrogen in industrial wastes is present in organic and inorganic form. Microbial transformation of nitrogenous compounds has been studied for over a century. The process of ammonia transformation to nitrate is known as nitrification while the nitrate conversion to gaseous nitrogen is called denitrification. Nitrification is an aerobic process in which ammonia is oxidized to nitrate.

1. Nitrification

Two types of nitrification have been universally recognized: chemoautotrophic and heterotrophic nitrification. The chemoautotrophs derive energy solely from the oxidation of ammonia to nitrite and then nitrate, not from the reduction of carbon compounds.[119] Two distinctly separate types of bacteria are involved in these steps. They are the ammonium and the nitrite oxidizers. Focht and Verstraete[120] listed four genera of chemoautotrophic ammonium oxidizers: *Nitrosomonas, Nitrosospira, Nitrosococcus*, and *Nitrosolobus*. The nitrite oxidizers listed by them belong to *Nitrobacter, Nitrospira*, and *Nitrococcus*.

Heterotrophic nitrification is a biochemical process in which an ammonium ion or amino nitrogen is transformed to the end products hydroxyl amine, nitrite, or nitrate. Organic carbon is used as the energy source in this case. Approximately 104 varied species belong to *Arthrobacter, Azotobacter, Alkalegens, Bacillus, Pseudomonas, Mycobacterium*, and *Nocardia; Aspergillus* and others have been listed in the literature.[119,120] These heterotrophs utilize sucrose, succinate, acetate, pyruvic acid, or acetaldoxime as the carbon source and oxidize ammonia to nitrate.[119] However, the rate of heterotrophic nitrification is reported to be 1000 to 10,000 times slower than autotrophic nitrification.[120]

Autotrophic nitrifiers are slow growing and the yield of the cells is very low in comparison with heterotrophs. Along with carbon dioxide carbonate or bicarbonate and ammonia or nitrite, a minimum concentration of dissolved oxygen, phosphate, iron, copper, and biotin are required for growth.[121] Nitrifiers grow over a wide range of temperature (4 to 45° C).

a. Biochemical Pathway of Autotrophic Nitrification

Oxidation of NH_3N to nitrite involves the participation of molecular oxygen. It serves two purposes: (1) direct incoproration to the substrate and (2) act as a terminal electron acceptor. Oxidation of ammonia to nitrite, according to Aleem[122] should follow the pathway given in Figure 6.

The first stage in ammonia oxidation was proposed to be the formation of hydroxylamine. A two-electron transfer to form an unstable nitroxyl was proposed. This compound then combined with nitrite to form nitrohydroxyl amine. Further oxidation of nitrohydroxyl amine was shown to yield 2 mol of nitrite, of which 1 mol could combine with the nitroxyl. In the absence of oxygen, nitrous oxide is evolved during the oxidation of hydroxylamine.

Oxidation of nitrite to nitrate is a single-step enzymatic reaction and is strictly confined to the electron transport. The oxygen atom of nitrate is reported to have been derived from water.

b. Biochemical Pathway of Heterotrophic Nitrification

An important difference in regard to autotrophic nitrification is that heterotrophic nitrification most often is not linked to cellular growth nor is proportional to total cellular biomass. In the organic pathway, the possible intermediate is an amine or amide in place of hydroxyl amine. These intermediates could then be oxidized to a nitro compound. Cleavage of the nitro group from the carbon moiety could give rise to nitrite and/or nitrate as shown below.

$$R-NH_2 \rightarrow R-NHOH \rightarrow R-NO \rightarrow \quad R-NO_2$$
$$R= NOH \qquad\qquad\qquad \swarrow \searrow$$
$$NO_2 \qquad NO_3^-$$

2. Denitrification

The health significance of nitrate in the environment was first recognized when Comly[124] reported that infantile cyanosis was caused by the oxidation of hemoglobin to methemoglobin with the presence of nitrate in well water. Since then, much attention has been paid to eliminate nitrate from the aqueous environment. Denitrification is a microbial process. It is presently common knowledge that denitrification is brought about by the same respiratory electron-transport chain which is present in aerobic bacteria. However, much of the current literature on denitrification is related to facultative bacteria. There are two types of nitrate reduction: assimilatory and dissimilatory. The former is distinct from the latter in that it has no involvement of electron transport. In pollution control the interest is mainly on dissimilatory nitrate reduction as by this process the nitrate can be removed from the liquid waste to nitrogen gas. Many facultative anaerobes can utilize nitrate as the terminal electron acceptor by the respiratory means to reduce it to nitrite only.[120] However, in a limited number of microorganisms, nitrite is reduced to N_2. Both heterotrophs and autotrophs have been reported to be able to reduce nitrate.

Acenitobacter, Achromobacter, Bacillus, Cytophaga, Gluconobacter, Flavobacterium, Halobacterium, Hypohomicrobium, Microcossus, Moraxella, Paracoccus, Pseudomonas, Propionebacterium, Rhodopseudomonas, Spirillum, Thiobacillus, and *Xanthomonas* have been reported to have the denitrifying capabilities.[120]

a. Biochemical Pathway of Denitrification

The initial step in the biochemical, denitrification reaction involves reduction of nitrate to nitrate. Payne[125] proposed a biochemical pathway for denitrification, which is inconclusively accepted and is given below.

$$NO_3^- \rightarrow NO_2^- \rightarrow NO \rightarrow N_2O \rightarrow N_2$$

The enzyme (nitrate reductase) in the case of dissimilatory process is reported to be particle-bound while, in the case of assimilatory rection, it is soluble. Many of the common cytochrome inhibitors block every step in the reaction series except the reduction of nitrite. The latter step has been reported to be coupled at the flavoprotein level[126] in the case of *T. denitrificans*, whereas in *Pseudomonas denitrificans* the phosphorylation activity has been reported to be associated with the reduction of nitric to nitrous oxide.[127] The enzymatic reduction of nitrous oxide to dinitrogen is mediated by particulate enzymes and is coupled with phosphorylation through the respiratory cytochrome chain.[125,127,128]

The denitrification process requires a carbon source, rarely molecular hydrogen, phosphorus sulfur, and the microminerals iron, copper, manganese, and molybdenum. Ammonia or amino acids are also needed in the case of nitrate-respiring species. Carbohydrates and organic and other cheap carbon sources can form suitable hydrogen donors. Methanol and nitrate, when provided, can enrich to obtain virtually pure culture.[119] Recently, several studies have shown that aromatic molecules can form the sole carbon source to denitrifying bacteria.[129,130] In *Moraxella* sp., in the absence of oxygen and presence of nitrate, the enzymes of aerobic aromatic pathways are suppressed and a reductive mechanism is operative. In the reductive pathway the addition of a water molecule into the aromatic ring has been suggested.[130]

The presence of dissolved oxygen in the environment tends to repress the synthesis of the enzymes of dissimilatory nitrate reduction pathway in nearly all species tested. Downey et al.[131] reported that nitrate sequentially induced the synthesis of nitrate and nitrite reductases as a result of which there has been a dramatic fall in the a-type cytochromes. In contrast to the nitrate reduction nitrite reduction is not sensitive to the presence of dissolved oxygen.

D. Biological Transformation of Sulfur-Containing Pollutants

Waste waters discharged from fossil fuel processing industries contain thiocyanate, thiosulfates, sulfides, thiols, mercaptans, and thiopenes as the sulfur-containing molecules. Organophosphorus pesticides also contain sulfur as the constituent. The ultimate stable end product of sulfur metabolism in an aqueous environment is the inorganic sulfate. The physiology of the oxidation of reduced sulfur compounds by chemolithotrophic bacteria is well known, particularly that of thiocyanate and thiosulfate. Happold et al.[132] isolated and characterized a thiocyanate-oxidizing bacteria *T. thiocyanoxidans*. *T. thioparus* was isolated from the laboratory activated sludge-treating carbonization effluent.[133] A heterotrophic bacterium, *P. stutzeri*, capable of utilizing thiocyanate has been described by Stafford and Callely.[134]

Youatt[135] proposed the breakdown mechanism of thiocyanate in which nitrogen and sulfur moieties are transformed to ammonia and sulfate. The overall chemical reaction sequence is given below.

$$2KCNS + 5H_2O + 3O_2 = K_2SO_4 + (NH_4)_2SO_4 + CO_2 + (CH_2O)$$

The formation of sulfate from thiocyanate could be stochiometrically demonstrated during thiocyanate degradation.[136] Stafford[137] proposed a modified pathway of inorganic sulfur metabolism including thiocyanate (Figure 7).

In the metabolism of sulfur-containing compounds, polythionates are formed as intermediates in thiocyanate oxidation under oxygen starvation. The accumulation of polythionates can produce the formation of elemental sulfur which behaves as a sulfur sink. Polythionates, thiosulfates, and tetrathionates can be oxidized to sulfate under unlimiting oxygen concentration. All the sulfur-containing compounds can be oxidized to sulfate when oxygen is present in excess. Growth substances such as pyruvate are effective in accelerating thiocyanate oxidation reactions. Tracer studies using labeled pyruvate indicated that about 40% of the radioactivity could be recovered from the carbon dioxide released during oxidation of pyruvate.

E. Biochemical Aspects of Cyanide Metabolism

A variety of microorganisms belong to fungi and bacteria have been found to metabolize cyanide;[138] some of these are listed in Table 11. However, the biochemical mechanisms of cyanide metabolism were different depending upon the type and enzyme systems of organisms. Representatives of each mechanism of cyanide metabolism is given in Figure 8.

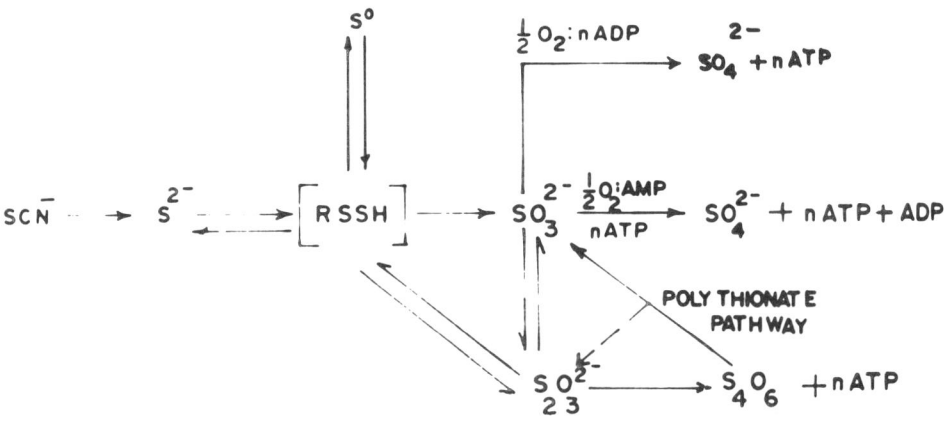

FIGURE 7. Metabolic pathways of inorganic sulfur metabolism. (From Stafford, D. A., *Coke Oven Managers Yearbook*, 1976, 1. With permission.)

Table 11
CYANIDE-DEGRADING
MICROORGANISMS

Bacteria	Fungi
Bacillus megatherium	*Fusarium solani*
B. pumilus	*Stemphilium loti*
B. subtilis	*Aspergillus niger*
Pseudomonas acidovorans	*Gleocercospora sorghi*
B. stereothermophilus	*F. novali*
Arthrobacter sp.	*Rhizopus nigricans*
Chromobacterium viola-	*Poliota* sp.
ceum	
Nocardia sp.	*Psychrophilic basidiomycetes*
	Rhizoctonia solani

Metabolic pathways leading to the formation of amino acids have been proposed. Cyanide could be traced by incorporation into serine to give aspartic acid via β-cyanoalanine.[139] Psychrophilic basidiomycetes was found to incorporate cyanide to acetaldehyde or succinic semialdehyde to finally give alanine or glutamic acid, respectively.[140,141] The fungus *Rhizoctonia solani* was found to give α-aminobutyric acid on coupling with propionaldehyde.[142] The fungi *Stemphylium loti* and *Gloeocercospora sorghi* converted cyanide to formamide.[143,144] Cyanide could also be detoxified to thiocyanate in the presence of thiosulfate by certain microorganisms.[145-149] The enzyme nitrogenase was found to carry out the reductive process of detoxifying cyanide in in vitro studies.[150]

VI. BIOLOGICAL TREATMENT OF TOXIC INDUSTRIAL WASTES

The performance of any waste treatment process such as biofilter, activated sludge, or anaerobic digester depends upon the activities of the communities of the living organisms present in the biological process. These communities are multispecies assemblages in which the component organisms interact with each other and live independently with the physical and chemical environment. The classic ecological theory postulates that a mixture of a wide

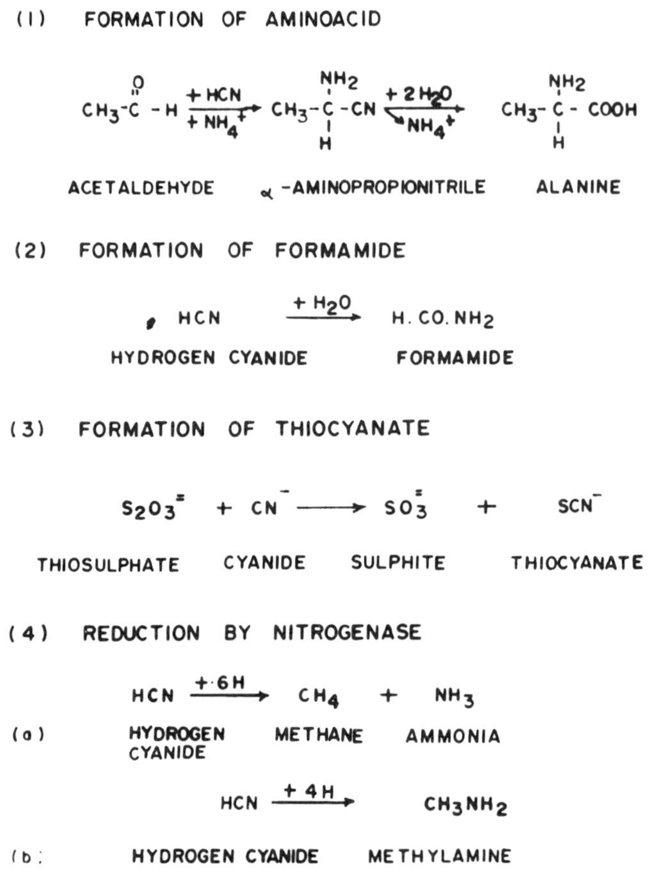

(1)   FORMATION OF AMINOACID

$$CH_3\text{-}\overset{O}{\overset{\|}{C}}\text{-}H \underset{+NH_4^+}{\overset{+HCN}{\rightleftharpoons}} CH_3\text{-}\overset{NH_2}{\underset{H}{\overset{|}{C}}}\text{-}CN \underset{NH_4^+}{\overset{+2H_2O}{\rightleftharpoons}} CH_3\text{-}\overset{NH_2}{\underset{H}{\overset{|}{C}}}\text{-}COOH$$

ACETALDEHYDE   α-AMINOPROPIONITRILE   ALANINE

(2)   FORMATION OF FORMAMIDE

$$\text{HCN} \xrightarrow{+H_2O} H.CO.NH_2$$

HYDROGEN CYANIDE   FORMAMIDE

(3)   FORMATION OF THIOCYANATE

$$S_2O_3^= + CN^- \longrightarrow SO_3^= + SCN^-$$

THIOSULPHATE   CYANIDE   SULPHITE   THIOCYANATE

(4)   REDUCTION BY NITROGENASE

$$\text{HCN} \xrightarrow{+6H} CH_4 + NH_3$$

(a)   HYDROGEN   METHANE   AMMONIA
CYANIDE

$$\text{HCN} \xrightarrow{+4H} CH_3NH_2$$

(b)   HYDROGEN CYANIDE   METHYLAMINE

FIGURE 8.   Biochemical pathways of cyanide metabolism.

variety of substrates shall support a rich and varied population while a single substrate shall, by selective pressure, give rise to a restricted population. Domestic sewage contains a variety of substrates and supports a mixed population, whereas most of the industrial wastes have few or even a single substrate which supports the growth of a selectively few species. It may seem surprising to find that effluents containing a single substrate give rise to a comparatively varied population. This is mainly due to the release of the cellular material of the primary population that metabolizes the substrate in question, either by high metabolic activity or autolysis of the starved cells and thus supports the growth of other heterotrophs. More than one species can utilize the same substrate. Several industrial wastes contain more than one pollutant — carbonaceous, nitrogenous, or others. Sometimes these wastes may have more than one constituent which is toxic to the activated sludge process. However, from the preceding sections, it could be seen that even the most potent poison such as cyanide could be detoxified and stabilized by microorganisms.

In the early history of industrial waste treatment, the only consideration given was what could be the dilution required for an effective treatment. Frankland and Silvester[151] used bacterial contact beds and trickling filters to satisfactorily treat a mixture of 9% gasworks ammonia-spent liquor along with municipal waste. In 1929, Mohlman[152] reported that ammoniacal liquor could be mixed with sewage at a proportion of 0.5 to 3% and treated in a biological process. The treatment criteria defined included the limitation of phenolics to 40 mg/ℓ in the mixture. Studies made by Mathews[153] and reported in 1953 indicated that the concentration of phenol in the mixture could be about 25 mg/ℓ. Mackinney et al.[154] in 1956, however, reported that phenol up to 500 mg/ℓ was not toxic to the activated sludge process. The existence of such literature pertaining to phenolic waste alone could be because destructive distillation of coal to tar in gasificators were the oldest industries to be established in the process of industrialization. Apart from this, coke ovens in steel mills also discharged phenolic wastes. The addition of other industrial wastes to the domestic waste-carrier system was not prominent as was the case with the coal conversion processes. Refinery, pesticides, metallo-organic compounds, aromatic hydrocarbons, pharmaceuticals, and fine chemicals have been added to this list during the past few decades. The ratio between industrial and domestic waste in the waste water carriage system was very high during the initial stages of industrial development. However, this proportion reduced as more industries were established in the peripheral area of the townships, followed by inclusion of the waste water into the combined waste water carriage and treatment. Today, the ratio between industrial and domestic waste in many of the waste water treatment facilities of sanitation districts falls between 35 and 50%.[155-157] The concern over the effect of these toxic pollutants on the performance of the treatment processes and the receiving fresh water courses such as rivers and lakes became a subject of investigation during the past decade. Simultaneously, the possibility of treating such wastes independently (without mixing with domestic sewage) was also studied, primarily for two reasons. When the sewage system had insufficient carrying capacity, the waste had to be treated at the site. Second, biological treatment of these wastes became a widely accepted method especially with coking plant wastes. One of the first waste treatment plants treating pehnolic waste was constructed in South Wales in 1954.[37] The Bethlehem Steel Corporation had constructed and put on stream an independent waste treatment plant in 1962.[158]

Toxic wastes which are amenable for biodegradation could be treated in the combined waste water treatment or in independent facilities, depending upon the situation, characteristics, and volume of the waste. It became a subject of investigation during the last 3 decades. These studies have been related to (1) monitoring by stream and lake surveys for the possible effect of these toxic materials on the flora and fauna of the receiving waters, (2) laboratory studies relating to biodegradability studies in bench-top treatment units and development of mathematical models using the kinetic constrants derived from laboratory experiments, (3) computer simulation studies for predictive purposes, and (4) field studies in the actual waste treatment plants.

From a microbiological point of view, the studies have been directed toward understanding the population dynamics of the complex system: pure culture studies in the laboratory using synthetic waste, and single or mixed (heterogenous) population. One of the recent developments in this aspect is the biochemical augmentation of the biological waste treatment plants with specific microbial cultures for effective degradation of the specific pollutants. These aspects will be discussed later.

Monitoring of streams and lakes must be carried out to gain information about the levels of harmful or potentially harmful pollutants that were discharged. Taken together with the best information available, including the effect of these pollutants on all living creatures, they give an insight into the condition of the environment and inhabitants. An assessment can provide an early warning system of the trouble ahead, when the trend of pollution concentration is clearly rising towards the level at which hazards or damage can occur. If the pollution levels have already reached undesirably high levels, control measures to reduce the pollution from the points of origin can be adopted.

River and lake monitoring programs have been going on in Britain since 1951[159] and in some localities from much earlier. BOD, nitrogen levels, and dissolved oxygen are the commonly employed basic parameters to monitor pollution. It is considered that these parameters can give valuable information about the capacity of the water and the degree to which organic matter is present in the streams as a substrate for bacterial action. The European

Economic Community (EEC) regulation demands that the water quality in the Rhine River must be monitored under the directives of the Commission on the Protection of the River Rhine. For this purpose, several automated monitoring stations are operating in the countries through which the river passes.[160]

As a result of stream, river, and lake water surveys conducted by different agencies in the U.S., several potentially dangerous chemicals have been observed in alarming concentration. This has resulted in the formation of a list of priority pollutants which consists of 129 names of various chemicals.[161] Many of these compounds listed are organic in nature and are biodegradable as evidenced from the discussion made under biochemical aspects.

A. Biodegradability of Toxic Pollutants

Three different kinds of approaches were made to study the biodegradability of the toxic pollutants. The most widely employed procedure is the classical "acclimation" of microorganisms to the waste. The sources for these organisms have been soil, sewage sludge, or activated sludge receiving such waste, if there is one already existing. The procedure can be simply explained; the microorganisms of a diverse population are exposed in a nutritive medium containing a complex carbon source so that all the species are first allowed to grow. These organisms are then fed with the particular waste along with deficient nutrients. The dose of the waste is kept low in the first instance followed by a slow increase until it reaches a point beyond which the concentration of the toxic pollutant becomes intolerable to the microbial population. In principle, the acclimation is nothing but a selective enrichment of a particular class of species for which the pollutant forms a carbon source of biosynthesis and energy. The experimenter in this instance generally has never been interested in verifying the microbial entity of the population present in such systems whether it is a single species or a group of organisms having diverse biochemical properties. The constraints could be that the complex nature of the waste water, having more than one pollutant, and selection of one pollutant would be considered as a misnomer. The objectives of any waste water treatment is to remove soluble organic materials from the feed streams. The degree of stabilization attained in these systems is assessed by monitoring the reduction of the pollutants. Since these processes are being aerobic in nature, the BOD test is used as an indicator for assessing the degree of stabilization achieved.

Much literature has accumulated on biological treatability studies of phenolics in coal conversion processes. This interest in microbial degradation of toxic or inhibitory carbon sources is due more to the presence of phenolics in coal conversion, petroleum refining, pharmaceutical, and a variety of other industries. Phenol has been used as a model toxic compound to conduct such studies as coal is predicted to play a major role in the energy market in the future.

The British Carbonization Research Association began investigations on the treatment of coking plant effluent in 1955.[162] It was reported that, of the many methods available, the biological process was considered to be the most promising avenue for investigation. Kabler et al.[163] studied the degradation of aromatic compounds by phenol-adapted bacteria. They used soil, compost, and mud from a catalytic cracking plant waste lagoon as the source of microorganisms. Prakasam and Dondero[164] used sewage treatment plant sludge in laboratory model units receiving synthetic waste containing 11 aromatic compounds to study the utilization of these compounds. They employed two procedures, the Confidence Interval method and Mountford Index method, to analyze the microbial populations. They observed that the activity of seven sludges correlated well with the population structure. However, they had not reported the biotyping of the population. Laboratory studies were not only directed toward testing the biodegradability of the pollutant, but also toward deriving biokinetic constants for predictive purposes, mathematical modeling, and computer simulation programs. These studies were also helpful in developing design criteria for upscale of the flow sheet and selecting the appropriate design type.

The application of kinetic parameters in design development for waste water treatment is a sound practice and safe approach and should be based on biochemical engineering experience. Until the 1950s, research in waste treatment was mostly focused within an empirical framework of analysis directed toward specific system characteristics of interest of practical problems to be solved. The kinetic parameter usually employed for analysis has been the BOD values. The proper kinetic description for the removal of the organic substrate has been a subject of discussion and will receive increased attention in the future. Although the objective is to achieve an overall removal of putrescible materials in waste, elimination of the major pollutants which contribute to the pollution load should be the key design parameter for the treatment plant. In this regard, kinetic studies with a single substrate with monoculture or heterogenous populations has gained importance in recent years. From batch culturing in flasks and continuous culturing in laboratory chemostats to field-scale pilot plants, a breakthrough in understanding the biodegradation of the pollutants has been made.

Venkata Rao and Bhat[165] conducted batch cultures in flasks to study the growth of *Candida tropicalis* and *Trichosporon cutaneum* in phenol and catechol. Both yeasts were found to require biotin, and *T. cutaneum* also required thiamin. They observed that both inorganic and organic nitrogenous compounds (except nitrite and nitrate) served as nitrogen sources. Phenol- and catechol-adapted cultures were able to grow on a variety of other aromatic compounds. Tatsuo et al.[164] studied the adaption of *C. tropicalis* for phenol degradation in batch culture. They reported that decomposition of phenol confirmed Monod's equation up to 500 mg/ℓ, beyond which the specific growth rate decreased with increase in phenol concentration. Kumaran and Parhad[66] observed that the substrate inhibition concentration, K_i value, for *C. tropicalis* isolated from soil was 387 mg/ℓ. Yang and Humphry[65] reported that during the microbial degradation of phenol, it should be possible to achieve phenol removal from waste water down to levels of 1 to 2 mg/ℓ in a single-stage system with a monoculture as well as with a mixture of two cultures. However, because of the effect of substrate inhibition on kinetic behavior, long-lasting transients occurred. This behavior of the system could not be used for determination of maximum specific growth rate (μ^{max}) and substrate saturation constant (K_s). A transient rise of the substrate and the response of the culture could be one of the possible considerations. Batch-culture experiments have been carried out by most of the investigators to calculate the biokinetic constants such as specific growth rate (μ), maximum specific growth rate (μ^{max}), substrate saturation constant (K_s), and also the substrate inhibition constant (K_i) after fitting with various equations. Several investigators have studied the kinetics of biodegradation of phenol from synthetic waste using heterogenous populations.

Batch and continuous culturing with pure cultures,[62,64,65,166] heterogenous populations, and synthetic[63,164,167] and live waste[29] have been used to collect biokinetic parameters and fit into various kinetic models developed on the basis of the theory of continuous cultivation.

Kinetic studies on bacterial growth have given rise to a considerable amount of literature, particularly in reference to the application of Monod's equation. The technique of continuous culturing theory and application was developed by Monod.[168] Novik and Slizard[169] experimented with a chemostat on spontaneous mutation of bacteria while Herbert et al.[170] made theoretical and experimental studies on the continuous culturing of bacteria. These equations have been developed in reference to noninhibitory substrates. Most design and predictive models developed and used in biological waste treatment plants had not considered the inhibitory model approach. The use of these generalized designs for treatment of toxic wastes have become a topic for discussion in the recent past. There are two schools of thought; one view is that the toxic substance (in this case, phenol) can be considered a noninhibitory compound and that degradation follows Monod's relationship.[82,167] The other considers that it should be treated on the basis of Haldane's equation.[63] However, these models have been developed considering a single toxicant. It is seldom found that a complex toxic industrial waste can have only a single pollutant which exerts toxicity. Even though phenolic waste contains phenolics as the major pollutant, it may contain other secondary toxicants also. As an example, coal conversion waste contains cyanides, sulfides, and ammonia. Any kinetic equations developed should also consider the influence of secondary toxicants on the biochemical activity of the population. The predictive models developed with synthetic waste data have not been found to hold true when it was applied to actual industrial wastes. Kumaran et al.[171] observed a very low K_s value (0.00025 mg/ℓ) for phenol with a heterogenous population with synthetic waste in pilot plant studies. When the same plant was operated with a live industrial waste (LTC waste)[129] the K_s was found several-fold high (13.18 $\pm$ 1.98 mg/ℓ). A change in the K_s value could have been due to the influence of other secondary toxicants on the conversion of the phenols.

In the continuous culturing systems where the population density is not too large and the organisms are well nourished, the specific growth rate and the velocity of the reaction (measured as the yield, Y) take constant values. When a well-balanced medium is thoroughly aerated and the substrate functions as a nonlimiting factor, the specific growth rate reaches maximum (μ^{max}) to yield a maximum product accumulation (Y^{max}). Waste water treatment systems are aimed at working under steady-state and feed-starve conditions, as the aim is to bring minimum yield and maximum oxidation. Thus, the batch-culture kinetic data obtained cannot be applied to predict the behavior of these systems when they are operated under steady-state, continuous conditions. However, these data are useful when the systems are operating under transient conditions and also predict the recovery rate of the process as well as the start-up of new treatment plants.

Biological waste treatment plants for the stabilization of industrial wastes are not designed and constructed today on the basis of the inhibitory kinetic model. Design engineers have not accepted these predictive models because, rather than having a clear idea about the inhibition of the pollutants, they are more confused with these models.[172] Mackinney[173-175] proposed a mathematical model for a CMAS. Eckenfelder[176] proposed another model. Goodman and Englande[177] suggested that these two models are identical and can be bridged to form a unified approach to activated sludge-process design. This model is called as the Eckenfelder-Mackinney model and is widely employed in the design and operation of biological waste treatment plants for industrial waste. According to this concept, the CMAS system is the appropriate method for the treatment of most of the industrial wastes. For a detailed discussion of these models the original articles may be referred.[172-177] The basic assumption of the CMAS is that the effluent concentration of the pollutant tends to reach a value nearer to zero irrespective of the undulations in the influent concentration. There are several factors which can influence this condition. One of the most important conditions is whether it is possible to achieve a complete mixing of the liquid content in large-scale waste water treatment plants. Even though it is possible to achieve complete mixing in bench-top or pilot-plant models, the full-scale treatment plants never attain complete mixing at all times. This can result in the plants operating under a pseudosteady state. Under such conditions, the influence of the toxic pollutants on a susceptible population can be predicted by taking into account the microbiological aspect of the treatment plants.

1. Microbiological Aspects of Toxic Waste Treatment

Practically very few references are available on the microbiological aspects of toxic waste water treatment. Even though many reports have accumulated on the existence of microorganisms in soil and other environments, no attempt has been made to qualitatively or quantitatively estimate the microbial population that is present in the treatment plant. Even attempts to identify the dominant flora have not been made so far. This void in the literature could be due to the lack of suitable methods for the estimation of the specific population.

The parameters which are generally accepted for assessing the biological activity of these processes are the mixed liquor suspended solids (MLSS), volatile suspended solids (VSS), BOD, and COD. The solids represent the biomass and BOD/COD as the pollutant load. The solids not only represent the living population but also consist of nonliving (inert) solids. VSS can differentiate between the organic and inorganic residues. Several investigators and design engineers believe that in the absence of a better assay method, VSS can form the basis for the design purpose with reference to the biomass.[178] Other methods such as the estimation of enzyme activity, DNA content, and ATP content of the sludge have also been proposed for assessing the biological activity. Stafford and Calley[179] expressed that the precise and absolute relationship between the viable part of the activated sludge biochemical activity must be related to a nonconservative component such as ATP. It is associated with all living cells. In a viable cell, the amount is maintained within a narrowly defined limit. Chappelle and Levin[180] estimated the ATP content of 20 bacterial strains and reported a concentration range of 0.28 to 8.9×10^{-10} mg per cell. Weddle and Jenkins[181] observed, assuming that a bacterial cell weighs 10^{-12} g, the ATP content per viable cell was constant over the growth range of 0.03 to 6.4 days as 10^{-8} to 10^{-9} μg per cell or 0.1 to 1% ATP on a dry-weight basis. These authors have quoted from Patterson et al. a value of 0.1 to 0.03 μg ATP per microgram of activated sludge. It must be realized that the ATP estimation of the sludge mass gives a measure of the total viable population, while the transformation of the soluble material is brought about by a part of the total viable mass (bacteria, yeast, and certain fungi) present in the sludge mass.

Another method employed for estimating the biological activity is the estimation of oxygen uptake rate by the sludge mass. To a greater extent, respirometric methods can provide more or less accurate results on the biological activity of the sludge. However, this procedure functions as a batch experiment whereas the waste treatment is a continuous process.

Stainer et al.[182] stated that the standard plate count was the most sensitive method for viable count determination in pure cultures. Weddle and Jenkins[181] observed several difficulties in estimating the viable counts in activated sludge mass as it is a mixed-culture system. The dispersion of sludge flocs to disengage the bacterial cells from other populations, inert material, and selection of proper medium were suggested to overcome these difficulties. However, it must be remembered that the standard plate count gives only the bacterial or, to some extent, the yeast population. The protozoan and other ciliates are lost in this method of estimation.

Studies on the viability relation to the VSS have been carried out by several investigators. Weddle and Jenkins[181] reported that the actual viable organism content in the activated sludge solids was usually <20% of the VSS. Curds,[183] based on computer simulation studies, estimated that 30% may be viable in the conventional activated sludge. With single substrate once flow through and a monoculture, Kumaran and Parhad[184] could demonstrate that the viable population in pure culture could be achieved between 68.5 and 85% of the cell mass. A comparison of these results can reveal that, depending upon the population, ciliates and protozoa can contribute to the rest of the population at least an equal proportion of the VSS. Because of these inherent drawbacks in the estimation procedures for a plant operator or design engineer, VSS can form the operational or design parameter of biological mass which can be safely relied upon for the complex waste waters.

a. Microbial Interactions

Several types of interactions between microbes themselves and with the environment are possible. A list of interactions and a close definition with suitable examples pertaining to toxic industrial wastes are given in Table 12.

Competition for the same substrate by two or more different groups has been illustrated on the basis of certain theoretical consideration by Harder.[185] When two groups of organisms A and B compete for a growth substrate in a chemostat, the outcome will depend upon the

Table 12
MICROBIAL INTERACTIONS

Interactions	Definition	Examples
Competition	A race for the same carbon source, nutrients of same nature	Several bacteria, yeasts, and fungi compete for the same carbon source (glucose or phenol)
Neutralism	No interaction and independent existence	Heterotrophs and chemoautotrophs exist together
Mutualism	Each benefiting from the activity of the other	Production of essential growth factors for the benefit of others
Commensalism	One group benefiting while others are unaffected	Bacteria capable of synthesizing vitamins which are utilized by others
Synergism	Metabolic products of one group form the substrate for the other group	Nitrification and nitrafication
Parasitism	One stealing food of other groups	Nematodes in biofilters
Antibiosis	Excretion of biological products which acts as toxic or biocidal	Toxins by bacteria or fungi
Predation	One feeding on others	Ciliates and other protozoa feeding on bacteria
Inhibition	Biochemical inhibition by chemical pollutants	Phenols, cyanides, and sulfides inhibit cyanide or ammonia oxidation
Cometabolism	Oxidation of a compound without being utilized as a carbon or energy source	Cyanides or pesticides metabolized along with other pollutants
In vivo gene transfer	Transformation of one strain in reference to acquiring new capabilities by natural processes of transfer of genetic material	Acquiring capability of transcribing messages for the synthesis of proteins for degradation
Environment	Fluctuations in pH, temperature, and dissolved oxygen	Low dissolved oxygen conc retard aerobic process

growth rate of each organism. If A has a higher specific growth rate, it will grow faster and outnumber the slow-growing B. The chemostat can be operated at any specific growth rates by hydraulic manipulation. However, if A and B have the same specific growth rates, both organisms can exist in equal numbers. Neutralism is a situation in which two different groups coexist without influencing the existance of each other. Heterotrophic bacteria utilize the carbonaceous material as the energy source while chemoautotrophic nitrifiers utilize inorganic carbon as the energy source. Mutalism is the behavior of the microorganisms in which two members are benefited by the activity of each other. Several classic examples can be given to this type of behavior. Quite often the behavior of *Lactobacillus plantarum,* which provides folic acid for *Streptococcus faecalis* which in turn reciprocates by furnishing phenylalanine to the former, is an example of mutualism. An appropriate example for the mutualistic behavior in toxic waste treatment has recently been demonstrated by us in coculturing phenol-degrading yeasts with cyanide-degrading bacteria in a mixed culture seeded with these cultures.[136] In a biological, completely mixed aeration system, phenol- and cyanide-degrading cultures were seeded and biodegradation of both toxic pollutants could be achieved. Phenol otherwise was toxic to cyanide-degrading bacteria and cyanide was toxic to phenol-degrading yeast. When they were cocultured, both detoxified the individual toxicants, thereby benefiting each other.

Commensalism is the behavior of two groups when grown together; growth of one group can benefit from the other. When phenol, cyanide, and thiocyanate were treated in the biological treatment process explained above, the thiocyanate-degrading population was benefited by the growth of both cyanide- and phenol-detoxifying populations.[136] Synergism has been explained in Table 12 in reference to waste treatment. Antibiosis may not exist in waste treatment.

Predation is an aspect which needs some investigation whether it is desirable in toxic waste treatment or it has any adverse influence on the population. It is an interaction between two populations in which one is dependant upon the other. Stalked and free-swimming ciliates must depend upon the bacterial or other primary population. The performance of the biological treatment plant is most often adjudged by the presence of ciliated protozoa. Microbial population dynamics had been studied with reference to domestic waste only; except in one case, it was with coke oven effluent.[186] In the latter case, the report mentioned that 14 different groups of bacteria were characterized. Two groups of ciliated protozoa and rotifers were often found in the coke oven biological treatment system. It was mentioned that changes in the protozoa and rotifer population could indicate the breakdown of the treatment plant before the chemical changes in the treated effluent were noted, using the normal methods of analysis. Curds[187] reported that ciliated protozoa feed on bacteria and carnivorous ciliates feed on bacteria-consuming ciliates. As long as the bacteria-consuming protozoa are absent or present in small numbers, the detoxifying population can exist and bring about an effective removal of the toxicants. When the population increases, the predators prey upon it, thereby bringing the bacterial population down. The carnivores then feed upon the ciliates. There can be oscillatory behavior between these populations. Because of the interaction between these populations, the system behaves as a pseudosteady state, in reference to the population.

The interaction between the chemical environment and the microbial population is one of the most important aspects to be considered as it has a direct effect on the quality of the treated effluent discharged. A detailed discussion on the inhibitory effect of the toxic pollutant on the population is necessary. The behavior of the microbial population in a biological treatment plant receiving toxic waste is mainly dependent upon the residual concentration of the pollutants. The chemical characteristics as explained earlier indicate that the wastes have a very complex nature. Most of them contain one or more toxic pollutants and are toxic to a variety of microbial entities. As long as these constituents are present in concentrations below the lethal dose for the susceptible flora, the flora will exist in the reactors. When the primary population that is responsible for the elimination of these toxicants are depleted by predation, an increase in the concentration of these toxicants can be expected. This in turn can deplete the predator population by the inhibitory effect of the toxic pollutants. In order to have satisfactory quality effluent, it is necessary to maintain the primary population in the biological reactor without having any influence by the predators.

A state of cometabolism can be observed in waste water treatment receiving toxic pollutants. It is a phenomenon in which microorganisms oxidize a compound without deriving any energy from the oxidation. Microbial degradation of hydrocarbons,[188] pesticides,[189,190] and surfactants are some examples.[191,192,118] However, in one of the recent publications, Golovleva and Skryabin[109] demonstrated that for the complete biodegradation of DDT *P. aeruginosa* 640K required only a succession of cosubstrates.

Temperature and hydrogen ion concentration are the two environmental factors which interact with the microbial population present in biological treatment systems. Even though microbes are classified into psychrophyls, mesophils, and thermophils on the basis of temperature optima, in waste water treatment processes, mesophils have a temperature tolerance between 15 and 45°C proliferate. The influence of diurnal variations on the biochemical activity of microbial populations had never received any attention.

In the early years of molecular biology and genetics, the subject received purely academic attention. However, in the recent past considerable interest was shown in applied genetics because of the introduction of thousands of new synthetic chemicals into the environment through trade wastes. The fact that these chemicals persist may be because they are so different from the ''acceptable'' carbon or energy source. The microorganisms need several new metabolic steps to channel these xenobiotic compounds into the existing metabolic pathways. Hegeman and Rosenberg[193] have summarized the possible mechanisms through which microorganisms can acquire the ability to degrade new compounds. It is either by the constitutive production of a previously inducible enzyme or by the acquisition of permeability to a degradable compound to which the parent compound is impermeable or the acquisition of decreased sensitivity to toxic metabolites from a new compound or by a change in the specificity of induction of enzymes which already possesses some small activity towards the novel compounds. If one accepts the retrograde theory of evolution of metabolic pathways proposed by Horowitz,[194] the nonspecific enzyme activities from the existing pathways can be recruited by borrowing the beneficial activities. The scattered genes for these enzymes are brought together as a result of recombination, transduction, or through plasmid transfer. If these mechanisms could be demonstrated in the laboratory by in vitro experiments, why could it not occur in the natural systems where the diverse organisms having diverse capabilities exist? The conditions of the laboratory are an artificial duplication of the natural systems. The extrachromosomal DNA to which these transcriptional messages are attributed can be transmitted from one organism to another by contact. It may not only provide the way by which the organisms maintain the ability to degrade the pollutants but also provide a way of improving the organisms in biological waste treatment.[195] Recently, the possibility of such a mechanism has been reported by Chatterjee and Chakrabarty.[196] *Klebsiella pneumoniae,* having PCB-degrading genes (PAC 21), was cocultured along with *P. putida* (PAC25) (specifying the biodgradation of 3-chlorobenzoate, TOL, and other degradative plasmids) in a chemostat and continued selection on 4-chlorobenzoate, led to the emergence of cells that harbor the plasmid PAC27 encoding the 4-chlorobenzoate pathway enzymes. The possibility of such type of in vivo gene transfer and construction cannot be ruled out in biological waste water treatment processes receiving xenobiotic or toxic pollutants. However, the probability of emergence of such a microbial population in a continuous culturing system or through natural selection would be low compared to such a population obtained through laboratory genetic manipulation techniques. Therefore, structuring of microorganisms to degrade specific xenobiotic compounds using genetic engineering techniques has acquired great significance in the industrial waste treatment technology.

b. Biochemical Augmentation

The conventional approach of building up the sludge in the biological treatment of toxic industrial waste is receiving secondary preference because of the application of specific, adapted inocula into the process streams. Arguments and counterquestions on the efficacy of such systems have been raised. A review of these systems, based purely on microbial principles, will be made through the following few pages.

The conventional concepts of natural population selection and the ubiquitous nature by which biological waste water treatment systems develop and maintain the appropriate flora will be discussed first. The improvements that could be achieved through the use of microbial inocula will then be explained.

The conventional concepts of natural population selection in waste water treatment systems are based on the process of acclimation followed by adaption of bacteria, yeast and certain fungi, actinomycetes, and *Nocardia* to the new environment. Depending upon the complex nature of the waste, the community also takes a complex shape. Each of the various strains has a functional role within the biological community of which it forms a part enabling it to survive and contribute towards the overall effectiveness of the community in removing the pollutional load from the waste water. According to the well-established notions, those organisms best able to survive the existing environmental conditions and compete most effectively for the available food sources will remain and be "selected" as a biological community.

The distribution of each strain or species in the community is ultimately a function of the types of pollutant present and the environmental conditions which exist in the waste water treatment systems. In some cases, neither the waste composition nor the environmental conditions are constant. Therefore, the distribution of organisms in the biological community will shift as a function of the change in the chemical composition of the waste. As discussed earlier, natural population selection undergoes shifts due to the oscillation of bacteria to protozoa-bacteria. The effectiveness of the natural population selection mechanisms in developing and maintaining the best biological community is dependent upon the appropriate "seed" of microbial strains. In the complete absence of such a seed the natural selection mechanism cannot provide the development of a proper community.

It is often assumed that bacteria are ubiquitous, all strains are continuously available at all times in the biological treatment process, and the natural population selection will develop the best possible community for the most effective possible treatment of a given waste water. This also leads to the assumption that the individual group of seeds will be present in sufficient numbers to be effective to act as a seed for the natural selection process. The development of a biological community capable of effective treatment of the waste component must occur within a period of time so that it would respond to frequent transient conditions of waste water characteristics. If the development is too slow, the waste components will accumulate in the aqueous environment and appear in the treated effluent. When this component is toxic it can disrupt the entire biological process in which the susceptible community will be eliminated and the resistant community will persist, but will enter into a dormant state. It is the only population which can degrade the constituents that will proliferate provided that they are present in enough numbers. Thus, the available concentration of a particular seed population has a significant role to play in the community. For the effective removal of the pollutants it should be present in sufficient numbers. Theoretically, it has a significant impact on the time it takes to develop the appropriate population once the seed is selected. The time taken for the development of enough population is called the response time and the resultant behavior of the process is known as the recovery period. The events are dependent upon (1) the maximum specific growth of the selected organisms, (2) the avaiable nutrient concentration, (3) the starting concentration of the selected seed, and (4) the residual concentration of the toxicants in the system.

It is claimed that by seeding with adapted, mutated, or genetically engineered bacteria, it is possible to reduce the response time considerably. In a recent paper, Christiansen[197] discussed three case studies where mutated bacteria were used in solving refinery waste water treatment upsets. In the first instance, a refinery experienced periodic problems with the sour water stripper. It was observed that the sulfide content in the waste was up to 150 mg/ℓ resulting in the decreased efficiency of phenol and ammonia removal. An aerated lagoon was processed to remove sulfides as per the design flow sheet, followed by an activated sludge system. The author used the product of Biochem© 1008 SF in the aerated lagoon. Another product, Biochem© 1010 N culture, containing organisms that oxidize ammonia to nitrite and nitrate was used for nitrification. The second culture was added after the 4th day of adding the culture to the aerated lagoon. The results presented showed that within 7 days the ammonia was reduced to the permissible concentration. In the second case study in which a two-stage, activated sludge process was employed, the first stage was meant to reduce the high BOD. The second stage was for nitrification. Addition of Biochem© 1003 FG increased the MLSS concentration in the first stage while the addition of Biochem©

1004 TX to the second stage brought down the ammonia concentration. In the third case study, phenol concentration was said to have increased in the effluent. The phenolic shocks could be overcome by the addition of another mutant culture.

The success of the process recovery can be explained on the basis of the biochemical mechanisms and the interaction that the chemical pollutants and the microorganisms present in the process. High concentrations of the inhibitors such as sulfides and phenols could not permit the proliferation of the nitrifying population. By introducing the specific cultures for detoxification of these toxicants, the recovery period has been enhanced. The introduction of the nitrifying organisms in the second unit had a positive influence on the recovery process because the inhibition of nitrification has ceased by the activity of other population.

The adapted- or mutant-culture technology, which is being developed by several private sector consultants claims surprisingly good success in the improved treatment of both domestic and varieties of industrial wastes. It is claimed that these products are obtained by cultivating and adapting selected naturally occurring strains to increasingly higher concentrations of the preferred substrate. From this population, individual microbial cells showing significantly increased capability to degrade the substrate in question are isolated. In certain cases, induced mutation is also carried out. Large-scale production of these strains is then carried out employing fermentation systems and finally they are preserved. A variety of individual strains of specialized microorganisms are then blended to create a broad-spectrum inoculum which, with the addition of other components, forms the final product. Such a product is said to contain strains capable of rapidly degrading aliphatic and aromatic hydrocarbons, phenols, halogenated phenols, cyanides, and other organic pollutants. It is difficult for microbiologists to believe these claims of the suppliers. For example, the technical information on a product mentions that it has the capability to "clean" by breaking down large chains of insoluble molecules of proteins, starch, cellulose, and fats. The composition of the product is: protease, 7000 U; amylase, 8000 U; cellulase, 4 U; lipase, 1000 U; and the organisms 7×10^7 U. (all Us are units per gram). It has a storage life of a minimum of 3 years. Similarly, some of the products are remarkably versatile having degradation capabilities; the substrate ranges are benzenes and phenolics, including chlorinated phenols, hydrocarbons, cyanides, and alcohols.

Genetic engineering technology has advanced to a greater extent in the laboratories in the in vitro synthesis of DNA and RNA. Construction of specific genes for the transcription of messages followed by fermentative production of proteins such as somatostatin, insulin, and interferon has been possible in the laboratory under controlled conditions of experimentation. The microorganisms responsible for such hyperactivity are "super bugs" in modern terminology. It is quite often argued that the release of such genetically engineered super bugs into an uncontrolled environment may prove disastrous as man will not have any control over them, once they are released. If one looks through a biochemical perspective, these "monsters" cannot survive in an environment where the substrate is depleted in time. Hyperdegradative activity persists until the substrate is present in a high concentration along with other nutrients. Second, the other nutrients may not be available in large concentrations for the organisms to utilize for proliferation.

It is already reviewed that the genetic information for such degradative activities for toxic substances and xenobiotics are stored in plasmids. When plasmid technology has been used successfully to improve the metabolic capabilities of a number of microbial strains, those lacking stability will not be effective in carrying out such tasks as they are likely to lose plasmids in a complex environment.[110,179,198] The selection of strains having a stable genetic trait is a primary criterion for environmental application. A strain having constitutive genes are preferred over the plasmid-borne, inducible enzyme systems. Biochemical augmentation based more on the sound scientific and engineering principle can bring about positive results in solving environmental problems. The application of a specific culture or a blend of cultures to specific needs requires other considerations such as an understanding of the conditions and proper maintenance required for these organisms to express themselves in the particular system. If these conditions are not provided, these cultures are likely to be eliminated during the competition between them and the natural population.

Laboratory-stored cultures are useful for biochemical augumentation when compared to the acclimatized seed, provided that these stored cultures are characterized by the ability to degrade the particular pollutant. It is also necessary to know the influence of other toxicants as well as substrates on the metabolism of the pollutant of interest. Before these cultures are used, the waste water characteristics and nutritional deficiencies of the pollutants should be known. A few advantages of laboratory-stored cultures are that they are easy to handle and can be mass-cultured to obtain high viable counts within short periods of time. Studies on these cultures should form a series of experiments starting from the isolation of the culture, batch culturing, continuous culturing with simulated wastes, and actual industrial wastes.

Kinetic studies in batch culture with synthetic waste can give valuable information on the behavior of the culture in a treatment system operating under an organic shocking (organic transient) state. Cultures for detoxification of phenols, cyanide, and thiocyanate were isolated in the authors's laboratory and were studied under the steps mentioned above.

C. tropicalis NCYC 1503 was isolated and characterized in batch-culture experiments. Phenol was provided as the sole carbon source to the yeast. Static and shake-flask experiments have shown that the phenol utilization rate for the normal cells range between 0.11 and 0.2 mg phenol per milligram cell mass per hour while actively growing cells metabolize phenol at 0.22 to 0.23 mg phenol per milligram cell mass per hour. The oxygen requirement for phenol oxidation is 0.652 mg O_2 per milligram cell mass per hour, including that required for respiration. Phenol about 387 mg/ℓ is inhibitory to utilization itself and above 5350 mg/ℓ is lethal. Ammonium chloride up to 10,000 mg/ℓ at neutral pH, free cyanide up to 10 mg/ℓ, and sulfide up to 20 mg/ℓ have no inhibitory effect on phenol biodegradation.[66] The interaction between secondary toxicants and the primary population play an important role in deciding the quality of the treated effluent. An understanding of these interactions can be useful in predicting the effluent quality in the event of fluctuating concentrations of the pollutants. It also needs a population that is not susceptible to the net residual concentration under the organic transient conditions.

Continuous cultivation of a culture with a single substrate and a monoculture can provide information on the optimum hydraulic and organic loading rate under which it can achieve maximum viability as well as the effective degradation of the substrate. *C. tropicalis* NCYC 1503 was cultivated in a chemostat as a monoculture at various hydraulic detention periods. The apparatus was operated as a once flow-through system in which, theoretically, the hydraulic detention time became the specific growth rate. Conclusions on these studies are that the yeast can have a viability between 68 and 85% at specific growth rates of 0.05 and 0.2 hr^{-1}. The substrate concentration can be maintained to below 1 mg/ℓ up to a specific growth rate of 0.1 hr^{-1}.

The interaction between a specific culture and the heterogenous population should be studied to assess the fate of such organisms in a mixed population because in biological waste water treatment, on several occasions, a particular culture can dominate others, depending upon the substrate that it receives. Continuous cultivation in open-bench model pilot plants are necessary to ascertain whether the results obtained in batch culture as well as chemostat experiments can be translated into field application. Such a study with synthetic phenolic waste and yeast as a starter seed was carried out in the laboratory. With proper recirculation of the biological mass it is possible to reach an organic loading of 2914 ± 50 mg/ℓ phenol at a hydraulic detention period of 10 ± hr^{-1} [199] The yeast population consisted of 78% of the total viable cells at an optimum loading rate of 1250 ± 100 mg/ℓ.

The methods developed with synthetic waste also need verification with actual industrial waste, in understanding the influence of other interfering substances of the waste and other secondary toxicants. Two different types of industrial wastes were experimented with the culture. Both had phenols as the major toxic pollutants. The conclusion drawn from both the cases was that the culture can be used for bioaugmentation. In the first instance, a producer gas plant effluent having the characteristics as shown in Table 5 was treated in laboratory model pilot plants. The phenol concentration could be brought down from 600—966 to 4—8 mg/ℓ, within a hydraulic detention period of 12 hr.[31] In the second case where LTC waste was used in bench model, field-scale pilot,[29] and full-scale treatment plants, the efficiency of phenol reduction remained nearly the same, as long as the concentrations of other secondary toxicants were maintained below the inhibitory concentrations for the culture.

Cyanide forms a secondary toxic pollutant in coal conversion processes. The elimination of cyanide from the waste water during biological treatment has been reported to be erratic.[200] However, it was possible to demonstrate the effective removal of alkali cyanide by biological treatment in a CMAS.[136] The presence of appropriate microbial population is essential, especially when the waste containing more than one toxic material is to be dealt with. The population must have the specific microorganisms capable of detoxifying both the pollutants. *P. acidovorans,* a cyanide-degrading culture, was cocultured with *Alcaligenes faecalis,* a phenol-degrading, cyanide-resistant culture, in batch experiments.[61] The results revealed that only after the residual phenol concentration reached below 5 mg/ℓ, cyanide degradation progressed.

Biological degradation of alkali cyanide was possible in a bench-top, continuously fed aerobic system.[201] Microbial sludge for this system was developed by the enrichment culture technique in a batch process; when cyanide degradation was evidenced the sludge was transferred to the CMAS and feeding with synthetic waste containing cyanide and other nutrient supplements was started. Studies on the microbial degradation of cyanide with glucose, acetate, peptone, or sewage showed that these sources could form carbon sources for cyanide detoxification. A microbial assay of the reactor sludge showed a cyanide resistant count ranging between 2.5 to 4.5 $\times$ 10^7/mℓ.

VII. FULL-SCALE BIOLOGICAL WASTE TREATMENT PLANTS FOR TOXIC WASTES

Cooper[162] concluded from the investigations at the British Carbonization Association Laboratory that the treatment of strong toxic wastes, with particular reference to those arising from the carbonization of coal, has led to the conclusion that biological treatment using ASP was the most efficient and successful method from the cost benefit point of view. In the U.K., ASP is the only full method of treatment which has been adopted by the carbonization industries. Of the several possible versions, the single-stage, completely mixed system with surface aerators has been in general use. Performance results of one such treatment plant receiving coke oven effluent show an efficiency between 98 to 99% reduction in phenols, 95% reduction in BOD$_5$, and 75 to 99% reduction in thiocyanate while ammonia nitrogen removal was practically negligible. (Results were obtained from the plant operators through personal communication.) The average levels in the untreated waste were phenols, 445 mg/ℓ; BOD$_5$, 738 mg/ℓ; thiocyanate, 139 mg/ℓ; and ammonia nitrogen 824 mg/ℓ. Abson and Todhunter[202] described a three-stage biological process for treatment of coke oven effluents. The first stage removed phenols and tars in the ASP; thiocyanates, thiosulfates, and cyanides were removed by the biofilter. Ammonia was removed by nitrification denitrification in the sludge tanks in the third stage. Kostenbader and Flecksteiner[158] reported the successful performance of an ASP for coke plant waste water. A two-stage biological treatment process consisting of a percolating tower followed by an ASP has been used for treating tar distillery waste in Germany.[203] The oxidation ditch has been widely accepted as a method for treating coking plant waste water. Adeema[204] described the largest oxidation ditch in the world which treated coking plant waste water and some other chemical plant waste. This has a volume of 30,000 m³ with a rotor length of 250 m. The hydraulic detention time is 3 days. Though there has been satisfactory removal of carbonaceous material, nitrogen stabilization has not been satisfactory. The explanation given by the author was that hydrogen released in the nitrification process reduced the alkalinity of water. Biczysko and Suschika[205] reported that crude phenolic waste from coking plants could be treated with a high degree of efficiency in an oxidation ditch equipped with a specific trickling filter. In addition to phenolic removal, cyanogen and sulfur compounds were also removed. However, nitrification was not reported in this paper. On a formal discussion during the conference, Ide presented the results from a CMAS process receiving phenolic wastes from a chemical treatment process of a stocking factory. The effluent contained phenol up to 6000 mg/ℓ. This waste was treated efficiently after 12-fold dilution. Coking plant effluent treatment in India has been done by any one of the processes: biological filters, activated sludge, or oxidation ditch. In an integrated steel mill where British technology is used, ASP has been in use, while a biofilter process is used in Germany. The oxidation ditch has been in use in a few industries where producer gas plants are being operated.

Process waste water from refineries is treated in a combination of physical and biological processes. In the EEC area, prior to the EEC directives of 1976, most of the refineries used gravity separators to remove oil. The waste was then let into receiving waters after dilution with ballast and cooling tower water.[40] The primary treatment process for refinery wastes is meant to remove the oil and grease that are not mixable with water. It is necessary to remove these hydrocarbon emulsions prior to biological treatment because the oil globules, if present in waste water, can form a barrier for oxygen transfer. Second, the oil can form a thin coating over the biomass, thus preventing the transport of the substrates and other nutrients. The modern refinery waste treatment flow diagram consists of an oil separator unit. This may be single or multiple units. The single unit generally employed is the API separator. In the multiple units in the modern refineries, an oil-water classifier followed by surge ponds and API separators are employed. Emulsified oil is flocculated by chemicals such as polyelectrolytes or iron salts and removed before the waste is allowed to enter into the biological process. A physical method such as steam stripping is employed for removing H_2S.[40] A biological method consisting of an aerated lagoon is also employed for removing H_2S.[197] Where low sulfur-containing oil is used for processing, this step is not applied. The waste water is taken to a combination of biofilter and activated sludge. This combination is either a biofilter followed by an activated sludge or the opposite. In one of the newest refineries in India, an activated sludge followed by biofilter is employed. In another, a roughing biofilter precedes a conventional ASP. These variations amply illustrate that there is no uniformity in the selection of the process-flow diagram. When a biofilter forms the first unit of the secondary treatment, it has often been observed to function as a sieve for the removal of oil and grease. While the ASP forms the first unit, the sludge solids contain much grease. These operational inefficiencies upset the refinery waste water treatment. Effluent from the biological processes is led to a balancing basin where other uncontaminated water from the boiler blows down and runoffs are mixed. This water is discharged into oceans or streams or is recycled through soil irrigation. A Concawe report[206] indicated that 60 to 70% of the oil refineries in the EEC are totally or partly equipped with secondary waste water treatment facilities. The CMAS system or biofilters are extensively used to treat sour water by the American petroleum industries.[207]

Biological oxidation of pollutants present in petrochemical wastes has not been fully discussed.[45] The problems are that, as no two factories wastes are similar, some form of pilot plant development is necessary to establish the critical parameters for the sizing of the biofilter or aeration basins. Petrochemical waste water is treated in biofilters or ASPs. The imperial chemical industries (ICI) in England employ the high-rate Floccor system to treat the petrochemical waste water. One of the three Indian petrochemical units treats the combined waste water in an extended aeration, biological ditch while another unit has plans to treat the waste in a CMAS process.

The application of biological processes in the pesticide industry has been accepted in regard to full scale treatment plants. Atkins[208] observed that chlorinated hydrocarbon herbicide waste was first diluted to a nontoxic level and treated in roughing filters. Parathion waste was treated in an activated sludge plant. Studies at the Monsanto parathion plant could prove that a single-stage activated sludge having high suspended solids (15,000 to 18,000 mg/ℓ) could treat the waste water.[209] These authors observed that waste blending, pH adjustment, and prechlorination for the control of odor were necessary. Full-scale ASP has been employed for the treatment in the industrial premises by the company. The treated waste is then discharged into the municipal sewer. Lue-Hing and Brady[210] employed a two-stage biological waste treatment (activated sludge) process for treating organic phosphorus pesticide waste water. The waste contained several types of pesticides.

VIII. CONCLUSIONS AND FUTURE PERSPECTIVES

An attempt has been made to critically review the causes for the deteriorating environmental conditions, as a resut of the indiscriminate discharge of toxic industrial effluents. The toxicity of the pollutants present in these wastes to aquatic and terrestrial life has been evaluated. The ability of certain microorganisms to detoxify and sometimes utilize these pollutants as carbon and energy sources has been traced through biochemical and genetic traits. The merits and demerits of application of biotechnology to the treatment of toxic industrial effluents through the ''classical approach'' have been discussed. Recent trends in bioengineering with special reference to biochemical augumentation of waste water treatment plants are highlighted. Now it is time to discuss the missing links and work out the needs in combating environmental pollution due to the discharge of toxic wastes.

If one tries to learn from the past experiences of others and avoid the mistakes committed by his predecessors, he can save time as well as money and the precious health of the present and future generations. Industrial waste treatment, especially toxic wastes in the combined waste water treatment plants, failed to eliminate these pollutants during the treatment. As a result, many of the toxic ''priority'' pollutants have accumulated and the persistence of these pollutants in the environment causes great concern. The application of the design and operational criteria developed for nontoxic or domestic waste to biological treatment of toxic industrial waste is questionable. The concepts regarding the limitations of these criteria to design engineers do not seem to be clear; especially those relating to the interactions between the microorganisms and the toxic pollutants. The physiological and biochemical properties of the population present in a biological waste treatment process depend upon the environment in which they exist. It is necessary to maintain a dynamic equilibrium between the population and the environment. An optimum viability ratio must be maintained in order to take care of the variations in organic as well as hydraulic shock loading.

Industrial waste water is complex in nature. It often may contain more than one toxic pollutant, in addition to others. The microbial community that is responsible for the degradation of one toxic pollutant should have the capacity to resist the toxic effect of the other (secondary) toxicants, only then will it be able to survive in treatment plant in the event of organic shocks. The gap in the present-day knowledge on biodegradation and application of toxic pollutants to environmental cleaning can be traced and filled by systematic studies as described in Figure 9.

Waste water characterization in reference to dominant and micropollutants of significant

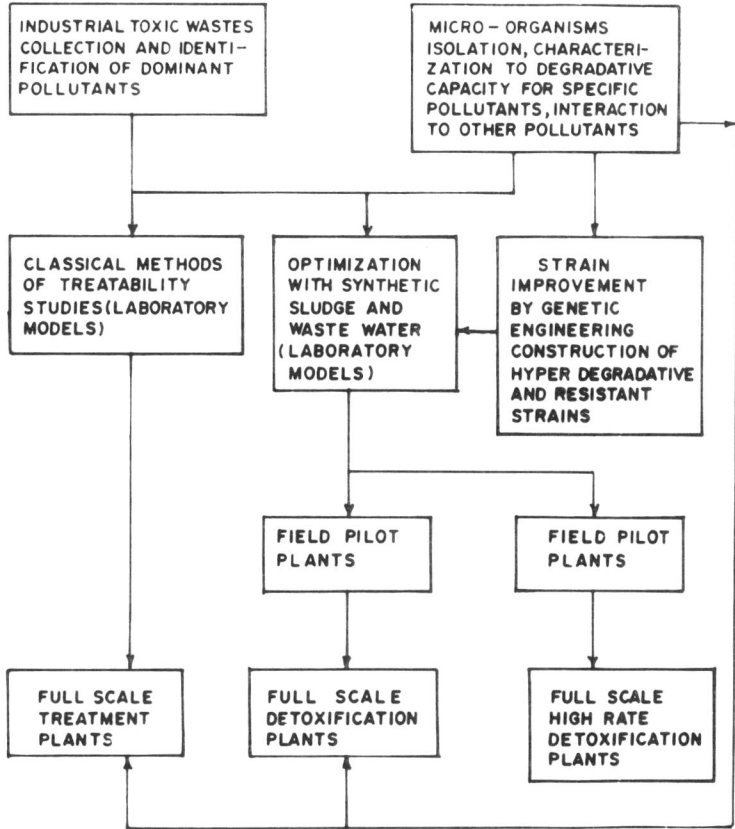

FIGURE 9. Process selection method for biological detoxification of toxic wastes.

nature is necessary as the ''surrogate tests'' have failed to indicate the elimination of these pollutants. Once the characteristics are known, a treatment flow sheet for the waste water incorporating the various treatment modes can be established. Biological treatment modes consisting of one or more units should be suitably incorporated in the flow diagram so that these modes are not affected by other processes. For example, if chemical methods precede biological methods it can reduce the efficiency of the latter by the presence of excess chemical additives.

Isolation of specific microorganisms for the degradation of the specific pollutants, characterization of these organisms with reference to biodegradation kinetics, and the influence of secondary toxicants on the biodegradation of the specific pollutant are the ideal steps in the systematic approach. The advantage of this approach over the classic acclimation procedure is that the former procedure has better reproducibility than the latter. In the acclimated sludge, the identity of the microbial population is unknown while it is possible to synthesize a sludge with known microbial entities. Optimization of biological systems, using the synthetic sludge with synthetic and live waste (if one is available) in the laboratory and field-scale pilot plants can produce design and operational parameters for full-scale treatment plants. Biological treatment processes with high rates of transformation are ideal for detoxification of toxic constituents. These high-rate systems with optimum specific growth rates can maintain high viability ratios. These plants can be used at the end of the pipeline in the industrial premises.

Genetic engineering for the construction of gene banks in the specific culture will be necessary to store the information. The bank should have information to be transcribed not only for the synthesis of degradative proteins but also for resistant proteins for secondary toxicants. The hyperdegradative strains can be used in capsule reactors for detoxification, possibly in a chain. Microbial technology can serve mankind in conserving energy for future use.

ACKNOWLEDGMENTS

We express our gratitude to Prof. B. B. Sundaresan, former Director of NEERI (presently Vice-Chancellor at Madras University, Madras), for the valuable suggestions made during the preparation of this chapter. We gratefully express our thanks to Mr. K. R. Bulusu, Director of NEERI, Nagpur, for the permission to present the review on the subject and for availing the secretarial facility. We received some literature from Prof. A. F. Gaudy Jr., Prof. C. W. Randall, and Concawe (Netherlands), who were very useful in the preparation of this review. We sincerely thank them.

REFERENCES

1. **Stanbridge, H. H.,** History of sewage purification in Britain, in *Water Pollution Control,* Maidstone, Kent, England, 1977, 4.
2. **Davis, J. G.,** Microbial aspects of pollution; some general development, in *Microbial Aspects of Pollution,* Sykes, G. and Skynner, F. A., Eds., Academic Press, London, 1971, 4.
3. Document for the United Nations Conference on the Human Environment, Stockholm, June 15 to 16, 1972, Part 1; State Department, Washington, D.C., March 1972, 348.
4. Shrubsale, V. J., Inland water pollution and disposal of solid wastes, in *Industrial Effluent Treatment,* Walter, J. K. and Coint, A., Eds., Applied Science, London, 1981, 325.
5. **Lester, W. F. and Chem, C.,** Implementation of the control of pollution Act 1974, *Water Pollut. Control,* 79, 165, 1980.
6. **Vaughn, W. A.** Improving the effectiveness of environmental management, in *Proc. 37th Industrial Waste Conference,* Purdue University, Bell, E. M., Ed. Ann Arbor Science, Ann Arbor, Mich., 1982, 885.
7. **Gaudy, A. F., Jr., Stein, M., Ettinger, M. G., Powers, T. J., Sawer, C. N., and Svore, J. H.,** Symp. joint vs. separate treatment of municipal and industrial wastes, *J. Water Pollut. Control Fed.,* 36, 345, 1964.
8. **Feliciano, D. V.,** Multipurpose projects stretching the federal dollars, *J. Water Pollut. Control Fed.,* 51, 2347, 1979.
9. **Bosselievre, E. B. and Schwartz, M.,** Role of federal government in pollution control, in *Treatment of Industrial Wastes,* McGraw-Hill, New York, 1976, chap. 9.
10. **Stoddart, F. W.,** *The Continuous Sewage Filter,* J. Wright and Co., Bristol, 1901.
11. **Bruce, A. M. and Macmillion, S. C.,** Research development in high rate biological filtration, *J. Inst. Public Health Eng. (India),* 178, 1970.
12. **Chipperfield, P. N. J., Askew, M. W., and Benton, J. A.,** Multistage plastic media treatment plants, in *Proc. 25th Industrial Waste Conference,* Purdue University, Ann Arbor Science, Ann Arbor, Mich., 1970, 214.
13. **Borchart, J. A.,** Biological waste treatment using rotating discs, *Biotechnol. Bioeng. Symp.,* (2nd Series), Canale, R. P., Ed., Wiley-Interscience, New York, 1971, 131.
14. **Stanbridge, H. H.,** History of sewage purification in Britain, 7- Activated sludge process, in *Water Pollution Control,* Maidstone, Kent, England, 1977, 9.
15. **Arden, E. and Lockett, W. T.,** Experiments on the oxidation of sewage without the aid of filters, *J. Soc. Chem. Ind.,* 33, 523, 1914.
16. **Caink, T.,** The activated sludge process for sewage purification. Experiments at Worcester, in *Proc. Assoc. Managers Sewage Disposal Works,* 1917, 34.
17. **Coomps, E. P.,** Air diffusers, their history and use in activated sludge processes, *J. Proc. Inst. Sew. Purif. (England),* 4, 304, 1955.
18. **Forster, C. F.,** Bio-oxidation, in *Treatment of Industrial Effluents,* Cally, A. G., Forster, C. F., and Stafford, D. A., Eds., Hodder and Stoughton, London, 1977, chap. 5.

19. **Lister, A. R. and Boon, A. G.,** Aeration in deep tank — an evaluation of a fine bubble diffused air system, *Water Pollut. Control,* 72, 590, 1973.
20. **Bolton, D. H., Bonchard, J. B., and Hines, D. A.,** Application of ICI deep shaft process for industrial effluents, in *Proc. 31st Industrial Waste Conference,* Purdue University, Bell, J. M., Ed., Ann Arbor Science, Ann Arbor, Mich., 1976, 344.
21. **Kessenger, H. J. N.,** Small bio-aeration plants, in *Proc. Assoc. Managers, Sewage Disposal Works,* England, 1930, 90.
22. **Pasveer, A.,** A contribution to the development of activated sludge process, *J. Proc. Inst. Sew. Purif. (England),* 4, 436, 1959.
23. **Knop, E. and Kalbokopf,** Energy and hydraulic test systems on mechanical aeration systems, in *Proc. 4th Int. Conf. Water Pollut. Research, Prague,* Jenkins, S. H., Ed., Pergamon Press, Elmsford, N.Y., 1969, 497.
24. **Forster, C. F.,** A comparison of the performance achieved by carrousel and mammoth rotors version in oxidation ditches, *Environ. Technol. Lett.,* 1, 366, 1980.
25. **Stover, E. L. and Kincannon, D. F.,** Biological treatability of specific organic compounds found in chemical industry waste water, in *Proc. 37th Industrial Waste Conference,* Purdue University, Bell, J. M., Ed., Ann Arbor Science, Ann Arbor, Mich., 1982, 1.
26. **Russel, L. L., Cain, and Jenkin, D. T.,** Impact of priority pollutants of POTW processes, in *Proc. 37th Industrial Waste Conference,* Purdue University, Bell, J. M., Ed., Ann Arbor Science, Ann Arbor, Mich., 1982, 876.
27. **Fritch, J. R.,** Energy 2000-2020, paper presented at *Third World Energy Conference Commission,* New Delhi, May 1, 1983, 1.
28. **Laska, R., Richer, D., Mayers, D., Ballerstero, J., Smith, L., and Lawrence, C.,** Research Highlights, Report 600/9-77-049, U.S. Environmental Protection Agency, Washington, D.C., 1977, 14.
29. **Kumaran, P., Kaul, S. N., Pandey, R. A., Choudhary, K. R., Swarnakar, N. G., Parhad, N. M., and Raman, V.,** Detoxification of phenol in LTC waste water, *Tech. Annu., Indian Assoc. Water Pollut. Control,* 10, 1, 1983.
30. **Luthy, R. G.,** Treatment of coal coking and coal gasification waste waters, *J. Water Pollut. Control Fed.,* 53, 325, 1981.
31. **Kumaran, P. and Parhad, N. M.,** Treatment of coal carbonization waste, *J. Inst. Public Health Eng. (India),* 1979, 10, 1979.
32. **Parhad, N. M., Kumaran, P., and Shivaraman, N.,** Microbial degradation of waste waters from the manufacture of metallurgical and domestic coke, in *Proc. Natl. Workshop on Microbial Degradation of Industrial Wastes,* Sundaresan, B. B., Kumaran, P., Joshi, S. R., and Shivaraman, N., Eds., NEERI, Nagpur, India, 1981, 78.
33. **Graham, P. W.,** Kinetic aspects of the treatment of phenolic wastes, in *Advances in Water Pollution Research,* Jenkins, S. H., Ed., Pergamon Press, London, 1969, 447.
34. **Cooper, R. L. and Wheatstone, E. C.,** Determination of phenolics in aqueous effluents, *Water Res.,* 7, 1375, 1973.
35. **Pearce, A. S. and Punt, S. E.,** Biological treatment of liquid toxic wastes, *Effluent Water Treat. J.,* 15, 32, 1975.
36. **Walters, R. W. and Luthy, R. G.,** Liquid/suspended solid phase partitioning of polycylic aromatic hydrocarbons in coal coking waste waters, *Water Res.,* 18, 795, 1984.
37. **Catchpole, J. R. and Stafford, D. A.,** *Treatment of Industrial Effluents,* Callely, A. G., Foster, C. F., and Stafford, D. A., Eds., Hodder and Stroughton, London, 1977, chap. 16.
38. **Ashmore, A. G., Catchpole, J. R., and Cooper, R. L.,** Biological treatment of coal carbonization effluent. II. Studies on the influence of liquor composition, *Water Res.,* 2, 555, 1968.
39. **Rodd, E. H.,** *Chemistry of Carbon Compounds,* Vol. 3, Part A, Elsevier, New York, 1954. 475.
40. **Paulins, C. D. A.,** The environmental impact of refinery effluent. II. Aqueous effluents discharged by oil refinery, in Concawe, The Hague, Netherlands, 1980, 3.
41. **Doctor, T. R.,** Microbial degradation of waste water from petroleum refinery, in *Proc. Natl. Workshop on Microbial Degradation of Industrial Wastes,* Sundaresan, B. B., Kumaran, P., Joshi, S. R., and Shivaraman, N., Eds., NEERI, Nagpur, India, 1981, 78.
42. Development Document for Effluent Limitations Guide Lines for Petroleum Refinery Point Source, Report 440/1-74-014-a, U.S. Environmental Protection Agency, Washington, D. C., 1984.
43. **Devos, L. L., Bridie, A. L., and Herberg, S.,** Analysis of oil contaminated ground water, H_2O, 10, 277, 1977.
44. **Chakraborti, R. N.,** Microbial degradation of petrochemical wastes, in *Proc. Natl. Workshop on Microbial Degradation of Industrial Wastes,* Sundaresan, B. B., Kumaran, P., Joshi, S. R., and Shivaraman, N., Eds., NEERI, Nagpur, India, 1981, 45.
45. **Cox, A. P.,** The petrochemicals and resins industry, in *Treatment of Industrial Effluents,* Callely, A. G., Forster, C. F., and Stafford, D. A., Eds., Hodder and Stroughton, London, 1977, chap. 13.

46. **Davey, D. C. and Harkness, N.,** Some effluents from the manufacture and use of synthetic resins and polymers, *Effluent Water Treat. J.,* 11, 320, 1971.
47. **Srivastava, B. P.,** Research and development activities of pesticide industry with particular reference to India, *Chem. Age India,* 32, 564B, 1981.
48. **Krishnamurti, C. R.,** India's bloom in chemicals, *World Health,* Aug.-Sept., 18, 1984.
49. **Sundaresan, B. B., Subrahamiam, P. V. R., and Bhide, A. D.,** Toxic and hazardous waste scene in India, paper presented at 2nd Int. Symp. on Solid and Hazardous Wastes, Philadelphia, March 6 to 11, 1983.
50. **Boltred, D. C. and Smith, R. F.,** Integrated pest management, *Environ. Sci. Technol.,* 16, 284A, 1982.
51. **Mrinalini, M.,** Pesticide hazards — a growing global problem, *Chem. Age India,* 35, 347, 1984.
52. **Wilson, D. C.,** Lessons from Seveso, *Chem. Ind. News,* 27, 353, 1982.
53. **Shrinde, R. T., Caskey, J. M., and Gillespie, C. K.,** Detection and quantitative estimation of synthetic organic pesticide by chromatography, *J. Am. Water Works Assoc.,* 54, 1407, 1962.
54. **Randal, C. W.,** Toxicity of organophosphate insecticides to fresh water microorganisms, in *Proc. 2nd Am. Water Resources Conference,* Chicago, November 20 to 22, 1966, 352.
55. **Davis, S. T., Wright, A. T., and Coates, H. A.,** Status of US-EPA program to control toxic industrial wastes, in *Proc. 32nd Industrial Waste Conference,* Purdue University, Bell, J. M., Ed., Ann Arbor Science, Ann Arbor, Mich., 1977, 438.
56. **Jackson, S. and Brown, V. M.,** Effect of toxic wastes on treatment processes and watercourses, *Water Pollut. Control,* 69, 292, 1970.
57. **Abson, J. W. and Todhunter, K. W.,** Factors effecting biological treatment of coal carbonization effluent, *Gas World,* 149, 61, 1959.
58. **Jones, G. L. and Carrington, E. G.,** Growth of pure and mixed cultures of microorganisms concentration in the treatment of coal carbonization liquors, *J. Appl. Bacteriol.,* 35, 395, 1972.
59. **Tomlinson, T. G., Boon, A. G., and Trotman, C. N. A.,** Inhibition of nitrification in the activated sludge process of sewage disposal, *J. Appl. Bacteriol.,* 29, 266, 1966.
60. **Stafford, D. A.,** The effect of phenol and heterocyclic bases on nitrification in the activated sludges, *J. Appl. Bacteriol.,* 37, 75, 1974.
61. **Shivaraman, N.,** Microbial Degradation of Cyanide in Cyanide Bearing Wastes, Ph.D. thesis, Nagpur University, Nagpur, India, 1981.
62. **Jones, G. L., Jansen, F., and Mekay, A. J.,** Substrate inhibition of growth of bacterium NCIB-8250 by phenol, *J. Gen. Microbiol.,* 74, 139, 1973.
63. **Powlowsky, U. and Howel, J. A.,** Mixed culture biooxidation of phenol. I. Determination of kinetic parameters, *Biotechnol. Bioeng.,* 15, 889, 1973.
64. **Hill, G. A. and Robinson, C. W.,** Substrate inhibition kinetics: phenol degradation by *Pseudomonas putida, Biotechnol. Bioeng.,* 17, 1599, 1975.
65. **Yang, R. D. and Humphry, A. E.,** Dynamic steady state studies of phenol biodegradation in pure and mixed cultures, *Biotechnol. Bioeng.,* 17, 1211, 1975.
66. **Kumaran, P. and Parhad, N. M.,** Isolation and characterization of a phenololytic yeast, NCYC 1503, *Indian J. Environ. Health,* 26, 112, 1984.
67. **Brown, V. M., Jordan, D. H. M., and Tittler, B. A.,** The acute toxicity to rainbow trout of fluctuating concentration of ammonia, phenol and zinc, *J. Fish Biol.,* 1, 1, 1969.
68. **Brown, V. M. and Dalton, R. A.,** The acute lethal toxicity to rainbow trout of mixtures of copper, phenol, zinc and nickel, *J. Fish Biol.,* 2, 211, 1970.
69. European Inland Fisheries Advisory Commission, Water quality criteria for European fresh water fish, in Report on Monohydric Phenols and Inland Fisheries. Tech. Paper 15, EIFAC, 1972, 18.
70. **Mann, H.,** Effect of flavour of fish by oils and phenols, in *Commun. Inter. Explor. Sci. Mer. Medil. Symp. Pollut. Mar. Par. Microorganism. Prod. Petrol.,* Monaco, Italy, 1984, 371.
71. **Barker, F. L., Landrigan, P. J., Bertozze, P. E., Patricia, H. F., and Skinner, H. G.,** Phenol poisoning due to contaminated drinking water, *Arch. Environ. Health,* 33, 89, 1978.
72. **Warren, K. S.,** Ammonia toxicity and pH, *Nature,* 195, 47, 1962.
73. **Huber, L.,** Environmental impact of refinery effluent, in Concawe, The Hague, Netherlands, 5/79, 1980, IV-1.
74. **Anthonison, A. S., Loeher, R. C., Prakasam, T. B. S., and Srinath, E. G.,** Inhibition of nitrification by ammonia and nitrous acid, *J. Water Pollut. Control Fed.,* 48, 835, 1976.
75. **Ford, D. L., Ehurchwall, R. L., and Kachtick, J. W.,** Comprehensive analysis of nitrification of chemical processing waste water, *J. Water Pollut. Control Fed.,* 52, 2726, 1980.
76. **Brebion, G., Gabridene, R., and Huriet, B.,** Biology of cyanide: possibilities of treatment of waste water containing cyanides — analytical technique, *Eau,* 53, 463, 1966.
77. **Parthasarathy, N. V.,** A survey of methods for treatment of effluents in electroplating industry, *Environ. Health,* 11, 358, 1969.

78. **Downing, A. L., Tomlinson, T. G., and Truesdale, G. A.,** Effect of inhibitors on nitrification in activated sludge process, VI, *J. Inst. Sewage Purif. (England)*, 537, 1964.
79. **Shivaraman, N., Kumaran, P., and Parhad, N. M.,** Phenol biodegradation by *Candida tropicalis* and influence of other toxicants, *Indian J. Environ. Health*, 20, 101, 1978.
80. **Zehnpfennig, R. G.,** Possible toxic effects of cyanates, thiocyanates, ferricyanides, ferrocyanides and chromates discharged to surface water, in *Proc. 22nd Industrial Waste Conference*, Purdue University, West Lafayette, Ind., 1967, 879.
81. **Mckee, J. E. and Wolf, H. N., Eds.,** *Water Quality Criteria, 2nd ed.*, Department of Health, Education and Welfare, 1963, 3-A, 137.
82. **Newfeld, R. D. and Valiknac, T.,** Inhibition of phenol biodegradation by thiocyanate, *J. Water Pollut. Control Fed.*, 51, 2283, 1979.
83. **Jones, J. R. E.,** *Fish and River Pollution*, Butterworths, London, 1964, 203.
84. **Bronzes, R. J. P.,** Fish toxicity with reference to pulp and paper industry — economic and technical review reports EPS-3WP-76-4, *Environ. Canada*, 81, 1976.
85. **Bertha, R. and Atlas, R. M.,** Microbiology of aquatic oil spills, *Adv. Appl. Microbiol.*, 22, 225, 1977.
86. **Lin, D. L. and Dutka, B. J.,** Biological oxidation of hydrocarbon in aquatic phase, *J. Water Pollut. Control Fed.*, 45, 232, 1973.
87. **Patrick, R. and Whipple, W.,** Petroleum industry in Delaware estuary, National Science Foundation, Ranne programme, in Concawe, The Hague, Netherlands, 1977, 440.
88. **Winters, K., Batterton, J. C., and Balen, C. V.,** Phanelen-1-one: occurrence in a fuel oil and toxicity to microalgae, *Environ. Sci. Technol.*, 11, 270, 1977.
89. **Swaminathan, T. and Shivaraman, N.,** Microbial degradation of waste waters generated in the manufacture and application of pesticides, in *Proc. Natl. Workshop on Microbial Degradation of Industrial Wastes*, Sundaresan, B. B., Kumaran, P., Joshi, S. R., and Shivaraman, N., Eds., NEERI, Nagpur, India, 1981, 259.
90. **Henderson, C., Pickering, Q. H., and Tarzwell, C. M.,** Relative toxicity of ten chlorinated hydrocarbon pesticides to four species of fish, *Trans. Am. Fish. Soc.*, 88, 23, 1959.
91. **Livingston, R. J.,** Review of current literature concerning the acute and chronic effects of pesticides on aquatic organisms, *Crit. Rev. Environ. Controls*, 4, 325, 1977.
92. **Sherrard, J. H., Sujarittanonta, S., Webber, A. S., and Bagby, M. M.,** The influence of cadmium and nickel on activated sludge performance, paper presented at the Am. Soc. Civ. Engineers Convention and Exposition, Atlanta, October 23 to 25, 1979.
93. **Shelly, C. R. and Sherrard, J. H.,** Cadmium shock nontoxic to organic removal, *Water Sewage Works*, 126(5), 50, 1979.
94. **Webber, A. S. and Sherrard, J. H.,** Effect of cadmium shocking on the completely mixed activated sludge process, *J. Water Pollut. Control Fed.*, 52, 2378, 1980.
95. **Sujarittanonta, S. and Sherrard, J. H.,** Activated sludge nickel toxicity studies, *J. Water Pollut. Control Fed.*, 53, 1314, 1981.
96. **Bagby, M. M. and Sherrard, J. H.,** Combined effects of cadmium and nickel on the activated sludge process, *J. Water Pollut. Control Fed.*, 53, 1609, 1981.
97. **Dagley, S.,** Catabolism of aromatic compounds by microorganisms, in *Advances in Microbial Physiology*, Vol. 6, Rose, A.-H. and Wilkins, J. F., Eds., Academic Press, London, 1971, 1.
98. **Daughton, C. G. and Hsieh, D. P. H.,** Parathion utilization by bacterial symbionts in a chemostat, *Appl. Environ. Microbiol.*, 34, 175, 1977.
99. **Barik, S., Siddaramappa, R., and Sethunathan, N.,** Metabolism of nitrophenol by bacteria isolated from parathion amented flooded soil, *Antonie van Leeuwenhoek; J. Microbiol. Serol.*, 42, 461, 1976.
100. **Evans, W. C.,** Oxidation of phenol and benzoic acid in soil bacteria, *Biochem. J.*, 41, 373, 1947.
101. **Newjahr, N. V. and Gaal, A.,** Phenol hydroxylase from yeast — purification and properties of the enzyme from *Trichosporon cutaneum*, *Eur. J. Biochem.*, 35, 386, 1973.
102. **Fewson, C. A.,** Biodegradation of aromatics with industrial relevance, in *Microbial Degradation of Xenobiotics and Recalcitrant Compounds*, (FEMs Symp. Ser. 12), Leisinger, T. Hütter, R., Cook, A. M., and Nüesch, J., Eds., Academic Press, London, 1981, 141.
103. **Kearney, D. C. and Kaufman, D. D.,** *Degradation of Pesticides*, Marcel Dekker, New York, 1969, 1.
104. **Torgson, D. C.,** *Fungicides*, Vol. 1, Academic Press, London, 1967, 1.
105. **Obrien, R. P.,** *Insecticides Action and Metabolism*, Academic Press, London, 1971.
106. **Cook, A. M., and Hütter, B.,** Biodegradation of striazene, in *Microbial Degradation of Xenobiotics and Recalcitrant Compounds*, (FEMs Symp. Ser. 12), Leisenger, T., Hütter, R., Cook, A. M., and Nüesch, J., Eds., Academic Press, London, 1981, 237.
107. **Eberspacher, J. and Lingens, F.,** Microbial degradation of herbicide chloridazon, in *Microbial Degradation of Xenobiotics and Recalcitrant Compounds*, (FEMs Symp. Ser. 12), Leisenger, T., Hütter, R., Cook, A. M., and Nüesch, J., Eds., Academic Press, London, 1981, 271.

108. **Munnecke, D. M.,** The use of microbial enzymes for pesticide detoxification, in *Microbial Degradation of Xenobiotics and Recalcitrant Compounds,* (FEMs Symp. Ser. 12), Leisenger, T., Hütter, R., Cook, A. M., and Nüesch, J., Eds., Academic Press, London, 1981, 251.

109. **Golovleva, L. A. and Skryabin, G. K.,** Microbial degradation of DDT, in *Microbial Degradation of Xenobiotics and Recalcitrant Compounds,* (FEMs Symp. Ser. 12), Leisenger, T., Hütter, R., Cook, A. M., and Nüesch, J., Eds., Academic Press, London, 1981, 291.

110. **Dagley, S.,** A biochemical approach to some problems of environmental pollution, *Essays Biochem.,* 11, 81, 1975.

111. **Wigmore, B. J. and Bayly, R. C.,** A partial order of genes determining enzymes of metacleavage pathway, *J. Gen. Microbiol.,* 100, 81, 1977.

112. **Williams, P. A. and Murray, K.,** Metabolism of benzoate and methyl benzoate in *Pseudomonas putida* (aruvilla) mt-2, *J. Bacteriol.,* 120, 416, 1974.

113. **Christopher, F., Franklin, H., Bagdasasarian, M., and Timmis, K. N.,** Manipulation of degradative genes in soil bacteria, in *Microbial Degradation of Xenobiotics and Recalcitrant Compounds,* (FEMs Symp. Ser. 12), Leisinger, T., Cook, A. M., Hütter, R., and Nüesch J., Eds., Academic Press, London, 1981, 109.

114. **Benson, S., Oppici, M., Shapiro, J., and Fennewald, M.,** Regulation of membrane peptides by *Pseudomonas* plasmid *alk.* regulon, *J. Bacteriol.,* 140, 754, 1979.

115. **Fiest, C. F. and Hegeman, G. D.,** Phenol and benzoate metabolism by *P. putida* — regulation of tangential pathways, *J. Bacteriol.,* 100, 869, 1969.

116. **Williams, P. A.,** Genetics of biodegradation, in *Microbial Degradation of Xenobiotics and Recalcitrant Compounds,* (FEMs Symp. Ser. 12), Leisinger, T., Cook, A. M., Hütter, R., and Nüesch, J., Eds., Academic Press, London, 1981, 97.

117. **Evans, W. C.,** Biochemistry of bacterial catabolism of aromatic compounds in anaerobic environment, *Nature,* 270, 17, 1977.

118. **Cain, R. B.,** Microbial degradation of surfactants and builder compounds, in *Microbial Degradation of Xenobiotics and Recalcitrant Compounds,* (FEMS Symp. Ser. 12), Leisinger, T., Cook, A. M., Hütter, R., and Nüesch, Jr., Eds., Academic Press, London, 1981, 325.

119. **Painter, H. A.,** Microbial transformation of inorganic nitrogen, *Prog. Water Technol.,* 8, 3, 1977.

120. **Focht, D. D. and Verstraete, W.,** Biochemical ecology of nitrification and denitrification, *Adv. Microbial Physiol.,* 135, 1977.

121. **Funk, B. B., Krulurich, T. A., and Tuttman, H. N.,** Effect of biotin on *Nitrobacter, agilis, Bacteriol,* in *Proc. 64th Annu. Symp. Soc. Microbiology,* May 3 to 7, 1964.

122. **Aleem, M. I. H.,** Oxidation of nitrogen compounds, *Annu. Rev. Plant Physiol.,* 21, 67, 1970.

123. **Aleem, M. I. H., Hock, G. E., and Vanner, J. E.,** Water as a source of oxidant and reductant in bacterial chemosynthesis, *Proc. Natl. Acad. Sci. U.S.A.,* 54, 869, 1965.

124. **Comly, H. H.,** Cyanosis in infants caused by nitrates in well water, *JAMA,* 129, 112, 1945.

125. **Payne, W. J.,** Reduction of nitrogenous oxides by microorganisms, *Bacteriol. Rev.,* 37, 409, 1973.

126. **Ishaque, M. and Aleem, M. I. H.,** Intermediates of denitrification in chemoautotrophic *Thiobacillus denitrificans, Arch. Microbiol.,* 94, 269, 1973.

127. **Terai, N. and Mori, T.,** Studies on phosphorylation coupled with denitrification and aerobic respiration in *Pseudomonas denitrificans, Bot. Mag. (Tokyo),* 88, 231, 1975.

128. **Cox, A. D., and Payne, W. J.,** Separation of soluble denitrifiying enzymes and cytochromes from *Pseudomonas perfectomarinus, Can. J. Microbiol.,* 19, 861, 1973.

129. **Taylor, B. F., Campbell, W. L., and Chinoy, I.,** Anaerobic degradation of benzene nucleus by facultative anerobic microorganisms, *J. Bacteriol.,* 102, 430, 1970.

130. **Williams, R. J. and Evans, W. C.,** The metabolism of benzoate by *Moraxella.* sp. through anaerobic nitrate respiration, *Biochem. J.,* 148, 1, 1975.

131. **Downey, R. J., Kiszkiss, D. F., and Nuner, J. H.,** Influence of oxygen on development of nitrate respiration in *Bacillus stearothermophilus, J. Bacteriol.,* 98, 1056, 1969.

132. **Happold, F.C., Johnstone, K. I., Rogers, H. J., and Yowatt, J. B.,** The isolation and characteristics of an organism oxidizing thiocyanate, *J. Gen. Microbiol.,* 10, 261, 1954.

133. **Jones, G. L. and Carrington, E. G.,** Growth of pure and mixed cultures of microorganisms in the treatment of coal carbonization liquors, *J. Appl. Bacteriol.,* 35, 395, 1972.

134. **Stafford, D. A. and Callely, A. G.,** The utilization of thiocyanate by a heterotrophic bacterium, *J. Gen. Microbiol.,* 55, 285, 1969.

135. **Yowatt, J. B.,** Studies on the metabolism of *Thiobacillus thiocyanoxidans, J. Gen. Microbiol.,* 11, 139, 1954.

136. **Shivaraman, N., Kumaran, P., Pandey, R. A., Chatterjee, S. K., Choudhary, K. R., and Parhad, N. M.,** Microbial degradation of thiocyanate, phenol and cyanide in a completely mixed aeration system, *Environ. Pollut. A.,* 39, 141, 1985.

137. **Stafford, D. A.,** The metabolic control of biooxidation of coal carbonization effluent in the activated sludge treatment system, *Coke Oven Managers Yearbook,* 1976, 1.

138. **Knowles, C. J.,** Microorganisms and cyanide, *Bacteriol. Rev.,* 40, 652, 1976.

139. **Castric, P. A. and Strobel, G. A.,** Cyanide metabolism by *Bacillus megatherium, J. Biol. Chem.,* 244, 4089, 1969.

140. **Strobel, G. A.,** The fixation of hydrocyanic acid by a phychrophilic basidiomycete, *J. Biol. Chem.,* 241, 2618, 1966.

141. **Strobel, G. A.,** 4-Amino-4-cyanobutyric acid as an intermediate in glutamate biosynthesis, *J. Biol. Chem.,* 242, 3265, 1967.

142. **Mundy, B. P., Lin, F. H. S., and Strobel, G. A.,** Alpha amino butyronitrile as an intermediate in cyanide fixation by *Rhizoctonia solani, Can. J. Biochem.,* 51, 1440, 1973.

143. **Fry, W. E. and Miller, R. L.,** Cyanide degradation by an enzyme from *Stemphytium loti, Arch. Biochem. Biophys.,* 151, 468, 1972.

144. **Fry, W. E. and Munch, D. C.,** Hydrogen cyanide detoxification by *Gleocercospora sorghi, Physiol. Plant Pathol.,* 7, 23, 1975.

145. **McChesney, C. A.,** Occurrence of rhodonese in species of *Thiobacillus, Nature,* 181, 347, 1958.

146. **Villarejo, M. and Westley, J.,** Sulphur metabolism of *Bacillus subtilis, Biochim. Biophys. Acta,* 117, 209, 1966.

147. **Atkinson A., Evans, C. G. T., and Yeo, R. G.,** Behaviour of *Bacillus stearothermophilus* grown in different media *J. Appl. Bacteriol.,* 38, 301, 1975.

148. **Yoch, D. C. and Lindstrom, E. S.,** Survey of the photosynthetic bacteria for rhodonese activity, *J. Bacteriol.,* 106, 700, 1971.

149. **Barton, C. P. and Akagi, J. M.,** Observations on the rhodonase activity of *Desulfotomaculum nigricans, J. Bacteriol.,* 107, 375, 1971.

150. **Hardy, R. W. F. and Knight, R., Jr.,** ATP-dependent reduction of azide and HCN by N$_2$-fixing enzymes of *Azotobacter vinalandi* and *Clostridium pasteurianum, Biochim. Biophys. Acta,* 139, 69, 1967.

151. **Frankland, R. E. and Silvester, H. J.,** The bacterial purification of sewage containing large proportion of spent gas liquor, *J. Soc. Chem. Ind.,* 26, 231, 1907.

152. **Mohlman, F. W.,** The biological oxidation of phenolic wastes, *Am. J. Public Health,* 19, 145, 1929.

153. **Mathews, W. W.,** Treatment of ammonia still waters by activated sludge process, *Sewage Ind. Waste,* 24, 164, 1952.

154. **McKinney, R. E., Tomlinson, H. D., and Wilcox, R. L.,** Metabolism of aromatic compounds by activated sludge, *Sewage Ind. Waste,* 28, 547, 1956.

155. **Kremer, J. G., Glasgow, D., and Dryden, F. D.,** Industrial waste surcharge program in Los Angeles County, *J. Water Pollut. Control Fed.,* 51, 2626, 1979.

156. **Vath, C. A. and Advani, R. K.,** Expanding a POTW to accommodate industrial waste waters — a case study, in *Proc. 33rd Industrial Waste. Conference,* Purdue University, Bell, J. M., Ed., Ann Arbor Science, Ann Arbor, Mich., 1979, 1.

157. **Hing, C. L., Lordi, D. T., Staneley, W., and Whiteblooms, S. W.,** Energy response to hazardous waste spills in metropolitan sanitary districts of Greater Chicago, in *Proc. 35th Industrial Waste Conference,* Purdue University, Bell, J. M., Ed., Ann Arbor Science, Ann Arbor, Mich., 1980, 746.

158. **Kostenbader, P. D. and Flecksteiner, J. W.,** Biological oxidation of coke plant weak ammoniacal liquor, *J. Water Pollut. Control Fed.,* 41, 199, 1969.

159. Monitoring of the Environment in United Kingdom — A Report by the Central Unit on the Environmental Pollution, Pollution Paper 1, Her Majesty's Stationery Office, London, 1974, 44.

160. **Kalweit, H.,** Water quality station — River Rhine, Mayence Wiesbaden, *Prog. Water Technol.,* 9(5/6), 167, 1977.

161. **Gaudy, A. F., Jr. and Kincannon, D. F.,** Toxic pollutants and POTW, paper presented at the 2nd Annu. Meet. Pollution Control Association of Oklahoma, Fountainhead Lodge, Okla., April 17 to 18, 1979, 1.

162. **Cooper, R. L.,** The treatment of toxic effluents: investigations at BCRA, *Effluent Water Treat. J.,* 17, 230, 1977.

163. **Kabler, P. W., Chamber, C. W., and Taback, H. H.,** Degradation of aromatic compounds by phenol adapted bacteria, *J. Water Pollut. Control Fed.,* 35, 1517, 1963.

164. **Prakasam, T. B. S. and Dondero, N. C.,** Aerobic heterotrophic bacterial population of sewage activated sludge. IV. Adaptation of sludge to utilization of aromatic compounds, *Appl. Microbiol.,* 19, 663, 1970.

165. **Venkata Rao, B. and Bhat, J. M.,** Characteristics of yeasts isolated from phenol and catechol adapted sludge, *Antonie van Leeuwenhoek; J. Microbiol. Serol.,* 37, 303, 1971.

166. **Tatsuo, S., Kazuyazu, A., Manabu, F., Nisaburo, N., and Kunisuke, I.,** Basic decomposition parameters of phenol by *Candida tropicalis, J. Ferment. Technol. (Japan),* 51, 803, 1973.

167. **Beltrame, P., Beltrame, P. L., Carniti, P., and Pitea, D.,** Kinetics of phenol degradation by activated sludge process in a continuous stirred reactor, *J. Water Pollut. Control Fed.,* 52, 126, 1980.

168. **Monod, J. M.,** La technique de culture continue: theorie et application, *Ann. Inst. Pasteur,* 79, 390, 1950.

169. **Novik, K. and Slizard, L.,** Experiments with chemostat on spontaneous mutations of bacteria, *Proc. Natl. Acad. Sci.,* 36, 708, 1950.

170. **Herbert, D., Elsworth, R., and Telling, R. C.,** The continuous culture of bacteria; a theoretical and experimental study, *J. Gen. Microbiol.,* 14, 601, 1956.

171. **Kumaran, P., Kaul, S. N., Pandey, R. A., Deshpandey, V. V., Shivaraman, N., Parhad, N. M., and Raman, V.,** Phenol removal by pilot plant activated sludge, *Tech. Annu. Indian Assoc. Water Pollut. Control,* 9, 16, 1982.

172. **Mckinney, R. E.,** The value and use of mathematical models for activated sludge systems, *Prog. Water Technol.,* 7(1), 17, 1975.

173. **Mckinney, R. E.,** Mathematics of complete mixing activated sludge. III, *Trans. Am. Soc. Civ. Eng.,* 128, 497, 1963.

174. **McKinney, R. E. and Ooten, R. J.,** Concepts of completely mixing inactivated sludge, in Trans. 19th Annu. Conf. of Sanitary Engineering, Bulletin of Engineering and Agriculture No. 60, University of Kansas, Lawrence, Kan., 1969.

175. **McKinney, R. E.,** Design and operation of completely mixed activated sludge, Report 1:3:1, Environmental Pollution Control Services Laboratory, Lawrence, Kan., 1970.

176. **Eckenfelder, W. W.,** *Activated Sludge and Extended Aeration-Process Design in Water Quality Engineering,* Vanderbilt University, Nashville, Tenn., 1971.

177. **Goodman, B. L. and Englande, J.A., Jr.,** A consolidated approach to activated sludge process design, *Prog. Water Technol.,* 7(1), 1, 1975.

178. **Grady, C. P. L., Jr. and Roper, R. E., Jr.,** A model for the bio-oxidation process which incorporates the viability concept, *Water Res.,* 8, 471, 1974.

179. **Stafford, D. A. and Callely, A. G.,** Microbiological and biochemical aspects, in *Treatment of Industrial Effluents,* Callely, A. G., Forster, C. F, and Stafford, D. A., Eds., Hodder and Stoughton, London, 1977, chap. 8.

180. **Chappelle, E. W. and Levin, G. V.,** Use of fire fly bioluminescent reaction for rapid detection and counting of bacteria, *Biochem. Med.,* 2, 41, 1968.

181. **Weddle, C. L. and Jenkins, D.,** The viability and activity of activated sludge, *Water Res.,* 5, 621, 1971.

182. **Stanier, R. Y., Donderoff, M., and Alderberg, E. A.,** *The Microbial World,* Prentice-Hall, Englewood, Cliffs, N.J., 1963.

183. **Curds, C. R.,** Computer simulations of microbial population dynamics in the activated sludge process, *Water Res.,* 5, 1049, 1971.

184. **Kumaran, P. and Parhad, N. M.,** Biodegradaton of phenol by *C. tropicalis* — studies in a chemostat, *Indian J. Environ. Health,* 26, 123, 1984.

185. **Harder, W.,** Enrichment and characterization of degrading organisms, in *Microbial Degradation of Xenobiotics and Recalcitrant Compounds,* (FEMS Symp. Ser. 12), Leisinger, T., Cook, A. M., Hütter, R., and Nüesch, J., Eds., Academic Press, London, 1981, 77.

186. Coal Carbonization Research Report — Microorganisms Occurring in the Activated Sludge Treatment of Carbonization Effluents — Preliminary Studies, British Carbonization Research Association, Chesterfield, England, Report No. 56, November 1978, 7.

187. **Curds, C. R.,** A theoretical study of factors influencing the microbial population dynamics of the activated sludge process. I, *Water Res.,* 7, 1269, 1973.

188. **Mckenna, E. J. and Kallio, R. E.,** The biology of hydrocarbons, *Annu. Rev. Microbiol.,* 19, 183, 1967.

189. **Wedemeyer, G.,** Biodegradation of dichlorodiphenyl, trichloroethane. Intermediates in dichlorophenyl acetic acid metabolism by *Aerobacter aerogenes, Appl. Microbiol.,* 15, 1494, 1967.

190. **Focht, D. D. and Alexander, M.,** Aerobic co-metabolism of DDT analogues by *Hydrogenomonas sp., J. Agric. Food Chem.,* 19, 20, 1971.

191. **Benarde, M. A., Koft, B. W., Horvath, R. S., and Shawlis, L.,** Microbial degradation of sulfonates and dodecyl benzene sulfonates, *Appl. Microbiol.,* 13, 103, 1965.

192. **Horvath, R. S. and Koft, B. W.,** Degradation of alkyl benzene sulfonate by *Pseudomonas sp., Appl. Microbiol.,* 23, 407, 1972.

193. **Hegeman, G. D. and Rosenberg, S. L.,** Evolution of bacterial enzyme systems, *Annu. Rev. Microbiol.,* 24, 429, 1970.

194. **Horowitz, N. H.,** The evolution of biochemical synthesis; retrospects and prospects, in *Evolving Genes and Proteins,* Bryson, V. and Vogel, H. J., Eds., Academic Press, New York, 1965.

195. **Waid, J. S.,** The possible importance of transfer factors in bacterial degradation of herbicides in natural ecosystems, *Residue Rev.,* 44, 65, 1972.

196. **Chatterjee, D. K. and Chakrabarty, A. M.,** Plasmids in the degradation of polychlorinated biphenyls and chlorobenzoates, in *Microbial Degradation of Xenobiotics and Recalcitrant Compounds,* (FEMS Symp. Ser. 12), Lesinger, L., Cook, A. M., Hütter, R., and Nüesch, J., Eds., Academic Press, London, 1981, 213.

197. **Christiansen J. A.,** Improving effluent quality of petrochemical waste waters with mutant bacterial cultures, in *Proc. 37th Industrial Waste Conference,* Purdue University, Bell, J. M., Ed., Ann Arbor Science, Ann Arbor, Mich., 1982, 567.

198. **Cain, R. B.,** Surfactant biodegradation in waste waters in *Treatment of Industrial Effluents,* Callely, A. G., Forster, C. F., and Stafford, D. A., Eds., Hodder and Stoughton, London, 1977, 318.

199. **Kumaran, P. and Sundaresan, B. B.,** Yeast technology in environmental pollution control paper, presented at the 7th Specialized Symposium on Yeasts, Bombay, India, January 24 to 28, 1983, A196.

200. **Barker, J. E. and Thompson, R. J.,** Biological Removal of Carbon and Nitrogen Compounds from Coke Plant Wastes, Report US-EPA, Ser. No. EPA-2-73-167, U.S. Environmental Protection Agency, Washington, D.C., April 1973.

201. **Shivaraman, N. and Parhad, N. M.,** Biodegradation of cyanide in a continuously fed aerobic system, *J. Environ. Biol.,* 5, 273, 1984.

202. **Abson, J. W. and Todhunter, K. H.,** Plant for continuous treatment of carbonization effluent, *Monograph,* 12, 147, *Soc. Chem. Ind.,* London, 1961, 1.

203. **Voigt, G. and Ranker, G.,** The biological treatment plant of the VEB tar distillery and chemical plant at Erkner, *Water Pollut. Abstr.,* 40 (Abstr. No. 133), 32, 1969.

204. **Adeema, D.,** The largest oxidation ditch in the world for treatment of industrial waste, in *Proc. 22nd Industrial Waste Conference,* Purdue University, West Lafayette, Ind., 1967, 717.

205. **Biczysko, J. and Suschka, J.,** Investigations on phenolic wastes treatment in an oxidation ditch, in *Advances in Water Pollution Research,* Vol. 2, Jenkins, S. H. and Mendia, L., Eds., Water Pollution Control Federation, Washington, D. C., 1967, 285.

206. **Boisvienx, P., Bonnier, P. E., Goethel, G. F., Jenkins, R. H., Lemlin, J. S., Levi, J. D., Marmin, A., Paulnis, C. D. A., Rotteri, S., Sibra, P., and Verschveren, K.,** The environmental impact of refinery effluent, (EIRE study), Concawe's assessment, Report No. 1/80, 9, *Concawe,* The Hague, Netherlands, February, 1980.

207. *Manual on Disposal of Refinery Waste Volume of Liquid Wastes,* 1st ed., American Petroleum Institute, New York, 1969, chap. 13.

208. **Atkins, P. R.,** The Pesticide Manufacturing Industry — Waste Treatment and Disposal, US-EPA Report No. 12020-FYE 61/72, U.S. Environmental Protection Agency, Washington, D.C., 1972.

209. **Coley, G. and Stutz., C. N.,** Treatment of parathion wastes and other organics, *J. Water Pollut. Control Fed.,* 38, 1345, 1966.

210. **Lue-Huig, C. and Brady, S. D.,** Biological treatment of organic phosphorous pesticide waste waters, in *Proc. 23rd Industrial Waste Conference,* Purdue University, West Lafayette, Ind., 1968, 1166.

Chapter 6

MICROBIAL, CHEMICAL, AND TECHNOLOGICAL ASPECTS OF THE ANAEROBIC DEGRADATION OF ORGANIC POLLUTANTS

Iman W. Koster

TABLE OF CONTENTS

I. Introduction..286
    A. Anaerobic Digestion..286
    B. Occurrence of Methanogenesis 286
    C. Historical Background..286
    D. Comparing Natural Ecosystems with Digesters 287

II. Anaerobic Degradation of Complex Organic Material 287
    A. Substrate Flows in Anaerobic Digestion.............................287
    B. Hydrolysis of Suspended Solids 290
    C. Fermentation of Amino Acids and Sugars.............................291
    D. Anaerobic Oxidaton of Long-Chain Fatty Acids.......................292
    E. Anaerobic Oxidation of Intermediary Products 292
    F. Nonmethanogenic Conversions of Acetic Acid, Methanol, and
        Hydrogen...294
    G. Methanogenesis 295

III. Chemical and Physical Requirements of Methane Fermentations 297
    A. Acidity and Buffering Capacity 297
    B. Temperature 299
    C. Nutrient Requirements..301
    D. Toxicity 303
        1. Introduction..303
        2. Effective Concentration of Potential Inhibitors...............304
        3. Antagonism and Synergism 304
        4. Adaptation 304

IV. Conclusions..306

References...307

I. INTRODUCTION

A. Anaerobic Digestion

In anaerobic digestion organic matter is transformed into methane via a series of successive bacteriological processes. In the first half of this chapter, the microbial, chemical, and technological aspects of the main processes occurring in anaerobic digestion of complex organic material will be described in detail. In the second part of this chapter, the main environmental factors which are important for a stable anaerobic digestion will be discussed. Since the methanogenic bacteria are the key organisms in anaerobic digestion, the discussion of the environmental factors focuses on methanogenesis, but the impact of the environmental factors on the other metabolic stages in the process of anaerobic digestion is also dealt with.

Anaerobic digestion is a general term for the process of degradation of organic material with the exclusion of oxygen which can take place in solid waste treatment, manure digestion, sludge digestion, waste water treatment, etc. It should be noted that in this text the term "digester" is used to describe any reactor in which anaerobic digestion occurs, including the manure- or sewage-sludge treating reactors for which the term is usually reserved.

B. Occurrence of Methanogenesis

Methane is one of the major end products in microbial life in anaerobic environments where organic matter is decomposed.[1] Methane production is common in nature and it even occurs in extreme ecosystems such as glacier ice,[2] acid tundra peat,[3] thermophilic (55 to 80°C) agricultural waste digesters,[4] sedimentary rocks from oil fields,[5] and possibly also in submarine hydrothermal vents at pressures of about 250 bar and temperatures exceeding 300°C.[6]

C. Historical Background

Scientific interest in the gases produced by decomposing animal and vegetable material goes back to the early scientific revolution in the writings of Robert Boyle and Denis Papin in 1682 and of Stephen Hales in 1727.[7] In 1776, Volta was the first who recognized the close relationship between the decaying vegetation in the sediments of lakes and streams and the appearance of combustible gas.[8] The gas was analyzed as methane in 1806 by Henry. In 1868, Béchamp, who advocated the theory of spontaneous generation, was the first who connected the formation of methane with microbial activity.[9] Popoff, in 1875, was the first to systematically investigate the formation of methane using various complex substrates.[10] In 1881, the first description of an anaerobic waste water treatment system appeared in a French journal.[11] Due to the fact that in those days microbial techniques and equipment were rather defective, especially in respect to the necessary exclusion of oxygen, nobody was able to isolate methanogenic bacteria in pure culture. Söhngen, in 1906, was able to enrich a coculture of two distinct acetotrophic bacteria: a sarcina and a rod.[12]

The next breakthrough in the development of the microbiology of methanogenesis was made by Schnellen, who, like Söhngen, was working within the framework of the "Delft School of Microbiology".[13] Schnellen was the first to obtain pure cultures of methanogenic bacteria: *Methanosarcina barkeri* and *Methanobacterium formicicum*.[14]

In 1967, Bryant et al.[15] discovered that the conversion of ethanol into methane by *Methanobacillus omelianskii*:

$$2CH_3CH_2OH + CO_2 \rightarrow 2CH_3COOH + CH_4 \ (\Delta G_0' = -132.6 \ kJ)$$

is only feasible because this "organism" is a mixed syntrophic culture of a nonmethanogenic bacterium which converts ethanol into acetic acid and hydrogen:

$$2CH_3CH_2OH + 2H_2O \rightarrow 2CH_3COOH + 4H_2 \ (\Delta G_0' = +6.3 \text{ kJ})$$

and a hydrogenotrophic methanogenic bacterium:

$$4H_2 + CO_2 \rightarrow CH_4 + 2H_2O \ (\Delta G_0' = -138.9 \text{ kJ})$$

which keeps the hydrogen concentration low enough to allow the nonmethanogenic bacterium to grow. This discovery made clear that the spectrum of substrates for methanogenic bacteria is rather limited, with acetate and carbon dioxide plus hydrogen as the most important substrates. The thermophilic methanogenic rod called *M. kuzneceovii,* which was reported to ferment methanol to methane and acetic acid, most probably is a mixed syntrophic culture of a hydrogenotrophic methanogenic bacterium and a hydrogen-producing acetogenic bacterium.[16]

The latest "major step forward" in the development of our knowledge of the microbiology of methanogenesis is the discovery that methanogenic bacteria are part of a unique genealogical group.[17] Apart from the eukaryotes (cells which have well-formed nuclei, e.g., fungi) and the prokaryotes (cells which do not have nuclei, the common bacteria) there is the primary group of the Archaebacteria, which includes three very different kinds of bacteria: extreme halophiles, thermoacidophiles, and methanogens. Archaebacteria differ from prokaryotes at the molecular level, especially in ribosomal RNA sequences. Methanogenic bacteria also differ from other bacteria in the composition of the cell wall.[18] Because muramic acid is not present in the cell walls of methanogenic bacteria, they are resistant to cell-wall active antibiotics such as penicillin, vancomycin, D-cycloserine, and cephalosporin.[19]

D. Comparing Natural Ecosystems with Digesters

Much information concerning methane-producing bacterial populations can be derived from natural anaerobic ecosystems such as sediments or the rumen. The rumen especially has been the subject of extensive microbial research.[20,21] It should be realized, however, that these natural ecosystems in many cases are not in all aspects comparable with anaerobic waste water treatment systems or sludge (or manure) digesters. Marine sediments are characterized by a sulfate pool which cannot be exhausted.[1] Under conditions of sulfate sufficiency, sulfate-reducing bacteria will win the competition for hydrogen and acetate (from the methanogenic bacteria.)[22,23]

The main difference between the rumen and a manure digester or an anaerobic wastewater treatment system is on the basis of the turnover time.[24] The rumen turnover time is about 1 day, whereas a manure digester turnover time is always longer, generally about 30 days. This implies that many bacteria whose growth rates are such that they are washed out of the rumen are able to occupy a habitat in a manure digester which in fact is fed with material that has passed the rumen.

II. ANAEROBIC DEGRADATION OF COMPLEX ORGANIC MATERIAL

A. Substrate Flows in Anaerobic Digestion

The effective conversion of complex organic material into methane depends on the combined activity of a miscellaneous microbial population consisting of various genera of obligate and facultative anaerobic bacteria. The activities of the mixed population present in an anaerobic digester can be summarized in seven distinct processes (Figure 1):

1. Hydrolysis or liquefaction of suspended solids
2. Fermentation of amino acids and sugars
3. Anaerobic oxidation of long-chain fatty acids

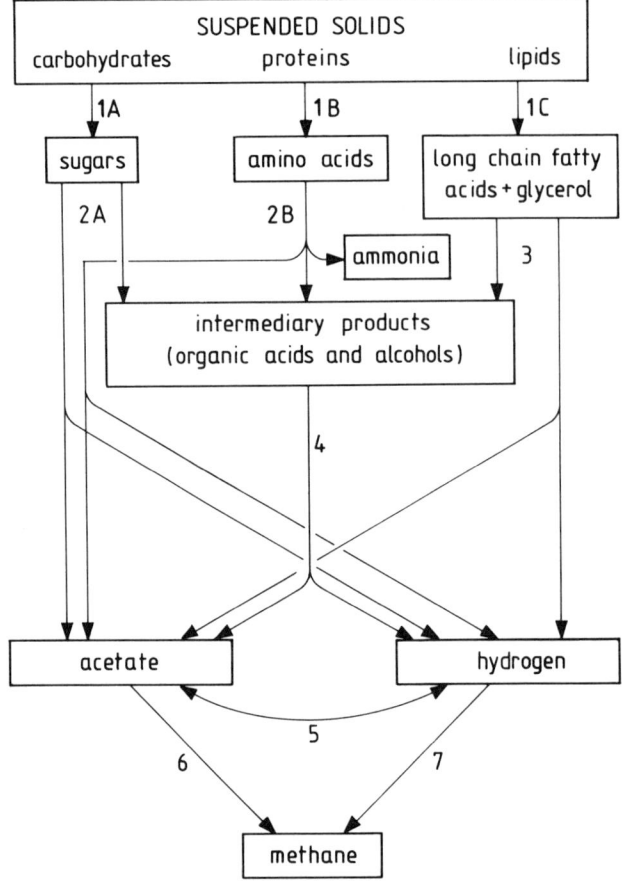

FIGURE 1.   Schematic representation of substrate flows in anaerobic digestion of complex organic material.

4.   Anaerobic oxidation of intermediary products, mainly volatile fatty acids such as propionic acid, butyric acid, etc.
5.   Nonmethanogenic conversions of acetate and hydrogen
6.   Acetoclastic or acetotrophic methanogenesis
7.   Hydrogenotrophic methanogenesis

These processes can be arranged into four distinct metabolic stages:[25]

1.   In hydrolysis or liquefaction, complex, nonsoluble organic compounds are solubilized by enzymes excreted by hydrolytic bacteria. Since these enzymes work outside the bacteria, they are called exoenzymes. In fact, hydrolysis is the conversion of polymers into monomers.
2.   In acidogenesis, soluble organic compounds, including the products of the hydrolysis, are converted into organic acids such as acetic acid, propionic acid, and butyric acid.
3.   In acetogenesis or intermediary acidogenesis, the products of the acidogenesis are converted into acetic acid, hydrogen, and carbon dioxide.
4.   In methanogenesis, methane is produced from acetic acid or from hydrogen plus carbon dioxide. Methane can also be formed directly from some other substrates, from which formic acid and methanol are the most important.

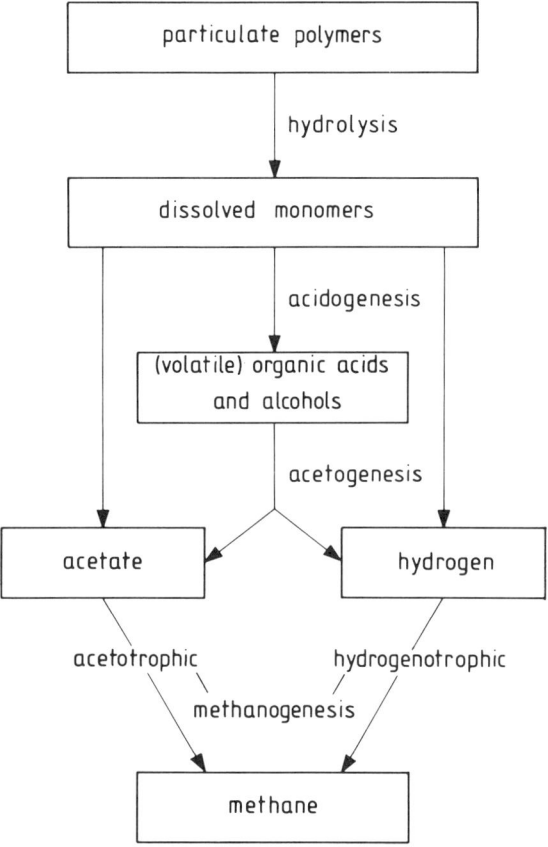

FIGURE 2. Metabolic stages and products in the anaerobic digestion of complex organic material.

A diagram of the four metabolic stages that can be distinguished in anaerobic digestion is shown in Figure 2. In well-balanced anaerobic digestion, all products of a previous metabolic stage are converted into the next one, so that the overall result is a nearly complete during conversion of the biodegradable organic material in the influent into end products such as methane, carbon dioxide, hydrogen sulfide, ammonia, etc. If the amount of organic material used for the growth of biomass and the amount of intermediary products leaving the system without being converted into biogas are negligible, the anaerobic digestion can be described by the following equation:[26]

$$C_nH_aO_bN_cS_dM_{ev} + \left(n - \frac{a}{4} - \frac{b}{2} + \frac{7c}{4} + \frac{d}{2} + \frac{3ev}{4}\right)H_2O \rightarrow$$

$$\left(\frac{n}{2} - \frac{a}{8} + \frac{b}{4} - \frac{5c}{8} + \frac{d}{4} - \frac{9ev}{8}\right)CO_2 + \left(\frac{n}{2} + \frac{a}{8} - \frac{b}{4} - \frac{3c}{8} - \frac{d}{4} - \frac{ev}{8}\right)CH_4$$

$$+ cNH_4HCO_3 + dH_2S + eM(HCO_3)_v$$

In this equation the letters C, H, O, N, and S have the usual chemical significance; M stands for any metal ion with a valence v; and all other subscripts stand for the number of the respective atoms.

B. Hydrolysis of Suspended Solids

Hydrolysis of complex organic compounds is a slow process, which depends on the action of extracellular enzymes such as cellulases, amylases, proteases, and lipases. The rate of hydrolysis is to a great extent influenced by pH and detention time.[27,28] In experiments concerning the addition of barley straw to piggery waste digesters, the straw was little-degraded at a detention time of 10 days, whereas approximately 35% of the straw was digested to biogas at a detention time of 20 days.[29] In experiments concerning the fermentation of rice straw to methane, the hydrolysis also varied with the detention time. At a detention time of 15.5 days, 59.3% of the hemicellulose and 34.8% of the cellulose was destructed, whereas at 100-days detention time these figures were 96.1 and 75.8%, respectively.[30]

The optimum pH for hydrolysis is different for various substrates. For easily degradable carbohydrates, hydrolysis and acidogenesis proceed at maximum rates in the pH range 5.5 to 6.5.[31,32] The optimum pH for hydrolysis of proteins is at pH 7 or even higher.[33,34] The optimum pH for hydrolysis of lipids has never been assessed. In an acid-phase sludge digestion operated at pH 5.15, lipids are not degraded at all.[35] In a conventional (one-phase) sludge digestion operated in the pH range 6.7 to 7.4 lipid hydrolysis is not the rate-limiting step in the anaerobic digestion of lipids.[36] The optimum pH for the hydrolysis of solid cannery vegetable wastes has been reported to be pH 6.5.[37] For hydrolysis of a mixture of garbage and sewage sludge the pH optimum is at pH 5.8.[27]

The rate of hydrolysis and the extent to which a substrate may be hydrolyzed is also influenced by the accessibility of the substrate to be hydrolyzed for the exoenzymes.[38] This is especially important with respect to anaerobic digestion of fibrous materials such as straw and wood, in which the cellulosic and hemicellulosic microfibrils are aggregated and embedded within the lignified cell wall matrix. The crystallinity and surface area of the fibers are the most important features which determine the accessibility for exoenzymes.[39] In order to enhance the hydrolysis of such materials, various pretreatments can be applied.[40-43] Physical pretreatment methods such as heating or milling or chemical pretreatment methods such as soaking in sodium hydroxide can be applied, as well as a combination of methods. Another promising method is microbial pretreatment based on the capability of "White Rot" fungi to degrade lignocellulose.[44]

The importance of accessibility for exoenzymes is reflected in the effect of particle size on the rate of hydrolysis. In a one-stage methane fermentation of tomato solid wastes operated at a detention time of 18 days, the rate of methane production increased 3.24 times when the particle size was decreased from 20 to 1.3 mm.[45]

When engineering decomposition of materials such as primary and secondary sewage sludges, the rate-limiting step at solids retention times on the order of 10 to 15 days will be the methane formation, whereas at solids retention times greater than 15 days the hydrolysis of the organic solids becomes the rate-limiting step.[46] This implies that, if maximum gas production per unit solids is the aim, the loading potential of digesters treating sewage sludge or manure is to a great extent limited by the slow rate of hydrolysis. In such cases, addition of concentrated soluble waste streams as auxiliary substrates for the anaerobic digestion process significantly increases maximum loading rates and therefore makes the process more economical.[47] In a separate acid-producing reactor fed with sewage sludges the hydrolysis is always the rate-limiting step.[35,48] Hydrolysis also is the rate-limiting step in the anaerobic degradation of proteins.[49] Lipids are hydrolyzed very slowly, with the result that hydrolysis might be the rate-limiting step in the anaerobic degradation of wastes containing considerable amounts of lipids.[50] At 20°C, the anaerobic degradation of lipids is nil, even though methanogenesis continues at a reduced rate.[51] The slow rate of hydrolysis appeared to be the limiting factor to the application of one-stage anaerobic treatment of slaughterhouse waste water, especially at low temperatures (approximately 20°C).[52,53] That particular (swine) slaughterhouse waste water contained 40 to 50% suspended solids, 5% of the total solids as grease.

C. Fermentation of Amino Acids and Sugars

Fermentations can be defined as those biological processes that occur in the dark and that do not involve respiratory chains with oxygen or nitrate as electron acceptors. The absence of respiratory chains causes a low ATP yield for the fermentative reactions. Consequently, the amount of biomass attained per mole of substrate is much smaller than with aerobes and, in addition to cell material, large amounts of fermentation end products are formed.[54]

Most bacteria present in anaerobic digesters are obligate anaerobes. However, a small fraction of the fermentative population is also able to use oxygen. In general, approximately 1% of the nonmethanogenic population in a digester consists of facultative anaerobic bacteria.[55]

The products of the fermentative population vary depending on environmental conditions applied. In the fermentation of glucose taking place in a separate acid-producing reactor, the main products are butyric acid, acetic acid, hydrogen, and carbon dioxide. Interruption of the feed supply for a period of 1 to 24 hr changes the fermentation pattern to an increased production of propionic acid and acetic acid. The same happens after a downshift in dilution rate.[56] In a similar reactor, the product pattern from the fermentative bacteria is influenced by the pH. At pH values below pH 6, the main product of the fermentation of glucose is butyric acid, but with increaing pH the product pattern changes, first to lactic acid and subsequently to acetic acid, formic acid, and ethanol.[31] Hydrolysis and subsequent fermentation of gelatin in a separate acid-producing reactor at pH values above pH 6 result in the following products: acetic acid, propionic acid, valeric acid, and minor amounts of other volatile fatty acids. However, at pH values below pH 6, the relative amount of acetic acid decreases and the relative amount of propionic acid increases.[33] The product pattern in the fermentation of sucrose in a separate acid-producing reactor is also influenced by the cell residence time. At short cell residence times (up to 1 day), ethanol and lactic acid were the main products, whereas at higher cell residence times (3 to 5 days), a mixture of volatile fatty acids was produced.[57] The product pattern of fermentation also depends on the type of substrate used, that is, the type of waste water to be treated.[58]

Removal of hydrogen by hydrogenotrophic bacteria (e.g., methanogenesis, sulfate reduction, or denitrification) can also significantly influence the kinds of products formed by fermentative bacteria.[59] When hydrogen is continuously consumed by hydrogenotrophic bacteria, the fermentative bacteria are able to produce further-oxidized products than they would be able to at increased hydrogen levels, which supplies more energy per unit of substrate to the bacteria.[60-62]

In fermentation, the oxidation of substrate is coupled with the reduction of metabolic intermediates.[63] Some fermentative bacteria are able to form H_2 from reduced pyridine nucleotides, but the accumulation of H_2 inhibits this reaction. Other major mechanisms for formation of H_2 include those from formate and pyruvate. These routes are not susceptible to inhibition by H_2. The importance of the H_2-inhibited and -uninhibited routes in the fermentation of glucose can be illustrated by the fermentation pathway of *Ruminococcus albus*.[61] In pure culture, this bacterium ferments glucose to ethanol, acetic acid, hydrogen, and carbon dioxide. However, in coculture with a hydrogenotrophic bacterium, it ferments glucose to acetic acid, hydrogen, and carbon dioxide; ethanol is not a product. The effect of instant removal of hydrogen by methanogenic bacteria on the products formed by other bacteria was first demonstrated during the studies on the nonmethanogenic bacterium isolated from *M. omelianskii*. In pure culture, this so-called S-organism ferments pyruvate to ethanol, acetic acid, carbon dioxide, and only a trace of hydrogen. In coculture with a hydrogenotrophic methanogen, acetic acid, carbon dioxide, and methane (but not ethanol) are formed. When cocultured with a methanogenic bacterium that keeps the hydrogen partial pressure low, the S-organism uses the electrons generated during pyruvate catabolism for H_2 production rather than for the production of ethanol.[64]

Table 1
SOME ACETOGENIC REACTIONS

$$CH_3CHOHCOO^- + 2H_2O \rightarrow CH_3COO^- + HCO_3^- + H^+ + 2H_2 \quad \Delta G_0' = -4.2 \text{ kJ/mol}$$

$$CH_3CH_2OH + H_2O \rightarrow CH_3COO^- + H^+ + 2H_2 \qquad\qquad \Delta G_0' = +9.6 \text{ kJ/mol}$$

$$CH_3CH_2CH_2COO^- + 2H_2O \rightarrow 2CH_3COO^- + H^+ + 2H_2 \qquad \Delta G_0' = +48.1 \text{ kJ/mol}$$

$$CH_3CH_2COO^- + 3H_2O \rightarrow CH_3COO^- + HCO_3^- + H^+ + 3H_2 \quad \Delta G_0' = +76.1 \text{ kJ/mol}$$

Note: Values of free energy changes are from Reference 75.

An important end product of the fermentation of amino acids is ammonium. Ammonium is the source of nitrogen for methanogenic bacteria.[65,66] On the other hand, at concentrations exceeding about 700 mg nitrogen per liter, ammonium inhibits methanogenesis.[67,68]

In anaerobic digestion, alkalinity is mainly provided by ammonium and bicarbonate ions. A reduction in the ammonium content of a digester leads to a decrease in alkalinity and perhaps pH, thus illustrating that, apart from being essential as a nutrient, ammonium is also essential for a buffering capacity.[69] The contribution of ammonium ions to the buffering capacity in an anaerobic digester will only be of considerable importance if the ammonium concentration is of magnitude of a few grams N per liter, such as in some industrial waste waters, manure, and concentrated sewage sludge.

D. Anaerobic Oxidation of Long-Chain Fatty Acids

Experiments with ^{14}C-labeled, long-chain fatty acids have revealed that the anaerobic degradation of long-chain fatty acids occurs by β oxidation.[70,71] Unsaturated fatty acids are hydrogenated before being degraded by β oxidation. The overall reaction for β oxidation of even-carbon-numbered fatty acids, including the formation of hydrogen (which is the main sink for electrons) is as follows:

$$CH_3(CH_2)_nCOO^- + nH_2O \rightarrow (^1/_2n + 1)CH_3COO^- + ^1/_2n \, H^+ + nH_2$$

In this equation n is an even number. Anaerobic β oxidation of odd-carbon-numbered fatty acids results in the same products plus propionate. The only isolated anaerobic bacterium which β oxidizes long-chain fatty acids is *Syntrophomonas wolfei*.[72,73] Anaerobic β oxidation of long-chain fatty acids is thermodynamically unfavorable unless the hydrogen partial pressure is maintained at a very low level.[74] For this reason, *S. wolfei* cannot be isolated in a pure culture, but syntrophic cocultures with hydrogenotrophic organisms such as *Desulfovibrio* sp. or methanogenic bacteria are possible.

E. Anaerobic Oxidation of Intermediary Products

The products of the acidogenic bacteria are converted into substrates for the methanogenic bacteria by the acetogenic bacteria, which produce acetic acid, hydrogen, and carbon dioxide. Some acetogenic reactions are shown in Table 1, together with the associated values for the standard free-energy change $(\Delta G'_0)$.[75] The standard conditions for the calculation of free-energy changes in biochemical reactions are the following: temperature of 25°C, pH 7.0, pressure of 1 atm, activity of all solutes is 1 mol/kg, and water is a pure liquid. At these standard conditions the acetogenic conversion of ethanol, butyrate, and propionate is not possible, because these reactions do not yield energy (i.e., the free-energy change is positive). The actual free-energy change for the reaction $aA + bB \rightarrow cC + dD$ may be calculated from the standard free-energy change by means of the following equation:[76]

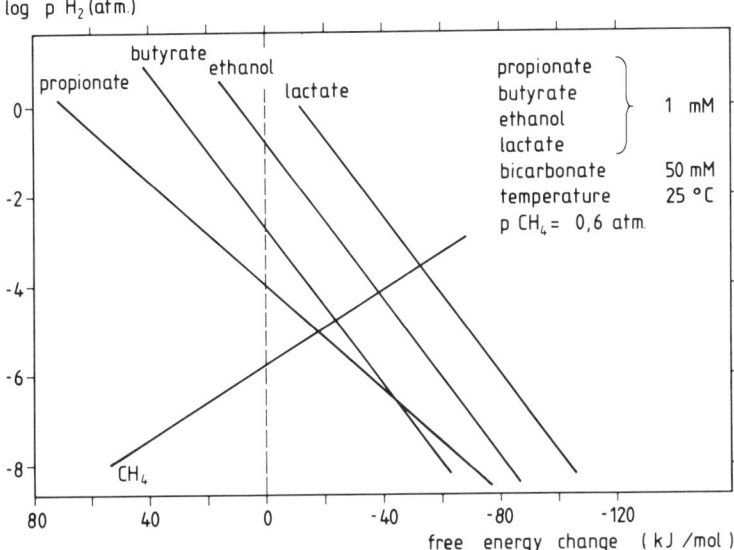

FIGURE 3. Effect of the partial pressure of hydrogen (pH$_2$) on the free energy change (ΔG′) for the acetogenic conversion of lactate, ethanol, propionate, and buryrate, and for the formation of methane from hydrogen and carbon dioxide.

$$\Delta G' = \Delta G_0' + R.T. \ln \frac{[A]^a \cdot [B]^b}{[C]^c \cdot [D]^d} \tag{1}$$

in which R is the molar gas constant (8.31 J.mol^{-1}.T^{-1}) and T is the absolute temperature. A low hydrogen partial pressure is necessary to make the acetongenic conversion of ethanol, butyrate, and propionate possible. This is illustrated in Figure 3.

The high substrate affinity of the hydrogen-consuming microorganisms in methanogenic ecosystems makes it possible to maintain sufficiently low hydrogen concentrations. The Michaelis-Menten half-saturation constant (K$_m$) for hydrogen has been reported to be 5.8 µM for rumen fluid, 6.0 µM for the contents of a sewage sludge digester, and 7.1 µM for sediment taken from the pelagic zone of a hypereutrophic lake.[77] These values are equivalent with a hydrogen partial pressure of 6.89 × 10^{-3}, 7.09 × 10^{-3}, and 7.82 × 10^{-3} atm, respectively.[78,79] The affinity for hydrogen exhibited by sulfate-reducing bacteria is higher than that of methanogenic bacteria. The sulfate reducer population in a certain lake sediment appeared to have a half-saturation constant for hydrogen uptake of 1.40 × 10^{-3} atm, which is much lower than the value of 5.91 × 10^{-3} atm found for the methanogenic population in the same sediment.[80] This higher substrate affinity is one of the reasons why hydrogen-otrophic methanogenic bacteria are out-competed by sulfate-reducing bacteria in environments where a sufficient amount of sulfate is present.[81]

In a well-balanced methane fermentation, the hydrogen partial pressure does not exceed 10^{-4} atm,[1] and is in most cases approximately 10^{-6} atm.[9] Such a low hydrogen partial pressure can only be maintained if all hydrogen produced by acidogenic and acetogenic bacteria is instantly and very effectively removed by hydrogen-consuming bacteria. It can be calculated that, in a well-balanced methane fermentation, a hydrogen molecule will be consumed within 0.5 sec after being produced, which means a maximum diffusion path of <0.1 mm.[82] This illustrates that the symbiotic relationship between hydrogen-producing and hydrogen-consuming bacteria is not only biochemical, but spatial as well. This symbiotic relationship makes it impossible to obtain pure cultures of acetogenic bacteria. Cocultures of acetogenic bacteria with hydrogen-consuming bacteria such as hydrogenotrophic methanogenic bacteria or sulfate-reducing bacteria are possible.[72,73,83,84]

F. Nonmethanogenic Conversions of Acetic Acid, Methanol, and Hydrogen

In anaerobic digestion, the main metabolic route for methanol and hydrogen (plus carbon dioxide) is the direct formation of methane. However, these substrates can also be utilized by nonmethanogenic bacteria which produce acetic acid and, in many cases, higher volatile fatty acids. The volatile fatty acids produced from hydrogen and methanol will be ultimately converted into methane, but this detour must be energetically unfavorable for the methanogenic bacteria. This means that, although in all cases methane will be the end product, there exists a competition for hydrogen and methanol between volatile fatty acid-producing and methanogenic bacteria.[85,86]

Synthesis of volatile fatty acids from hydrogen plus carbon dioxide has been found to occur in anaerobic digester sludge, cistern mud, and lake sediments,[87] as well as in the cecum of some rodents.[88] Synthesis of volatile fatty acids from hydrogen and carbon dioxide does not occur in the rumen;[87,88] apparently the hydrogenotrophic acetonogenic bacteria are outcompeted by the hydrogenotrophic methanogenic bacteria in this specific habitat. In anaerobic digesters, the population size of the hydrogenotrophic acetogenic bacteria is approximately 1% of the population size of the hydrogenotrophic methanogenic bacteria.[87] At present, only four species of the hydrogenotrophic acetogenic bacteria have been properly described and are in a laboratory pure culture: *Clostridium aceticum*,[89-92] *Acetobacterium woodii*,[93] *Acetogenium kivui*,[94] and *Acetoanaerobium notarae*.[236] There also is a pure culture of a hydrogenotrophic acetogenic bacterium that is morphologically different but physiologically similar to *C. aceticum*.[95]

The hydrogenotrophic acetogenic bacteria are also able to ferment organic substrates such as sugars. *Acetobacterium woodii* is able to use organic substrates and hydrogen plus carbon dioxide simultaneously, whereas *C. aceticum* is unable to take advantage of the simultaneous presence of an organic substrate and hydrogen plus carbon dioxide, only the organic substrate being utilized under such conditions.[96]

The conversion of acetic acid and hydrogen plus carbon dioxide to propionic acid takes place in pure cultures of the sulfate-reducing bacterium *Desulfobulbus propionicus* (Lindhorst), which originates from sediments of brackish coastal waters.[97] This conversion has also been observed in enrichment cultures of digested sewage sludge.[98] The formation of propionic and butyric acid by a mixed culture of bacteria originating from digested sewage sludge, which produces mainly acetic acid, has also been reported.[99]

The ecological significance of microorganisms which convert hydrogen and carbon dioxide into volatile fatty acids in methanogenic ecosystems is not well understood. They might play a role in keeping the hydrogen concentration low enough to permit acetogenic reactions such as propionic acid degradation. In competition with other hydrogen utilizers, such as sulfate-reducing bacteria and hydrogenotrophic methanogenic bacteria, some of the hydrogenotrophic acetogenic bacteria might take advantage of the ability to utilize organic substrates and hydrogen plus carbon dioxide simultaneously (e.g., *A. woodii*); others might take advantage of the slightly alkaline pH optimum (e.g., *C. aceticum*).

The formation of volatile fatty acids (mainly acetic and butyric acid) from methanol occurs in UASB reactors treating aqueous solutions of methanol[100] or fusel oil,[101] which is a mixture of methanol and higher alcohols dissolved in water. This production of volatile fatty acids from methanol occurred only if a sufficient amount of trace metals was added to the reactor feed. The fraction of methanol that was converted to volatile fatty acids depended on the bicarbonate concentration. Except when the acid production gave rise to pH values lower than pH 4.5, acetic acid production was predominant. From the sludge used in these experiments, an enrichment culture was started from which an obligatory anaerobic bacterium of the genus *Clostridium* producing acetic acid from methanol could be isolated.[102]

Table 2
METHANOGENIC REACTIONS

Reactions		ΔG_0 (kJ/mol CH_4)
$CH_3COO^- + H_2O$	$\rightarrow CH_4 + HCO_3^-$	-28.2
$4CH_3OH$	$\rightarrow 3CH_4 + CO_2 + 2H_2O$	-102.5
$4HCOO^- + 2H^+$	$\rightarrow CH_4 + CO_2 + 2HCO_3^-$	-126.8
$4H_2 + CO_2$	$\rightarrow CH_4 + 2H_2O$	-139.2
$4CO + 2H_2O$	$\rightarrow CH_4 + 3CO_2$	-185.1
$4CH_3NH_2 + 2H_2O + 4H^+$	$\rightarrow 3CH_4 + CO_2 + 4NH_4^+$	-101.6
$2(CH_3)_2NH + 2H_2O + 2H^+$	$\rightarrow 3CH_4 + CO_2 + 2NH_4^+$	-86.3
$4(CH_3)_3N + 6H_2O + 4H^+$	$\rightarrow 9CH_4 + 3CO_2 + 4NH_4^+$	-80.2

Note: Values of free energy changes are from Reference 9.

Other pure culture isolates of methanol utilizing nonmethanogenic anaerobic bacteria are *C. formicoaceticum*,[92,103] *Eubacterium limosum*,[104] and *Butyribacterium methylotrophicum*.[105]

The direct splitting of acetic acid in methane and carbon dioxide is the most important process of methane formation from acetic acid. However, a syntrophic coculture has been described in which one organism oxidizes acetic acid into carbon dioxide and hydrogen, while hydrogen is subsequently used by the other organism to reduce carbon dioxide to methane.[106] Acetic acid oxidation prior to methanogenesis has also been proved to occur in a two-step anaerobic digestion process fed with carbohydrates (glucose, starch, and cellulose).[107] In this particular process, labeling experiments revealed that methanogenesis from acetic acid by cleavage and via oxidation of acetic acid took place at similar order.

That one should be very careful in drawing conclusions concerning the way methane is formed from acetic acid if only enrichment cultures were used is illustrated by the fact that a strictly hydrogenotrophic methanogenic bacterium (a *Methanobacterium* strain) could be isolated from an acetate enrichment culture of sewage sludge.[108] In this culture, the hydrogenotrophic methanogen probably grew on hydrogen produced during fermentation of lysed acetotrophic methanogenic bacteria.

G. Methanogenesis

In methanogenic environments supplied with a complex organic substrate, approximately 70% of the methane is produced by acetate cleavage. The methyl group of acetate goes to methane; the carboxyl group goes to carbon dioxide.[109-112] Approximately 30% of the methane is produced by carbon dioxide reduction with hydrogen. Some methanogenic bacteria are able to utilize other substrates, such as formic acid, methanol, or methylamines, but, except for the case of some methanolic waste waters, these substrates play only a minor role in anaerobic waste and waste water treatment. The methanogenic reactions and free energy changes are shown in Table 2. *Methanosarcina barkeri* by far is metabolically the most versatile of all methanogenic bacteria isolated in pure culture so far. It has been shown to form methane by all the reactions compiled in Table 2, except for the reaction with formate.[113] A list of all methanogenic bacteria isolated in pure culture and sufficiently described to determine genus and species, together with substrate spectrum is shown in Table 3.[9,114-133,230-235]

Table 3

ISOLATED METHANOGENIC BACTERIA AND SUBSTRATE[a] SPECTRUM

Bacteria	Spectrum	Bacteria	Spectrum
Methanobacterium bryantii	H_2	*Methanogenium cariaci*	H_2 and HCOOH
M. formicicum	H_2 and HCOOH	*M. marisnigri*	H_2 and HCOOH
M. thermoautotrophicum	H_2	*M. tatii*	H_2 and HCOOH
M. alcaliphilum	H_2	*M. olentangyi*	H_2
		M. thermophilicum	H_2 and HCOOH
Methanobrevibacter arboriphilus	H_2	*M. bourgense*	H_2 and HCOOH
M. ruminantium	H_2 and HCOOH	*M. aggregans*	H_2 and HCOOH
M. smithii	H_2 and HCOOH		
		Methanococcoides methylutens	CH_3NH_2 and CH_3OH
Methanococcus vannielii	H_2 and HCOOH		
M. voltae	H_2 and HCOOH	*Methanothrix soehngenii*	CH_3COOH
M. deltae	H_2 and HCOOH	*M. concilii*	CH_3COOH
M. maripaludis	H_2 and HCOOH		
M. jannaschii	H_2	*Methanothermus fervidus*	H_2
M. thermolithoautotrophicus	H_2 and HCOOH		
M. frisius	H_2, CH_3OH, CH_3NH_2, and $(CH_3)_3N$	*Methanolobus tindarius*	CH_3OH, CH_3NH_2, $(CH_3)_2NH$, and $(CH_3)_3N$
Methanomicrobium mobile	H_2 and HCOOH	*Methanosarcina barkeri*	CH_3OH, CH_3COOH, H_2, CO, CH_3NH_2, $(CH_3)_2NH$, and $(CH_3)_3N$
M. paynteri	H_2		
Methanospirillum hungatei	H_2 and HCOOH	*Methanosarcina thermophila*	CH_3OH, CH_3COOH, H_2, CH_3NH_2, $(CH_3)_2NH$, and $(CH_3)_3N$
Methanoplanus limicola	H_2 and HCOOH		
M. endosymbiosus	H_2		

[a] Only electron donor for methane formation and growth.

The cleavage of acetate into methane and bicarbonate does not release enough energy to form ATP. Together with the fact that it is rather difficult to envision a mechanism of energy production due to a simple decarboxylation, this has made the conclusion that acetotrophic methanogenic bacteria generate energy directly by the splitting of acetate controversial.[134] Current evidence concerning acetotrophic methanogenesis suggests an absence of substrate-level phosphorylation and synthesis of ATP via an ATPase coupled directly or indirectly to a proton-motive force.[135,136] Proton-motive, force-driven synthesis of ATP might also play an important role in methanogenic bacteria utilizing other substrates.[137]

The relatively slow growth rate of acetotrophic methanogenic bacteria (which grow 5 to 10 times slower than hydrogenotrophic methanogenic bacteria) might be a consequence of the fact that energetically acetic acid is such a poor substrate.

The acetotrophic methanogenic bacteria can be divided into two groups: (1) the *Methanosarcina* group, characterized by a maximum specific growth rate (μ_m) of 0.3 day^{-1} and a substrate saturation constant (K_s) of 200 mg/ℓ and (2) the *Methanothrix* group, characterized by a maximum specific growth rate of 0.1 day^{-1} and a saturation constant of 30 mg/ℓ.[82] The facultative acetotrophic methanogenic bacterium *Methanococcus mazei*, which was reported in 1980,[138] is now believed to belong to the genus *Methanosarcina*. Because of a high substrate affinity (i.e., small K_s) *Methanothrix* sp. will outcompete other acetotrophic methanogenic bacteria at detention times above approximately 15 days (at 35°C). For this reason, in most digesters *Methanothrix* sp. will be the predominant acetate-cleaving organism.

In general, of the total bacterial population involved with methane production from a complex substrate, the methanogenic bacteria are the most sensitive with respect to pH, temperature, toxic compounds, etc. A stable anaerobic digestion process can only be accomplished if the methanogenic bacteria function well, because they are needed to remove acids formed in earlier metabolic stages and they play an important role in maintaining a low hydrogen concentration. Therefore, monitoring of the concentrations of methanogenic substrates is a very suitable way of controlling the performance of a methane-producing system. An increase in the hydrogen partial pressure is an indication of an instantaneous upset of the process, whereas the augmentation of acetic acid is an early warning for a possible forthcoming pH drop.[97] Since hydrogen is rather difficult to analyze in the extremely low concentration range present in anaerobic digestion processes, the propionic acid concentration is often used as an indirect parameter for the activity of the hydrogenotrophic population.[139]

III. CHEMICAL AND PHYSICAL REQUIREMENTS OF METHANE FERMENTATIONS

A. Acidity and Buffering Capacity

A crucial factor in the operation of anerobic digestion processes is the acid tolerance of the fermentative bacteria. Apart from methanogenesis in acid bogs, which might be explained by the existence of neutral microenvironments,[9] the methanogenic activity severely drops at pH values below pH 6. Methanogenesis is optimal in the pH range 6.7 to 7.4.[1,140] Many fermentative bacteria are still active at pH values below pH 6.[141] In the fermentation of green crops as silage, bacterial activity only stops at pH 4.[142,143] The same happens in the acidification of other organic material, e.g, waste tomatoes.[144] It should be noted that methanogenesis from methanol can occur at relatively low pH. In experiments with a UASB reactor treating solutions of methanol in a mineral medium, methanol was found to be still fermented at pH values as low as 3.5. The pH range 5.5 to 6.0 appeared to be optimal.[100]

Under balanced digestion conditions, the biochemical reactions tend to automatically maintain the pH in the proper range. The acidogenic reactions alone would result in a reduction of the pH caused by the production of organic acids, but this effect is counteracted

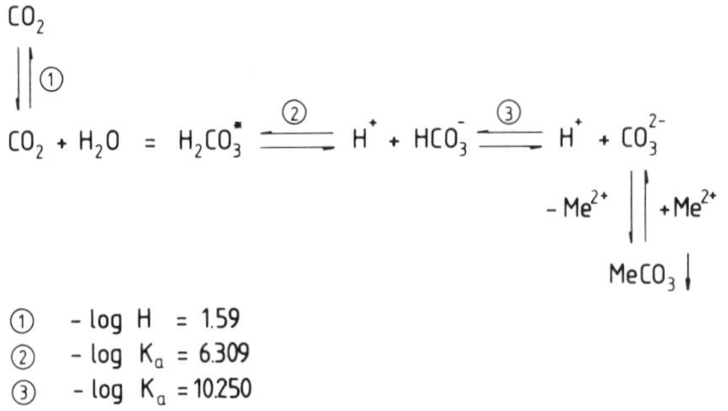

$$CO_2$$

$$\Updownarrow \text{①}$$

$$CO_2 + H_2O = H_2CO_3^* \xrightleftharpoons{\text{②}} H^+ + HCO_3^- \xrightleftharpoons{\text{③}} H^+ + CO_3^{2-}$$

$$-Me^{2+} \Updownarrow +Me^{2+}$$

$$MeCO_3 \downarrow$$

① $-\log H = 1.59$
② $-\log K_a = 6.309$
③ $-\log K_a = 10.250$

FIGURE 4. Physical and chemical reactions influencing the bicarbonate buffer system. Equilibrium constants adapted from Reference 148.

by the degradation of these acids and the concomitant reformation of the bicarbonate buffer during the methanogenic reactions. However, if an imbalance of the digestion process occurs, for example, by a sudden change in temperature, the introduction of a toxic compound, or an overloading with degradable organic material, the acid-producing bacteria will outpace the acid-consuming bacteria, therefore causing a buildup of organic acids in the system. If a balanced digestion is not restored, the buffering capacity of the system is overcome, followed by a decrease in pH. Since a decreasing pH hinders methanogenesis (= acid consumption) more than it hinders acidogenesis, a ''sour'' digester will be the result. Once the digester has become sour, the time needed to restore a methanogenic bacterial population after the environmental stress has been removed will be very long, since the viable methanogenic population left in the digester will be small and methanogenic bacteria are very slow growing. In fact, the best way to achieve a balanced anaerobic digestion again will be, in many cases, to start with a new seed.[145]

In anaerobic digesters where methanogenic bacteria are the main hydrogen-consuming organisms, even a relatively small pH drop may be the start of a severe upset of the digestion process, because it influences the hydrogen metabolism by inhibiting the hydrogenotrophic methanogenic bacteria, thus leading to a buildup of hydrogen in the digester environment. Hydrogen buildup is a very rapid process, and therefore is an almost immediate warning for digester instability.[146,147] Since the product pattern of the fermentative bacteria is influenced by the pH and to an even greater extent by the hydrogen concentration in the digester environment, a pH drop followed by a buildup of hydrogen will lead to a change in the composition of the fermentation products which are the substrates for the acetogenic and methanogenic bacteria. This substrate change might cause a decrease in the conversion rate of volatile acids, thus accelerating the pH drop which initially was very small.

With respect to pH control in an anaerobic digester, the alkalinity or acid-neutralizing capacity of the reactor contents is of vital importance. The most important buffering system in anaerobic digestion is the equilibrium between dissolved carbon dioxide and bicarbonate, which has a pK value of 6.31 at 35°C.[148] This buffer is influenced by the partial pressure of carbon dioxide and the precipitation of carbonate salts, as shown in Figure 4. The alkalinity of anaerobic digester contents can be determined by a simple titration method.[149,150] To maintain a stable digestion process, the molar bicarbonate concentration should be at least one order of magnitude higher than the total molarity of volatile fatty acids.[145,151] In sewage sludge digesters, the bicarbonate concentration normally varies between 0.04 to 0.1 mol/ℓ.[150] If the buffer capacity of the digester contents tends to be inadequate, addition of alkalinity becomes important. For this purpose the following chemicals can be used: quick lime or calcium oxide, hydrated lime or calcium hydroxide, anhydrous ammonia, sodium hydroxide, sodium bicarbonate, sodium carbonate, ammonium carbonate or ammonium hydroxide.[152,153]

Although it is the cheapest of the above-mentioned chemicals, the use of lime has a few disadvantages. Lime becomes insoluble at pH over 6.5, so that it can only be applied via slurry handling equipment. Lime also contains calcium ions, part of which will precipitate as calcium carbonate in the digester. In certain circumstances the carbonate precipitation might cause such a removal of carbon dioxide from the gas head space in the digester that it results in a partial vacuum (or negative pressure) during which atmospheric oxygen may gain access to the digester.[154] However, I have never heard of any practical scale operation in which this phenomenon occurred. Increased calcium concentrations in anaerobic digesters characterized by a high biomass retainment factor can also cause scaling of the active biomass (present as flocs, films, or granules) by calcium carbonate precipitates. As any precipitation in the biomass, this results in decreasing specific acitivities, because the bacteria are "diluted" with inactive solids.[155] Moreover, the calcium carbonate scales can become so thick that the diffusion of substrate into the biofilm or granule, instead of the microbial activity, becomes the rate-limiting factor of the process. An example of the detrimental effect calcium carbonate precipitation can have is shown in Plate 1.*[156]

The pH can also indirectly influence the rate of anaerobic digestion processes. The proteins present in dairy waste water precipitate at pH values below pH 6.5, which causes a drastic drop in the rate of hydrolysis and therefore results in a limitation of the maximum loading rates that can be applied in the treatment of such waste waters.[157]

For many compounds with a potential toxicity for bacteria active in anaerobic digestion, the molecular form is much more toxic than the ionized form. This has been shown to be the case for acetic acid,[158-160] ammonia,[161,162] and hydrogen sulfide.[163] The effect of pH on the dissociation equilibrium of ammonium is shown in Figure 5.

The precipitation of toxic heavy-metal ions such as carbonate salts is also influenced by the pH. In situations where heavy-metal toxicity occurs, it might be advisable to operate the digester in the pH range 7.4 to 8.0 instead of operating in the pH range optimal for uninhibited methanogenesis.[164]

B. Temperature

Since the early research efforts concerning the effect of temperature on the anaerobic digestion of sewage sludge, which were performed in the 1930s,[165-170] it is clear that in anaerobic digestion at least two different temperature ranges, corresponding with two different groups of bacteria, exist. Mesophilic methanogenic bacteria grow at temperatures up to 40°C, with an optimum at about 35°C. Thermophilic methanogenic bacteria grow at temperatures over 50°C, with an optimum between 55 and 75°C, depending on the particular species. It is generally assumed that in anaerobic digesters operated at psychrophilic temperatures (i.e., temperatures below 20°C) the methanogenic population is basically similar to the one in digesters operated at mesophilic conditions, but recently evidence for the existence of psychrophilic methanogenic bacteria has been presented.[171] Methane formation from acetate in peat samples from an acid-subarctic mire in Sweden was found to be optimal at 20°C. This population was not capable of consuming hydrogen.

Between the temperature ranges for mesophilic and thermophilic anaerobic digestion lies an intermediary temperature range in which neither the mesophilic nor the thermophilic population work well. Many mesophilic methanogenic bacteria are even killed at temperatures exceeding 42°C, so in operating a mesophilic anaerobic digestion process the intermediary temperature range should be avoided.

The response to short-term temperature variations will be a temporarily decreasing methanogenesis, whereas the acidogenesis is less sensitive. Hence, short-term temperature var-

---

* Plate 1 appears after page 304.

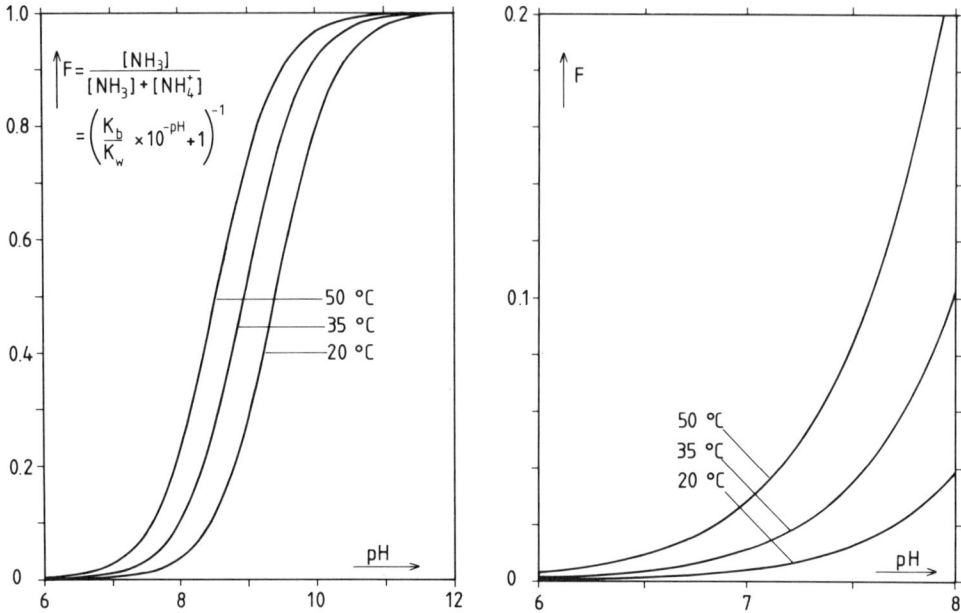

FIGURE 5.   The effect of pH on the dissociation of ammonium at various temperatures.

iations might result in a process breakdown caused by a fatal pH drop if the process loading rate is initially high and is not decreased simultaneously with the change in temperature.[172] Low-loaded anaerobic digestion processes are not very sensitive to temperature shocks. This has been demonstrated for fixed-film reactors in the mesophilic and psychrophilic temperature ranges,[173] as well as for suspended growth digesters operated at mesophilic and thermophilic temperatures.[174]

Although the rate of the anaerobic digestion process will decrease with a decreasing temperature, operation of the process at psychrophilic temperatures might be economically feasible for the treatment of cold liquid wastes in regions having a low ambient temperature. The energy required to maintain the reactor temperature at the microbial optimum represents by far the largest proportion of the total energy requirement of a digester treating cold liquid wastes, so operation at a lower (suboptimal) temperature will increase the total net energy production. The development of modern high-rate anaerobic waste water treatment systems, characterized by the retention of a high concentration of viable biomass,[175-179] makes it possible to treat waste water anaerobically at psychrophilic temperatures without needing extremely large reactors. In the temperature range of 8 to 20°C, treatment of raw domestic sewage in a granular bed UASB reactor with a hydraulic retention time of 12 hr results in a chemical oxygen demand (COD) reduction of 65 to 85%.[180] In a flocculant bed UASB reactor operated at 20°C with a hydraulic retention time of 8 hr, raw domestic sewage could be treated with a COD reduction of 75%.[181] Even slaughterhouse waste water, which contains a large fraction of suspended matter, can be treated successfully at psychrophilic temperatures. A flocculant sludge UASB reactor was reported to be able to handle organic space-loading rates up to 3.5 kg COD per cubic meter per day at a hydraulic retention time of 8 hr at temperatures as low as 20°C.[53] When granular, instead of flocculant, sludge was applied, organic space-loading rates up to 6 kg COD per cubic meter per day could be achieved at 20°C.[182]

Undoubtedly, thermophilic anaerobic digestion will become a very attractive alternative to mesophilic anaerobic digestion in the near future. At thermophilic temperatures higher loading rates can be achieved, higher methane yields can be obtained from complex wastes,

Table 4
AVERAGE ELEMENTAL COMPOSITION OF
METHANOGENIC BACTERIA

Element	% w/w	Element	% w/w	Element	ppm
Carbon	42.0	Potassium	1.6	Zinc	244
Nitrogen	6.3	Sodium	1.0	Nickel	101
Phosphorus	1.7	Magnesium	0.31	Cobalt	50
Sulfur	1.0	Calcium	0.25	Molybdenum	38
Hydrogen	11.3	Iron	0.15	Manganese	11

Adapted from Scherer, P., Lippert, H., and Wolff, G., *Biol. Trace Element Res.*, 5, 149, 1985.

and the killing of pathogens is much more effective.[183] In various cases, waste water temperatures are in the thermophilic range or thermophilic temperatures can be applied by using waste heat available from the production process generating the waste water. Important drawbacks of thermophilic anaerobic digestion are the high costs for heating in case waste heat is not available and the fact that thermophilic anaerobic digestion presumably is more sensitive (compared with mesophilic anaerobic digestion) to toxicants. Thermophilic digestion has been reported to be more sensitive to high ammonium concentrations than mesophilic digestion, which presumably is caused by an increase in the fraction of free ammonia (NH_3) with temperature.[184-186] (The effect of temperature on the dissociation equilibrium of ammonium is shown in Figure 5). At present, thermophilic anaerobic digestion is applied in full scale only for sewage sludge digestion.[187-190]

Thermophilic anaerobic digestion of agricultural wastes is undergoing extensive investigation.[4,191,237,238] Research concerning thermophilic anaerobic waste water treatment is receiving growing attention.[192,193,239-243]

C. Nutrient Requirements

One of the major advantages of anaerobic digestion as a waste water treatment process is the low growth yield of the bacteria involved. This results in a relatively small production of excess sludge, and the requirement of only low concentrations of nutrients. While municipal waste waters usually contain all nutrients in sufficient amounts, many industrial waste waters are not nutritionally balanced and often lack nitrogen, phosphorus, and/or several trace elements. Additions to the waste water of any of these nutrients should be made in the form of salts that can be readily consumed.

The elemental composition of the methanogenic bacteria gives an indication of the required waste water composition, but it should be realized that the affinity of the bacteria is not the same for all nutrients and trace elements. A very thorough and extensive study concerning the elemental composition of ten methanogenic species (with emphasis on *Methanosarcina barkeri* which was represented by five strains and cultivated on various substrates) was published by Scherer et al.[194] The values shown in Table 4 have been adapted from this study. The greatest variations were found with respect to sodium and potassium, which both seem to have important physiological functions. Although it is unknown whether zinc is an essential trace element for methanogenic bacteria, all species investigated contained remarkably high zinc contents. The high nickel content of methanogenic bacteria reflects the fact that they belong to the unique Archaebacteria and not to the common prokaryotes. Nickel is generally not essential for growth of bacteria, but it is essential for methanogenic bacteria because they contain nickel tetrapyrroles (e.g., factor F_{430}) which are involved as catalysts in methane formation.[195,196]

It has been established that ammonia is the nitrogen source for methanogenic bacteria.[65,66]

Studies with *M. barkeri* made clear that, although the amino acid cysteine stimulates growth, sulfide is always necessary as a sulfur source.[197] If the waste water to be treated lacks sulfur but contains organic material from which hydrogen will be produced during digestion, sulfur may be added as sulfate. Sulfate will be reduced to sulfide in due course, but does not form unsoluble precipitates in supply lines as sulfide might do. Phosphorus should be added as soluble phosphate salts.

It is not possible to give a general value for the appropriate COD to nutrients ratio, because it is subjective and depends on the required COD removal, the detention time, and the allowable nutrient leakage in the effluent.[198] Apart from these considerations, the COD to nutrients ratio required for a stable anaerobic digestion process depends mainly on the yield coefficient of the biomass, that is, the amount of biomass produced per amount of digested organic material.

Based on an extensive literature review, Henze and Harremoës[50] concluded that the yield coefficient decreases with a decreasing organic loading rate, resulting in a required COD to nitrogen ratio of 1000:7 or more at organic loading rates lower than 0.5 kg COD per kilogram VSS per day, whereas at organic loading rates of 1.0 kg COD per kilogram VSS per day or higher a COD to nitrogen ratio of 400:7 would be necessary. This relationship between the yield coefficient and organic loading rate is deduced from results obtained in experiments with a variety of different waste waters, so a considerable effect of waste water composition on the yield coefficient should have been accounted for. Nevertheless, the resulting relationship between the COD to nitrogen ratio and organic loading rate might be useful in certain cases. Lettinga and co-workers[101] account for the different yield coefficients of methanogenic plus acetogenic bacteria (assumed biomass yield on a waste water containing mainly volatile fatty acids: 0.05) and acidogenic bacteria (assumed biomass yield on a waste water containing mainly carbohydrates: 0.15). They advise a COD to nitrogen to phosphorus ratio of 1000:5:1 if volatile fatty acids serve as the main substrate, as is the case when the waste water is already acidified in the sewer system, and a COD to nitrogen to phosphorus ratio of 350:5:1 if complex, not-yet-acidified material serves as the substrate.

The addition of nutrients to enhance anaerobic digestion has been reported for several cases. Methanogenesis in a UASB reactor treating waste water from a factory producing liquid sugars dropped to an intolerable low level before phosphate was applied.[199] Phosphate addition also appeared to be necessary in the anaerobic treatment of rendering wastes.[200] In the presence of nickel, the specific acetate utilization rate of an acetate-enriched methanogenic culture (in which *Methanosarcina* was the predominant organism) was two to five times higher than in the absence of nickel. This stimulatory effect of nickel was enhanced by concomitant addition of iron and cobalt.[201] Growth of a pure culture of *M. barkeri* on methanol was found to be dependent on cobalt and molybdenum, whereas nickel and selenium had a stimulatory effect on growth on methanol.[202] Apart from these reports on enriched or pure cultures, there are also some reports concerning trace-element additions in anaerobic digestion of complex wastes. In 1964, McCarty[203] advised to add iron to some industrial waste waters in order to improve anaerobic treatment of these waste waters. The addition of iron, cobalt, and nickel greatly enhanced the performance of an anaerobic film-expanded bed reactor treating a powdered, whey-based solution. In this study, the methanogenic bacteria were more affected by nutrient limitation (i.e., shortage of iron, cobalt, and nickel) than were the nonmethanogenic bacteria.[204] The conversion of acetic acid to methane and carbon dioxide by the mixed microbial population from an anaerobic fixed-film digester treating bean-blanching waste water was stimulated by the addition of nickel and cobalt, especially if these elements were added in combination. Molybdenum addition was only slightly stimulatory when added in combination with nickel and cobalt.[205] Addition of cobalt to cattle manure has been reported to result in an increased biogas production per unit weight of volatile solids added during digestion. This effect was found to be at least partly due to increased cellular synthesis induced by the cobalt addition.[206]

It is very difficult, if not impossible, to give exact figures for the amount of nutrients and trace elements that should be added to secure a stable digestion process, even if the theoretically required amount has been determined, because it is the *available* amount of nutrients and trace elements that counts. The precipitation of metals with carbonate or sulfide is described in the paragraph concerning toxicity. Such precipitation processes can cause serious metal deficiencies. Precipitation and chelation of metal ions play a very important role in anaerobic digestion, determining the availability of metals as nutrients.[207,208] Phosphorus deficiency can occur in the presence of excessive amounts of iron, aluminum, and calcium ions that cause precipitation of insoluble phosphate salts.[209] The precipitation of calcium salts is discussed in more detail in the section concerning pH and buffering capacity.

D. Toxicity

1. Introduction

Since the susceptibility of methanogenic bacteria to toxic compounds in the environment is much greater than that of the acidogenic bacteria, the presence of toxic (inhibitory) compounds in the waste water might be the start of a fatal upset of the digestion process. The introduction of a certain compound that inhibits the methanogenesis will be followed by a buildup of organic acids which might result in a sour digester. The aspect of acidity is dealt with in Section III.A. In this section, the effect of toxic compounds that might occur in waste water on methane-producing bacterial populations will be discussed.

The potential inhibitors of methanogenesis can be divided in two categories:

1. Compounds that are not normally present in biological systems, and that even at very low concentrations disrupt the cell metabolism. Examples of compounds to be placed in this category are chloroform, cyanide, and monensin (an additive to cattle fodder).
2. Compounds that normally are present in anaerobic microbial environments. Usually these compounds are either inert to methanogenic bacteria or even necessary for growth. Substrates (e.g., acetic acid), nutrients (e.g., ammonium), and trace elements (e.g., nickel) among others are part of this category.

For compounds belonging to the first category, an increasing concentration will always result in an increasing inhibition. For compounds belonging to the second category, three concentration ranges, each characterized by a different effect of the particular compound on the methanogenic bacteria, can be distinguished.[210] Initially, an increasing concentration will result in an increasing bacterial activity. This concentration range is followed by a range in which the concentration is optimal (an increasing concentration neither stimulates nor inhibits the bacterial activity). Increasing the concentration further than the optimal range will result in inhibition of the bacterial activity. In general, this inhibition increases as the concentration increases. Ammonium-nitrogen may serve as an example of a compound belonging to the second category. McCarty reported a stimulation of methanogenic activity in the concentration range 50 to 200 mg/ℓ,[210] whereas others have found a stimulatory effect up to 700 mg/ℓ.[211] However, at concentrations exceeding 700 mg/ℓ, an increasing concentration results in a decreasing methanogenic activity.[67,68]

Some of the toxicants encountered in specific industrial wastewaters are[51]

- Heavy-metal catalysts from chemical processes
- Pharmaceuticals supplemented to animal feeds
- Detergents and disinfectants used in food processing equipment cleanup
- Solvents from degreasing operations
- Inhibitors formed as secondary products, e.g., cyanide in coking operations
- Toxic process stream leakage, e.g., formaldehyde
- Chemical inhibitor treatments for food preservation

Since it would be rare to find an industrial waste water completely devoid of all potential toxicants, there is a commonly held belief that anaerobic digestion is not appropriate for treatment of most industrial waste waters. However, this misbelief is based mostly on results obtained in the early days of scientific interest in anaerobic digestion as a waste water treatment process. At that time, experiments concerning toxicity problems in anaerobic digestion were performed for periods too short to allow any adaptation of the bacteria to occur, or they were performed with systems characterized by a relatively short sludge retention time such as in the anaerobic contact process. At present many mechanisms such as adaptation and antagonism that increase the tolerance of the anaerobic digestion process are known and more or less understood.[211] Moreover, it should be realized that in many cases the actual *effective* concentration of a potential toxicant in the digester environment is much less than the concentration in the waste water to be treated. Adaptation, antagonism, synergism, and effective concentration will be discussed further in this section.

2. Effective Concentration of Potential Inhibitors

In many cases, the extent of the toxic effect of a compound is influenced by the pH. Examples of compounds showing a pH-related toxicity are ammonium,[161,162] acetic acid,[158-160] and sulfide.[163] The unionized forms of these compounds are supposed to be the toxic agents, whereas the ionized forms have much less effect. It has been shown that the pH-controlled operation and a suitable choice of temperature can prevent the inhibitory effect of ammonia in manure digestion.[212]

Another mechanism that decreases the effective concentration of potential toxicants is the formation of insoluble precipitates. Precipitation of metal sulfides is probably the best example, since sulfide is the end product of the anaerobic conversion of sulfur-containing organic material. The precipitation of metal sulfides can be used to prevent sulfide toxicity, but the mechanism is better known in relation to toxicity of heavy metals.[213] The precipitation of metal carbonates also plays a role in the prevention of heavy metal toxicity.[214,215] but it needs a relatively high pH to be important. For the formation of cadmium carbonate, a pH above pH 7.2 is necessary; for the formation of zinc carbonate the pH should be even higher than pH 7.7. In cases of potential heavy-metal toxicity, it is advised to add sulfate (which will be reduced to sulfide) to the waste water and operate the digester at a somewhat higher pH than normal.[164,216]

Another example of precipitation reducing toxicity is the interaction between calcium ions and long-chain fatty acids such as oleic, palmitic, and stearic acid.[74]

3. Antagonism and Synergism

Antagonism is the reduction of the toxic effect of one compound by the presence of another. Syngerism is an increase in the apparent toxicity of one compound caused by the presence of a second compound. As with toxicity, antagonism and synergism are concentration dependent. Antagonism and synergism have been studied in detail using acetate-consuming enrichment cultures exposed to various cations.[217-219] Table 5 gives an overview of antagonistic and synergistic relationships in cation toxicity adapted from the work of Kugelman and Chin.[219] They reported that peak antagonism is achieved at 0.01 *M* with monovalent cations as antagonists and at 0.005 *M* when divalent cations serve as antagonists. They also found that the concentration of the synergist at which the synergistic effect begins is below the concentration at which the synergistic cation will produce inhibition when present alone.

Although synergism and antagonism undoubtedly play an important role in toxicity in anaerobic digestion, the addition of antagonistic chemicals to suppress toxicity is never applied in practice.

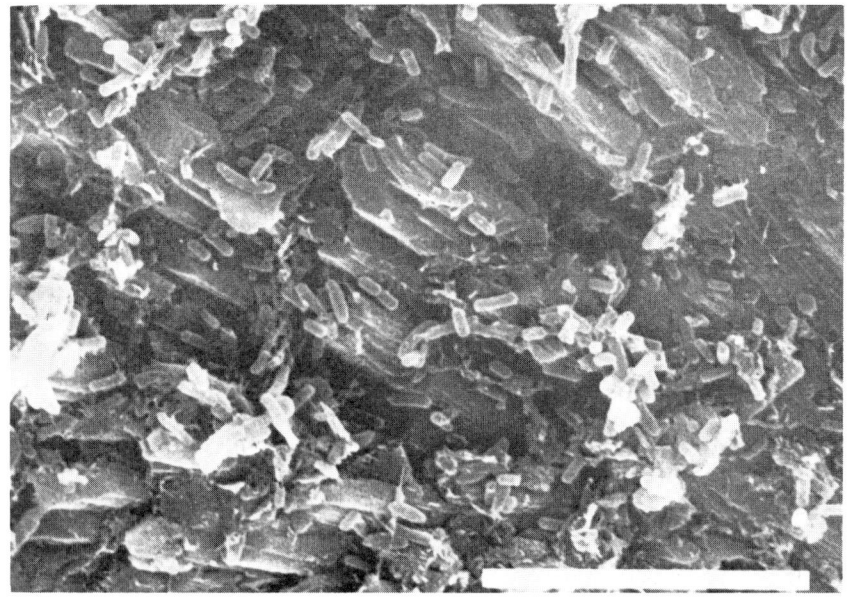

A

B

PLATE 1. Electron micrographs of granular methanogenic sludge after prolonged cultivation at conditions favoring calcium carbonate precipitation. (A): detail of outside surface area, bar = 10 μm and (B): scale around whole granule (biomass was lost during preparation), bar = 1000 μm.

Table 5
ANTAGONISM AND SYNERGISM IN CATION TOXICITY IN
ANAEROBIC DIGESTION

Primary toxicant	Antagonist	Synergist
Na^+	K^+	NH_4^+, Ca^{2+}, and Mg^{2+}
K^+	Na^+, Ca^{2+}, Mg^{2+}, and NH_4^+	
Ca^{2+}	Na^+ and K^+	NH_4^+ and Mg^{2+}
Mg^{2+}	Na^+ and K^+	NH_4^+ and Ca^{2+}
NH_4^+	Na^+	K^+, Ca^{2+}, and Mg^{2+}

Adapted from Kugelman, I. J. and Chin, K. K., *Anaerobic Biological Treatment Processes,*
(Advances in Chemistry Series, Vol. 105), Gould, R. F., Ed., American Chemical Society,
Washington, D.C., 1971, chap. 5.

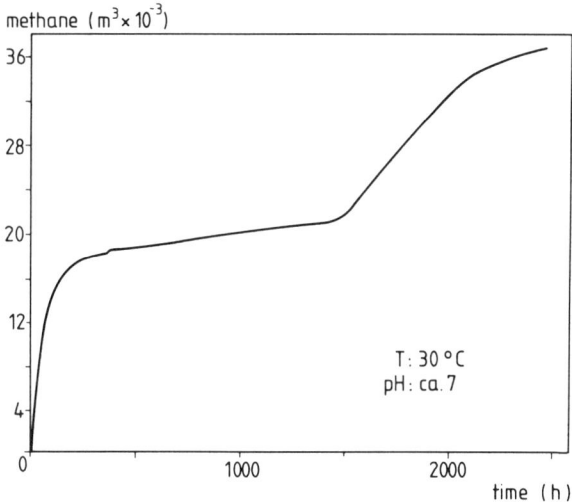

FIGURE 6.   Cumulative methane production as a function of time
in the anaerobic digestion of twice-diluted potato juice.

4. Adaptation

One of the most important aspects of toxicity in anaerobic digestion is the capacity of the
methane-producing bacterial population to adapt to the presence of toxic compounds. Adaptation (or acclimatization) takes place during application of gradually increasing concentrations of toxicants,[210,220] but also after a period of apparent death (i.e., zero gas production
caused by the toxic compound) recovery of the bacterial activity by adaptation can occur.[67,221,222] Such a recovery of methane production after an apparent death phase during
which adaptation occurred is shown in Figure 6, which is based on results of a batch
experiment in which diluted potato juice (approximately 2.3 g nitrogen per liter) is digested
by unadapted granular methanogenic sludge.[244] In the initial stage of the digestion, methane
production was possible because a large fraction of the nitrogen was still bound in proteins,
but as soon as these proteins had been broken down and the nitrogen had been released as
ammonium, the methane production stopped. The recovery of the methane production after
the inhibition is too fast to be accounted for by growth of mutants that can stand toxicity.
Apparently, during inhibition the methanogenic bacteria are able to rearrange or rebuild
metabolism in such a way that they can stand the initially inhibitory concentration of the
toxicant.

Apart from adaptation as a result of metabolic changes in the bacteria already present before the exposure to the inhibitory compound started, the population as a whole can also become adapted to the potential toxicant as a result of growth of bacteria that are able to decompose that particular compound. Examples are the reduction of sulfite used as a bleacher and bactericide in maize-starch production by sulfate-reducing bacteria in an UASB reactor,[223] the biodegradation of chlorophenols and chlorocatechols present in kraft-bleaching effluent,[224] and the biodegradation of pentachlorophenol which allows anaerobic treatment of waste water containing 5 mg/ℓ pentachlorophenol, whereas 0.2 mg/ℓ is the toxicity level for unacclimatized methane-producing populations.[225]

The acclimization potential of methanogenic bacterial populations is described in more detail by Parkin and Miller.[226]

As can be seen in Figure 6, adaptation may take some time. It is obvious that the methanogenic bacterial population in an anaerobic digester can become adapted to toxicant exposure only if the solids retention time at which the digester is operated allows for the adaptation period.[227,228] The reason, in the older studies concerning toxicity in anaerobic digestion, adaptation could not be noticed is that the systems then available (continuously stirred tank reactor and anaeobic contact process) did not offer a sufficiently long solids retention time. In cases where periodic shock loads with toxicants occur, a high solids retention time is needed to prevent washout of the biomass during the period of reduced growth rates directly after the shock exposure.[222]

Modern waste water treatment systems such as the anaerobic filter and the upflow anaerobic sludge bed reactor combine very high solids retention times with relatively low liquid retention times, which makes these systems able to handle toxic substances at least as well, if not better, than aerobic activated sludge systems.[229]

IV. CONCLUSIONS

In this chapter the microbiology, (bio)chemistry, and technology of the anaerobic degradation of organic pollutants have been discussed. From the viewpoint of the engineer who must design and operate anaerobic waste treatment equipment, it may be concluded that, at present, enough is known of the basics of anaerobic digestion to be able to design reliable and economically feasible methane-producing anaerobic waste or waste water treatment systems. Research concerning the physiological requirements of the various metabolic groups that play a role in anaerobic digestion has made it possible to optimize the design and operation of treatment systems for any specific situation. This approach will, in some cases, result in the application of two-phase systems (ie., hydrolysis and acidification in one reactor, acetogenesis and methanogenesis in the other reactor) instead of having all metabolic stages in one reactor. In a two-phase system, it is possible to optimize hydrolysis as well as methanogenesis. Two-phase waste water treatment is not always applied because it is much more expensive than the single-reactor treatment in which the situation is optimized for methanogenesis, often resulting in suboptimal conditions for hydrolysis and acidogenesis. For many industrial waste waters, there is no real need to optimize hydrolysis and acidogenesis, because the rates are sufficiently high even at suboptimal conditions. In fact, the sewer of the factory in such cases acts as a kind of first phase reactor.

Research concerning acetogenic bacteria has generated a still-growing understanding of the dynamics of anaerobic digestion, especially with respect to the hydrogen turnover. The understanding of the importance of extremely low hydrogen levels has opened ways to develop more sophisticated control systems which will make it possible to monitor a digester upset within seconds after this upset has started.

Research concerning the chemical and physical requirements of methane fermentations has made clear that anaerobic digestion is appropriate for treatment of a much wider variety of wastes and waste waters than was believed some 20 years ago. In retrospect, it appears that an inherent lack of trace metals and/or other nutrients in many past treatability studies may have been the cause of negative results. A similar story can be told concerning the sensitivity of anaerobic digestion for exposure to toxicants. In the past, anaerobic digestion was regarded as a very sensitive process, which could easily become unbalanced, resulting in a sour digester not able to produce biogas anymore. This (wrong) opinion was based on experiences with digesters characterized by a short cell-residence time. The present high-rate anaerobic treatment systems are characterized by a very long cell-residence time, which allows for adaptation of the bacterial population to compounds which initially act as toxicants, but which are harmless after adaptation has taken place. Therefore, the presently available anaerobic technologies are as feasible as aerobic technologies for the treatment of most kinds of organic waste and waste water.

Further research concerning the basic microbiology and biochemistry of anaerobic digestion is required, because even after the rapid increase in research efforts in this field during the last decade anaerobic digestion, in many cases, is still something that is put into operation as a "black box". The future development of anaerobic digestion therefore requires an extensive combined effort of researchers in biochemistry, microbiology, and technology. Research must be directed towards anaerobic digestion of substrates that at present are regarded as hard to degrade (e.g., fats and lignocellulosics) and to the operation of anaerobic digestion at suboptimal conditions, including psychrophilic and thermophilic anaerobic digestion and anaerobic digestion in the presence of inhibitory compounds.

The overall conclusion of this chapter is that anaerobic digestion is a form of biotechnology which has an enormous potential in solving the environmental problem of pollution with organic wastes and waste waters. Future research will even increase this potential.

REFERENCES

1. **Zehnder, A. J. B.,** Ecology of methane formation, in *Water Pollution Microbiology,* Vol. 2, Mitchell, R., Ed., John Wiley & Sons, New York, 1978, chap. 13.
2. **Berner, W., Bucher, P., Oescher, H., and Stauffer, B.,** AISH-AIMPA Symp. on Isotopes and Impurities in Snow and Ice, Grenoble, France, 1975.
3. **Svensson, B. H.,** Carbon Fluxes from Acid Peat of a Subarctic Mire with Emphasis on Methane, Ph.D. thesis, Swedish University of Agricultural Sciences, Uppsala, Sweden, 1983.
4. **Varel, V. H.,** Characteristics of bacteria in thermophilic digesters and effect of antibiotics on methane production, in *Fuel Gas Developments,* Wise, D. L., Ed., CRC Press, Boca Raton, Fla., 1983, 19.
5. **Belyaev, S. S. and Ivanov, M. V.,** Bacterial methanogenesis in underground waters, in *Environmental Geochemistry,* Hallberg, R., Ed., Forskningsradsnaemnden Foerlagstjaensten, Stockholm; *Ecol. Bull.,* 35, 273, 1983.
6. **Baross, J. A., Lilley, M. D., and Gordon, L. I.,** Is the CH$_4$ venting from submarine hydrothermal systems produced by thermophilic bacteria?, *Nature,* 298, 366, 1982.
7. **Pine, M. J.,** The methane fermentations, in *Anaerobic Biological Treatment Processes,* (Advances in Chemistry Series, Vol. 105), Gould, R. F., Ed., American Chemical Society, Washington, D. C., 1971, chap. 1.,
8. **Priestly, J.,** *Experiments and Observations on Different Kinds of Air,* Birmingham, 1790. Available from Kraus Reprint & Periodicals, Millwood, N.Y.,
9. **Zehnder, A. J. B., Ingvorsen, K., and Marti, T.,** Microbiology of methane bacteria, in *Anaerobic Digestion 1981,* Hughes, D. E., Stafford, D. A., Wheatley, B. I., Baader, W., Lettinga, G., Nyns, E. J., Verstraete, W., and Wentworth, R. L., Eds., Elsevier, Amsterdam, 1982, 45.
10. **Popoff, L.,** Ueber die Sumpfgasgärung, *Pfluegers Arch. Gesamte Physiol.,* 10, 113, 1875.
11. **McCarty, P. L.,** One hundred years of anaerobic treatment, in *Anaerobic Digestion 1981,* Hughes, D. E., Stafford, D. A., Wheatley, B. I., Baader, W., Lettinga, G., Nyns, E. J., Verstraete, W., and Wentworth, R. L., Eds., Elsevier, Amsterdam, 1982, 3.
12. **Söhngen, N. L.,** Het Ontstaan en Verdwijnen van Waterstof en Methaan Onder den Invloed van het Organisch Leven, Dissertation, Delft Technical University, J. Vis Jr., Delft, Netherlands, 1906.

13. **van Niel, C. B.**, The "Delft School" and the rise of general microbiology, *Bacteriol. Rev.*, 13, 161, 1949.
14. **Schnellen, Ch. G. T. P.**, Onderzoekingen over de Methaangisting, Dissertation, Delft Technical University, Rotterdam, 1947.
15. **Bryant, M. P., Wolin, E. A., Wolin, M. J., and Wolfe, R. S.**, *Methanobacillus omelianskii*, a symbiotic association of two species of bacteria, *Arch. Microbiol.*, 59, 20, 1967.
16. **Mah, R. A., Ward, D. M., Baresi, L., and Glass, T. L.**, Biogenesis of methane, *Annu. Rev. Microbiol.*, 31, 309, 1977.
17. **Woese, C. R.**, Archaebacteria, *Sci. Am.*, 244, 94, 1981.
18. **Kandler, O. and König, H.**, Chemical composition of the peptidoglycan-free cell walls of methanogenic bacteria, *Arch. Microbiol.*, 118, 141, 1978.
19. **Sahm, H.**, Microbiology and biochemistry of anaerobic wastewater treatment, convenor's address presented at the Eur. Symp. Anaerobic Waste Water Treatment, Noordwijkerhout, The Netherlands, November 1983.
20. **Hobson, P. N. and Wallace, R. J.**, Microbial ecology and activities in the rumen. I, *Crit. Rev. Microbiol.*, 9(3), 165, 1982.
21. **Hobson, P. N. and Wallace, R. J.**, Microbial ecology and activities in the rumen. II, *Crit. Rev. Microbiol.*, 9(4), 253, 1982.
22. **Winfrey, M. R. and Zeikus, J. G.**, Effect of sulphate on carbon and electron flow during microbial methanogenesis in freshwater sediments, *Appl. Environ. Microbiol.*, 33, 275, 1977.
23. **Laanbroek, H. J. and Veldkamp, H.**, Microbial interactions in sediment communities, *Philos. Trans. R. Soc. London, Ser. B*, 297, 533, 1982.
24. **Hobson, P. N.** The bacteriology of anaerobic sewage digestion, *Process Biochem.*, 8(1), 19, 1973.
25. **McInerney, M. J., Bryant, M. P., and Stafford, D. A.**, Metabolic stages and energetics of microbial anaerobic digestion, in *Anaerobic Digestion*, Stafford, D. A., Wheatley, B. I., and Hughes, D. E., Eds., Applied Science, London, 1980, 91.
26. **Symons, G. E. and Buswell, A. M.**, The methane fermentation of carbohydrates, *J. Am. Chem. Soc.*, 55, 2028, 1933.
27. **Verstreate, W., De Baere, L., and Rozzi, A.**, Phase separation in anaerobic digestion: motives and methods, *Trib. Cebedeau*, 34, 367, 1981.
28. **Ghosh, S. and Klass, D. L.**, Two phase anaerobic digestion, *Process Biochem.*, 13(4), 15, 1978.
29. **Hobson, P. N.**, Straw as feedstock for anaerobic digesters, in *Straw Decay and its Effect on Disposal and Utilization*, Grossbard, E., Ed., John Wiley & Sons, London, 1979, 217.
30. **Lequerica, L., Vallés, S., and Flors, A.**, Kinetics of rice straw methane fermentation, *Appl. Microbiol. Biotechnol.*, 19, 70, 1984.
31. **Zoetemeyer, R. J., Van den Heuvel, J. C.,and Cohen, A.**, pH influence on acidogenic dissimilation of glucose in an anaerobic digestor, *Water Res.*, 16, 303, 1982.
32. **Zoetemeyer, R. J.**, Acidogenesis of Soluble Carbohydrate Containing Wastewaters, Ph.D. thesis, University of Amsterdam, Amsterdam, 1982.
33. **Breure, A. M. and Van Andel, J. G.**, Hydrolysis and acidogenic fermentation of a protein, gelatin, in an anaerobic continuous culture, *Appl. Microbiol. Biotechnol.*, 20, 40, 1984.
34. **Breure, A. M. and Van Andel, J. G.**, Hydrolysis and acidification of gelatin, in Proc. Eur. Symp. Anaerobic Waste Water Treatment, Van den Brink, W. J., Ed., TNO Corporate Communication Department, The Hague, Netherlands, 1983, 31.
35. **Eastman, J. A. and Ferguson, J. F.**, Solubilization of particulate organic carbon during the acid phase of anaerobic digeston, *J. Water Pollut. Control Fed.*, 53, 352, 1981.
36. **O'Rourke, J. T.**, Kinetics of Anaerobic Waste Treatment at Reduced Temperatures, Ph.D. thesis, Stanford University, Stanford, California, 1968.
37. **Roy, F., Verrier, D., and Florentz, M.**, Importance of the liquefaction step in a two stage methanisation process of solid cannery vegetables wastes, in Proc. Eur. Symp. Anaerobic Waste Water Treatment, Van den Brink, W. J., Ed., TNO Corporate Communication Department, The Hague, Netherlands, 1983, 175.
38. **Hobson, P. N.**, Production of biogas from agricultural wastes, in *Advances in Agricultural Microbiology*, Subba Rao, N. S., Ed., Oxford & IBH, New Delhi, 1982, chap. 20.
39. **Fan, L. T., Lee, Y. H., and Beardmore, D. H.**, Mechanism of the enzymatic hydrolysis of cellulose: effects of major structural features of cellulose on enzymatic hydrolysis, *Biotechnol. Bioeng.*, 22, 177, 1980.
40. **Kelsey, R. C. and Shafizadeh, F.**, Enhancement of cellulose accessibility and enzymatic hydrolysis by simultaneous wet milling, *Biotechnol. Bioeng.*, 22, 1025, 1980.
41. **Hashimoto, A. G.**, Microbial hydrolysis of thermochemically treated and un-treated manure-straw mixtures, *Agric. Wastes*, 4, 345, 1982.
42. **Gharpuray, M. M., Lee, Y. H., and Fan, L. T.**, Structural modification of lignocellulosics by pretreatments to enhance enzymatic hydrolysis, *Biotechnol. Bioeng.*, 25, 157, 1983.

43. **Ibrahim, M. N. M. and Pearce, G. R.,** Effects of chemical pretreatments on the composition and *in vitro* digestibility of crop by-products, *Agric. Wastes,* 5, 135, 1983.
44. **Wicklow, D. T., Detroy, R. W., and Jessee, B. A.,** Decomposition of lignocellulose by *Cyathus stercoreus* (Schw.) de Toni NRRL 6473, a "White Rot" fungus from cattle dung, *Appl. Environ. Microbiol.,* 40, 169, 1980.
45. **Hills, D. J. and Nakano, K.,** Effects of particle size on anaerobic digestion of tomato solid wastes, *Agric. Wastes,* 10, 285, 1984.
46. **Pfeffer, J. T.,** Increased loading on digesters with recycle of digested solids, *J. Water Pollut. Control Fed.,* 40, 1920, 1968.
47. **De Baere, L. A., Rozzi, A.,and Verstraete, W.,** Solubilization of particulate organic matter as the rate-limiting step in anaerobic digestion loading, *Trib. Cebedeau,* 37, 75, 1984.
48. **Ghosh, S.,** Kinetics of acid-phase fermentation in anaerobic digestion, *Biotechnol. Bioeng. Symp.,* 11, 301, 1981.
49. **Nagase, M. and Matsuo, T.,** Interactions between amino-acid degrading bacteria and methanogenic bacteria in anaerobic digestion, *Biotechnol. Bioeng.,* 24, 2227, 1982.
50. **Henze, M. and Harremoës, P.,** Anaerobic treatment of wastewater in fixed film reactors — a literature review *Water Sci. Technol.,* 15(8/9), 1, 1983.
51. **Speece, R. E.,** Anaerobic biotechnology for industrial wastewater treatment, *Environ. Sci. Control.,* 17, 416A, 1983.
52. **Sayed, S. K. I.,** *The Anaerobic Purification of the Wastewater from Swine Slaughterhouses* (transl.), Department of Water Pollution Control, Agricultural University, Wageningen, Netherlands, 1982.
53. **Sayed, S. K. I., De Zeeuw, W. J., and Lettinga, G.,** Anaerobic treatment of slaughterhouse waste using a flocculant sludge UASB reactor, *Agric. Wastes,* 11, 197, 1984.
54. **Gottschalk, G.,** *Bacterial Metabolism,* Springer-Verlag, New York, 1979.
55. **Toerien, D. F. and Hattingh, W. H. J.,** Anaerobic digestion. I. The microbiology of anaerobic digestion, *Water Res.,* 3, 385, 1969.
56. **Cohen, A.,** Optimization of Anaerobic Digestion of Soluble Carbohydrate Containing Wastewaters by Phase Separation, Ph.D. thesis, University of Amsterdam, Amsterdam, 1982, chap. 7.
57. **Pipyn, P. and Verstraete, W.,** Lactate and ethanol as intermediates in two-phase anaerobic digestion, *Biotechnol. Bioeng.,* 23, 1145, 1981.
58. **Cohen, A.,** Two-phase digestion of liquid and solid wastes, in *Proc. 3rd Int. Symp. on Anaerobic Digestion,* Wentworth, R. L., Ed., Third International Symposium on Anaerobic Digestion, Cambridge, Mass., 1983, 123.
59. **Wolin, M. J. and Miller, T. L.,** Interspecies hydrogen transfer: 15 years later, *ASM News,* 48, 561, 1982.
60. **Wolin, M. J.,** Interactions between H_2-producing and methane producing species, in *Microbial Formation and Utilization of Gases,* Schlegel, H. G., Gottschalk, G., and Pfennig, N., Eds., E. Goltze, Göttingen, West Germany, 1976, 141.
61. **Wolin, M. J.,** The rumen fermentation: a model for microbial interactions in anaerobic ecosystems, *Adv. Microb. Ecol.,* 3, 49, 1979.
62. **Mah, R. A.,** Interactions of methanogens and non-methanogens in microbial ecosystems, in *Proc. 3rd Int. Symp. on Anaerobic Digestion,* Wentworth, R. L., Ed., Cambridge, Mass., 1983, 13.
63. **Wolfe, R. S.,** Fermentation and anaerobic respiration in anaerobic digestion, in *Proc. 3rd Int. Symp. on Anaerobic Digestion,* Wentworth, R. L., Ed., Cambridge, Mass., 1983, 3.
64. **Reddy, C. A., Bryant, M. P., and Wolin, M. J.,** Characteristics of S-organism isolated from *Methanobacillus omelianskii, J. Bacteriol.,* 109, 539, 1972.
65. **Bryant, M. P., Tzeng, S. F., Robinson, I. M., and Joyner, A. E., Jr.,** Nutrient requirements of methanogenic bacteria, in *Anaerobic Biological Treatment Processes,* (Advances in Chemistry Series, Vol. 105), American Chemical Society, Washington, D.C., 1971, 23.
66. **Kenealy, W. R., Thompson, T. E., Schubert, K. R., and Zeikus, J. G.,** Ammonia assimilation and synthesis of alanine, aspartate and glutamate, in *Methanosarcina barkeri* and *Methanobacterium thermoautotrophicum, J. Bacteriol.,* 150, 1357, 1982.
67. **Koster, I. W. and Lettinga, G.,** Ammonium-toxicity in anaerobic digestion, in *Proc. Eur. Symp. Anaerobic Waste Water Treatment,* Van den Brink, W. J., Ed., TNO Corporate Communication Department, The Hauge, Netherlands, 1983, 553.
68. **Koster, I. W. and Lettinga, G.,** The influence of ammonium-nitrogen on the specific activity of pelletized methanogenic sludge, *Agric. Wastes,* 9, 205, 1984.
69. **Kotzé, J. P., Thiel, P. G., and Hattingh, W. H. J.,** Anaerobic digestion, II. The characterization and control of anaerobic digestion, *Water Res.,* 3, 459, 1969.
70. **Jeris, J. S. and McCarty, P.L.,** The biochemistry of methane fermentation using ^{14}C tracers, *J. Water Pollut. Control Fed.,* 37, 178, 1965.
71. **Weng, C.-N. and Jeris, J. S.,** Biochemical mechanisms in the methane fermentation of glutamic and oleic acids, *Water Res.,* 10, 9, 1976.

72. **McInerney, M. J., Mackie, R. I., and Bryant, M. P.,** Syntrophic association of a butyrate-degrading bacterium and *Methanosarcina* enriched from bovine rumen fluid, *Appl. Environ. Microbiol.*, 41, 826, 1981.

73. **McInerney, M. J., Bryant, M. P., Hespell, R. B., and Costerton, J. W.,** *Syntrophomonas wolfei* gen. nov., sp. nov., an anaerobic syntrophic fatty acid degrading bacterium, *Appl. Environ. Microbiol.*, 41, 1029, 1981.

74. **Hanaki, K., Matsuo, T., and Nagase, M.,** Mechanism of inhibition caused by long-chain fatty acids in anaerobic digestion process, *Biotechnol. Bioeng.* 23, 1591, 1981.

75. **Thauer, R. K., Jungermann, K., and Decker, K.,** Energy conservation in chemotrophic anaerobic bacteria, *Bacteriol. Rev.*, 41, 100, 1977.

76. **Chang, R.,** *Physical Chemistry with Applications to Biological Systems*, MacMillan, New York, 1977.

77. **Robinson, J. A. and Tiedje, J. M.,** Kinetics of hydrogen consumption by rumen fluid, anaerobic digestor sludge and sediment, *Appl. Environ. Microbiol.*, 44, 1374, 1982.

78. **Schumpe, A., Quicker, G., and Deckwer, W.-D.,** Gas solubilities in microbial media, in *Advances in Biochemical Engineering*, Vol. 24, Fiechter, A., Ed., Springer-Verlag, Berlin, 1982, 1.

79. **Wilhelm, E., Battino, R., and Wilcock, R. J.,** Low-pressure solubility of gases in liquid water, *Chem. Rev.*, 77, 219, 1977.

80. **Loveley, D. R., Dwyer, D. F., and Klug, M. J.,** Kinetic analysis of competition between sulfate reducers and methanogens for hydrogen in sediments, *Appl. Environ. Microbiol.*, 43, 1373, 1982.

81. **Robinson, J. A. and Tiedje, J. M.,** Competition between sulfate-reducing and methanogenic bacteria for H_2 under resting and growing conditions, *Arch. Microbiol.*, 137, 26, 1984.

82. **Gujer, W. and Zehnder, A. J. B.,** Conversion processes in anaerobic digestion, *Water Sci. Technol.*, 15(8/9), 127, 1983.

83. **McInerney, M. J., Bryant, M. P., and Pfennig, N.,** Anaerobic bacterium that degrades fatty acids in syntrophic association with methanogens, *Arch. Microbiol.*, 122, 129, 1979.

84. **Boone, D. R. and Bryant, M. P.,** Propionate-degrading bacterium, *Syntrophobacter wolnii* sp. nov., gen. nov., from methanogenic ecosystems, *Appl. Environ. Microbiol.*, 40, 626, 1980.

85. **Adamse, A. D., Velzeboer, C. T. M., and Janssen, W. B.,** Competition between anaerobic acidogenic and methanogenic bacteria, in *Proc. Eur. Symp. Anaerobic Waste Water Treatment*, Van den Brink, W. J., Ed., TNO Corporate Communication Department, The Hague, Netherlands, 1983, 36.

86. **Adamse, A. D., Velzeboer, C. T. M., and Janssen, W. B.,** Competition between anaerobic acidogenic bacteria and methanogens, *Antonie van Leeuwenhoek; J. Microbiol. Serol.*, 50, 75, 1984.

87. **Braun, M., Schobert, S., and Gottschalk, G.,** Enumeration of bacteria forming acetate from H_2 and CO_2 in anaerobic habitats, *Arch. Microbiol.*, 120, 201, 1979.

88. **Prins, R. A. and Lankhorst, A.,** Synthesis of acetate from CO_2 in the cecum of some rodents, *FEMS Microbiol. Lett.*, 1, 255, 1977.

89. **Wieringa, K. T.,** The formation of acetic acid from carbon dioxide and hydrogen by anaerobic spore-forming bacteria, *Antonie van Leeuwenhoek; J. Microbiol. Serol.*, 6, 251, 1940.

90. **Karlsson, J. L., Volcani, B. E., and Barker, H. A.,** The nutritional requirements of *Clostridium aceticum*, *J. Bacteriol.*, 56, 781, 1948.

91. **Adamse, A. D.,** New isolation of *Clostridium aceticum* (Wieringa), *Antonie van Leeuwenhoek; J. Microbiol. Serol.*, 46, 523, 1980.

92. **Braun, M., Mayer, F., and Gottschalk, G.,** *Clostridium aceticum* (Wieringa), a microorganism producing acetic acid from molecular hydrogen and carbon dioxide, *Arch. Microbiol.*, 128, 288, 1981.

93. **Balch, W. E., Schoberth, S., Tanner, R. E., and Wolfe, R. S.,** *Acetobacterium*, a new genus of hydrogen-oxidizing, carbon dioxide-reducing, anaerobic bacteria, *Int. J. Syst. Bacteriol.*, 27, 355, 1977.

94. **Leigh, J. A., Mayer, F., and Wolfe, R. S.,** *Acetogenium kivui*, a new thermophilic hydrogen-oxidizing acetogenic bacterium, *Arch. Microbiol.*, 129, 275, 1981.

95. **Ohwaki, K. and Hungate, R. E.,** Hydrogen utilization by Clostridia in sewage sludge, *Appl. Environ. Microbiol.*, 33, 1270, 1977.

96. **Braun, K. and Gottschalk, G.,** Effect of molecular hydrogen and carbon dioxide on chemo-organotrophic growth of *Acetobacterium woodii* and *Clostridium aceticum*, *Arch. Microbiol.*, 128, 294, 1981.

97. **Laanbroek, H. J., Abee, T., and Voogd, I.,** Alcohol conversions by *Desulfobulbus propionicus* Lindhorst in the presence and absence of sulphate and hydrogen, *Arch. Microbiol.*, 133, 178, 1982.

98. **Zehnder, A. J. B. and Koch, M. E.,** Thermodynamic and kinetic interactions of the final steps in anaerobic digestion, in *Proc. Eur. Symp. Anaerobic Waste Water Treatment*, Van den Brink, W. J., Ed., TNO Corporate Communication Department, The Hague, Netherlands, 1983, 86.

99. **Goldberg, I. and Cooney, C. L.,** Formation of short-chain fatty acids from H_2 and CO_2 by a mixed culture of bacteria, *Appl. Environ. Microbiol.*, 41, 148, 1981.

100. **Lettinga, G., Van der Geest, A. Th., Hobma, S. W., and Van der Laan, J. B. R.,** Anaerobic treatment of methanolic wastes, *Water Res.*, 13, 725, 1979.

101. **Lettinga, G., De Zeeuw, W. J., and Ouborg, E.,** Anaerobic treatment of wastes containing methanol and higher alcohols, *Water Res.,* 15, 171, 1981.

102. **Adamse, A. D. and Velzeboer, C. T. M.,** Features of a *Clostridium,* strain CV-AA1, an obligatory anaerobic bacterium producing acetic acid from methanol, *Antonie van Leeuwenhoek; J. Microbiol. Serol.,* 48, 305, 1982.

103. **Andreesen, J. R., Gottschalk, G., and Schlegel, H. G.,** Clostridium formicoaceticum nov. spec., Isolation, description and distinction from *C. aceticum* and *C. thermoaceticum, Arch. Mikrobiol.,* 72, 154, 1970.

104. **Sharak Genther, B. R., Davis, C. L., and Bryant, M. P.,** Features of rumen and sewage sludge strains of *Eubacterium limosum,* a methanol and H_2-CO_2 utilizing species, *Appl. Environ. Microbiol.,* 42, 12, 1981.

105. **Zeikus, J. G., Lynd, L. H., Thompson, T. E., Krzycki, J. A., Weimer, P. J., and Hegge, P. W.,** Isolation and characterization of a new, methylotrophic, acidogenic anaerobe, the Marburg strain, *Curr. Microbiol.,* 3, 381, 1980.

106. **Zinder, S. H. and Koch, M. E.,** Non-aceticlastic methanogenesis from acetate: acetate oxidation by a thermophilic syntrophic coculture, *Arch. Microbiol.,* 138, 263, 1984.

107. **Weber, H., Kulbe, K. D., Chimel, H., and Trösch, W.,** Microbial acetate conversion to methane:kinetics, yields and pathways in a two-step digestion process, *Appl. Microbiol. Biotechnol.,* 19, 224, 1984.

108. **Archer, D. B.,** Hydrogen-using bacteria in a methanogenic acetate enrichment culture, *J. Appl. Bacteriol.,* 56, 125, 1984.

109. **Stadtman, T. C. and Barker, H. A.,** Studies on the methane fermentation. VII. Tracer experiments on the mechanism of methane formation, *Arch. Biochem.,* 21, 256, 1949.

110. **Pine, M. J. and Barker, H. A.,** Studies on the methane fermentation. XII. The pathway of hydrogen in the acetate fermentation, *J. Bacteriol.,* 71, 644, 1956.

111. **Mah, R. A., Smith, M. R., and Baresi, L.,** Studies on an acetate-fermenting strain of *Methanosarcina, Appl. Environ. Microbiol.,* 35, 1174, 1978.

112. **Zinder, S. H. and Mah, R. A.,** Isolation and characterization of a thermophilic strain of *Methanosarcina* unable to use H_2-CO_2 for methanogenesis, *Appl. Environ. Microbiol.,* 38, 996, 1979.

113. **Shapiro, S. and Wolfe, R. S.,** Methyl-coenzyme M, an intermediate in methanogenic dissimilation of C_1 compounds by *Methanosarcina barkeri, J. Bacteriol.,* 141, 728, 1980.

114. **Zeikus, J. G. and Wolfe, R. S.,** Methanobacterium thermoautotrophicus sp. n., an anaerobic, autotrophic, extreme thermophile, *J. Bacteriol.,* 109, 707, 1972.

115. **Zeikus, J. G. and Wolfe, R. S.,** Methanobacterium thermoautotrophicus should be Methanobacterium thermoautotrophicum, *Int. J. Syst. Bacteriol.,* 22, 395, 1972.

116. **Ferry, J. G., Smith, P. H., and Wolfe, R. S.,** Methanospirillum, a new genus of methanogenic bacteria, and characterization of *Methanospirillum hungatii* sp. nov., *Int. J. Syst. Bacteriol.,* 24, 465, 1974.

117. **Rivard, C. J. and Smith, P. H.,** Isolation and characterization of a thermophilic marine methanogenic bacterium, *Methanogenium thermophilicum* sp. nov., *Int. J. Syst. Bacteriol.,* 32, 430, 1982.

118. **Paynter, M. J. B. and Hungate, R. E.,** Characterization of *Methanobacterium mobilis,* sp. n., isolated from the bovine rumen, *J. Bacteriol.,* 95, 1943, 1968.

119. **Smith, P. H. and Hungate, R. E.,** Isolation and characterization of *Methanobacterium ruminantium,* n. sp., *J. Bacteriol.,* 75, 713, 1958.

120. **Stadtman, T. C. and Barker, H. A.,** Studies on the methane fermentation. X. A new formate-decomposing bacterium, *Methanococcus vannielii, J. Bacteriol.,* 62, 269, 1951.

121. **Whitman, W. B., Ankwanda, E., and Wolfe, R. S.,** Nutrition and carbon metabolism of *Methanococcus voltae, J. Bacteriol.,* 149, 852, 1982.

122. **Huber, H., Thomm, M., König, H., Thies, G., and Stetter, K. O.,** Methanococcus thermolithoautotrophicus a novel thermophilic lithotrophic methanogen, *Arch. Microbiol.,* 132, 47, 1982.

123. **Huser, B. A., Wuhrmann, K., and Zehnder, A. J. B.,** Methanothrix soehngenii gen. nov., sp. nov., a new acetotrophic non-hydrogen-oxidizing methane bacterium, *Arch. Microbiol.,* 132, 1, 1982.

124. **Wildgruber, G., Thomm, M., König, H., Ober, K., Ricchiuto, T., and Stetter, K. O.,** Methanoplanus limicola, a plate-shaped methanogen representing a novel family, the Methanoplanaceae, *Arch. Microbiol.,* 132, 31, 1982.

125. **Corder, R. E., Hook, L. A., Larkin, J. M., and Frea, J. I.,** Isolation and characterization of two new methane producing cocci: Methanogenium olentangyi sp. nov., and *Methanococcus deltae* sp. nov., *Arch. Microbiol.,* 134, 28, 1983.

126. **Jones, W. J., Paynter, M. J. B., and Gupta, R.,** Characterization of *Methanococcus maripaludis* sp. nov., a new methanogen isolated from salt marsh sediment, *Arch. Microbiol.,* 135, 91, 1983.

127. **Jones, W. J., Leigh, J. A., Mayer, F., Woese, C. R., and Wolfe, R. S.,** Methanococcus jannaschii sp. nov., an extremely thermophilic methanogen from a submarine hydrothermal vent, *Arch. Microbiol.,* 136, 254, 1983.

128. **Zabel, H. P., König, H., and Winter, J.,** Isolation and characterization of a new coccoid methanogen, *Methanogenium tatii,* spec. nov. from a solfataric field on mount Tatio, *Arch. Microbiol.,* 137, 308, 1984.
129. **Sowers, K. R. and Ferry, J. G.,** Isolation and characterization of a methylotrophic marine methanogen, *Methanococcoides methylutens* gen. nov., sp. nov., *Appl. Environ. Microbiol.,* 45, 684, 1983.
130. **Rivard, C. J., Heuson, J. M., Thomas, M. V., and Smith, P. H.,** Isolation and characterization of *Methanomicrobium paynteri* sp. nov., a mesophilic methanogen isolated from marine sediments, *Appl. Environ. Microbiol.,* 46, 484, 1983.
131. **Stetter, K. O., Thomm, M., Winter, J., Wildgruber, G., Huber, H., Zillig, W., Janecovic, D., König, H., Palm, P., and Wunderl, S.,** *Methanothermus fervidus* sp. nov., a novel extremely thermophilic methanogen isolated from an icelandic hot spring, *Zentralbl. Bakteriol., Int. Z. Mikrobiol. Hyg. 1. Abteilung, Originale C,* 2, 166, 1981.
132. **König, H. and Stetter, K. O.,** Isolation and characterization of *Methanolobus tindarius* sp. nov., a coccoid methanogen growing only on methanol and methylamines, *Zentralbl. Bakteriol., Int. Z. Mikrobiol. Hyg. 1. Abteilung, Originale C,* 3, 478, 1982.
133. **Patel, G. B.,** Characterization and nutritional properties of *Methanothrix concilii* sp. nov., a mesophilic, aceticlastic methanogen, *Can. J. Microbiol.,* 30, 1383, 1984.
134. **Smith, M. R., Zinder, S. H., and Mah, R. A.,** Microbial methanogenesis from acetate, *Process Biochem.,* 15(3), 34, 1980.
135. **Daniels, L., Sparling, R., and Sprott, G. D.,** The bioenergetics of methanogenesis, *Biochim. Biophys. Acta,* 768, 113, 1984.
136. **Vogels, G. D. and Visser, C. M.,** Interconnection of methanogenic and acetogenic pathways, *FEMS Microbiol. Lett.,* 20, 291, 1983.
137. **Blaut, M. and Gottschalk, G.,** Protonmotive force driven synthesis of ATP during methane formation from molecular hydrogen and formaldehyde or carbon dioxide in *Methanosarcina barkeri, FEMS Microbiol. Lett.,* 24, 103, 1984.
138. **Mah, R. A.,** Isolation and characterization of *Methanococcus mazei, Curr. Microbiol.,* 3, 321, 1980.
139. **Kennedy, K. J. and Van den Berg, L.,** Anaerobic digestion of piggery waste using a stationary fixed film reactor, *Agric. Wastes,* 4, 151, 1982.
140. **Clark, R. H. and Speece, R. E.,** The pH-tolerance of anaerobic digestion, in *Advances in Water Pollution Research,* Vol. 1, Pergamon Press, Oxford, 1971, II-27/1.
141. **Russell, J. B. and Dombrowski, D. B.,** Effect of the pH on the efficiency of growth by pure cultures of rumen bacteria in continuous culture, *Appl. Environ. Microbiol.,* 39, 604, 1980.
142. **McDonald, P.,** *The Biochemistry of Silage,* John Wiley & Sons, London, 1981,
143. **McDonald, P.,** Silage fermentation, *Trend Biochem. Sci.,* 7, 164, 1982.
144. **Koster, I. W.,** Liquefaction and acidogenesis of tomatoes in an anaerobic two-phase solid waste treatment system, *Agric. Wastes,* 11, 241, 1984.
145. **Sahm, H.,** Anaerobic wastewater treatment, in *Advances in Biochemical Engineering/Biotechnology,* Vol. 29, Fiechter, A., Ed., Springer-Verlag, Berlin, 1984, 83.
146. **Heyes, R. and Hall, R. J.,** Anaerobic digestion modelling — the role of H_2, *Biotechnol. Lett.,* 3, 431, 1981.
147. **Mosey, F. E.,** Mathematical modelling of the anaerobic digestion process: regulatory mechanisms for the formation of short-chain volatile acids from glucose, *Water Sci. Technol.,* 15(8/9), 209, 1983.
148. **Stumm, W. and Morgan, J. J.,** *Aquatic Chemistry, an Introduction Emphasizing Chemical Equilibria in Natural Waters,* Wiley-Interscience, New York, 1970.
149. **Van der Laan, J. B. R. and Hobma, S. W.,** Determination of short chain fatty acids and bicarbonate alkalinity by means of titration (in Dutch), H_2O, 11, 465, 1978.
150. **Jenkins, S. R., Morgan, J. M., and Sawyer, C. L.,** Measuring anaerobic sludge digestion and growth by a simple alkalimetric titration, *J. Water Pollut. Control Fed.,* 55, 448, 1983.
151. **Kaspar, H. F. and Wuhrmann, K.,** Product inhibition in sludge digestion, *Microb. Ecol.,* 4, 241, 1978.
152. **Brown, G. J., Lin, K.-C., Landine, R. C., Cocci, A. A., and Viraraghavan, T.,** Lime use in anaerobic filters, *J. Environ. Eng. Div., Am. Soc. Civ. Eng.,* 106, 837, 1980.
153. **Li, A. and Sutton, P. M.,** Determination of alkalinity requirements for the anaerobic treatment process, in *Proc. 38th Purdue Industrial Waste Conference,* Bell, J. M., Ed., Ann Arbor Science, Ann Arbor, Mich., 1983, 603.
154. **Brovko, N. and Chen, K. Y.,** Optimizing gas production, methane content and buffer capacity in digester operation, *Water Sewage Works,* 124, 54, 1977.
155. **Lettinga, G., Roersma, R., Grin, P., De Zeeuw, W., Hulshoff Pol, L., Van Velsen, A., Hobma, S., and Zeeman, G.,** Anaerobic treatment of sewage and low strength waste waters, in *Anaerobic Digestion 1981,* Hughes, D. E., Stafford, D. A., Wheatley, B. I., Baader, W., Lettinga, G., Nyns, E. J., Verstraete, W., and Wentworth, R. L., Eds., Elsevier, Amsterdam, 1982, 271.
156. **Pereboom, J.,** *Behaviour and Effect of Calcium in Anaerobic Wastewater Treatment* (in Dutch), Department of Water Pollution Control, Wageningen Agricultural University, Wageningen, Netherlands, 1984.

157. **Lettinga, G. and Van Velsen, A. F. M.,** Application of methane fermentation for treatment of low strength wastewater (in Dutch), H_2O, 7, 281, 1974.

158. **Andrews, J. F.,** Dynamic model of the anaerobic digestion process, *J. Sanit. Eng. Div., Am. Soc. Civ. Eng.,* 95, 95, 1969.

159. **Andrews, J. F. and Graeff, S. P.,** Dynamic modelling and simulation of the anaerobic digestion process, in *Anaerobic Biological Treatment Processes,* (Advances in Chemistry Series, Vol. 105), Gould, R. F., Ed., American Chemical Society, Washington, D. C., 1971, chap. 8.

160. **Duarte, A. C. and Anderson, G. K.,** Inhibition modelling in anaerobic digestion, *Water Sci. Technol.,* 14, 749, 1982.

161. **De Baere, L. A., Devocht, M., Van Assche, P., and Verstraete, W.,** Influence of high NaCl and NH_4Cl salt levels on methanogenic associations, *Water Res.,* 18, 543, 1984.

162. **McCarty, P. L. and McKinney, R. E.,** Salt toxicity in anaerobic digestion, *J. Water Pollut. Control Fed.,* 33, 399, 1961.

163. **Kroiss, H. and Plahl-Wabnegg, F.,** Sulfide toxicity with anaerobic waste water treatment, in *Proc. Eur. Symp. Anaerobic Waste Water Treatment,* Van den Brink, W. J., Ed., TNO Corporate Communication Department, The Hague, Netherlands, 1983, 72.

164. **Yang, J., Parkin, G. F., and Speece, R. E.,** Recovery of anaerobic digestion after exposure to toxicants, final report prepared for the U.S. Department of Energy, Drexel University, Philadelphia, 1979.

165. **Rudolfs, W. and Heukelekian, H.,** Thermophilic digestion of sewage solids, *Ind. Eng. Chem.,* 22, 96, 1930.

166. **Fair, G. M. and Moore, E. W.,** Heat and energy relations in the digestion of sewage solids. III. Effect of temperature of incubation upon the course of digestion, *Sewage Works J.,* 4, 589, 1932.

167. **Fair, G. M. and Moore, E. W.,** Time and rate of sludge digestion, and their variation with temperature, *Sewage Works J.,* 6, 3, 1934.

168. **Imhoff, K.,** Temperatur und Schlammfaulung, *Tech. Gemeindebl.,* 39, 162, 1936.

169. **Fair, G. M. and Moore, E. W.,** Observations on the digestion of a sewage sludge over a wide range of temperatures, *Sewage Works J.,* 9, 3, 1937.

170. **Imhoff, K.,** Die Temperatur in Schlammfaulräumen, *Gesund. Ing.,* 62, 310, 1939.

171. **Svensson, B. H.,** Different temperature optima for methane formation when enrichments from acid peat are supplemented with acetate or hydrogen, *Appl. Environ. Microbiol.,* 48, 389, 1984.

172. **Speece, R. E. and Kem, J. A.,** The effect of short-term temperature variations on methane production, *J. Water Pollut. Control Fed.,* 42, 1990, 1970.

173. **Jewell, W. J. and Morris, J. W.,** Influence of varying temperature, flowrate and substrate concentration on the anaerobic-attached film expanded-bed process, in *Proc. 36th Purdue Industrial Waste Conference,* Bell, J. M., Ed., Ann Arbor Science, Ann Arbor, Mich., 1981, 655.

174. **Kandler, O., Temper, U., Steiner, A., and Winter, J.,** Efficiency and stability of methane fermentation of wastes at mesophilic and thermophilic temperatures, paper presented at the Symp. on Recent Advances in Biotechnology during ChemTech + Ort 1983 Conference, New Delhi, February 1983.

175. **Lettinga, G.,** The prospects of anaerobic waste water treatment, in *Anaerobic Digestion and Carbohydrate Hydrolysis of Waste,* Ferrero, G. L., Ferranti, M. P., and Naveau, H., Eds., Elsevier, London, 1984, 262.

176. **Lettinga, G.,** Anaerobic wastewater treatment and its potentials, *Eur. Water Sewage, Water Serv. J.,* 89, 22, 1985.

177. **Lettinga, G., Hulshoff Pol, L. W., Koster, I. W., Wiegant, W. M., De Zeeuw, W. J., Rinzema, A., Grin, P. C., Roersma, R. E., and Hobma, S. W.,** High-rate anaerobic waste-water treatment using the UASB reactor under a wide range of temperature conditions, in *Biotechnology and Genetic Engineering Reviews,* Vol. 2, Russell, G. E., Ed., Intercept, Newcastle-on-Tyne, 1984, chap. 9.

178. **Van den Berg, L.,** Developments in methanogenesis from industrial waste water, *Can. J. Microbiol.,* 30, 975, 1984.

179. **Callander, I. J. and Barford, J. P.,** Recent advances in anaerobic digestion technology, *Process Biochem.,* 18(4), 24, 1983.

180. **Lettinga, G., Roersma, R. E., and Grin, P. C.,** Anaerobic treatment of raw domestic sewage at ambient temperatures using a granular bed UASB reactor, *Biotechnol. Bioeng.,* 25, 1701, 1983.

181. **Grin, P. C., Roersma, R. E., and Lettinga, G.,** Anaerobic treatment of raw sewage at lower temperatures, in *Proc. Eur. Symp. Anaerobic Waste Water Treatment,* Van den Brink, W. J., Ed., TNO Corporate Communication Department, The Hague, Netherlands, 1983, 335.

182. **Lettinga, G., Hobma, S. W., Hulshoff Pol, L. W., De Zeeuw, W. J., De Jong, P., Grin, P. C., and Roersma, R. E.,** Design operation and economy of anaerobic treatment, *Water Sci. Technol.,* 15(8/9), 177, 1983.

183. **Wiegant, W. M. and Lettinga, G.,** Starting up of a thermophilic anaerobic digestion, in *Energy from Biomass, Solar Energy R & D in the European Community,* Series E, Vol. 1, Chartier, P. and Palz, W., Eds., D. Reidel, Dordrecht, Netherlands, 1981, 126.

184. **Van Velsen, A. F. M., Lettinga, G., and Den Ottelander, D.,** Anaerobic digestion of piggery waste. III. Influence of temperature, *Neth. J. Agric. Sci.,* 27, 255, 1979.

185. **Van Velsen, A. F. M. and Lettinga, G.,** Digestion of animal manure, *Stud. Environ. Sci.,* 9, 55, 1981.

186. **Wiegant, W. M. and Lettinga, G.,** Maximum loading rates and ammonia toxicity in thermophilic digestion, in *Energy from Biomass, Solar Energy R & D in the European Community,* Series E, Vol. 3, Grassi, G. and Palz, W., Eds., D. Reidel, Dordrecht, Netherlands, 1982, 238.

187. **Popova, N. M. and Bolotina, O. T.,** The present state of purification in town sewage and the trend in research work in the city of Moscow, *Int. J. Air Water Pollut.,* 7, 145, 1963.

188. **Garber, W. F.,** Certain aspects of anaerobic digestion of wastewater solids in the thermophilic range at the Hyperion treatment plant, *Progress Water Technol.,* 8(6), 401, 1977.

189. **Garber, W. F.,** Operating experience with thermophilic anaerobic digestion, *J. Water Pollut. Control Fed.,* 54, 1170, 1982.

190. **Rinkus, R. R., Ryan, J. M., and Cook, E. J.,** Full scale thermophilic digestion at the West Southwest Sewage Treatment Works, Chicago, Illinois, *J. Water Pollut. Control Fed.,* 54, 1447, 1982.

191. **Shelef, G., Kimchie, S., and Grynberg, H.,** High-rate thermophilic anaerobic digestion of agricultural wastes, *Biotechnol. Bioeng. Symp.,* 10, 341, 1980.

192. **Schraa, G. and Jewell, W. J.,** Conversion of soluble organics with the thermophilic anaerobic attached film expanded bed process, in *Proc. Eur. Symp. Anaerobic Waste Water Treatment,* Van den Brink, W. J., Ed., TNO Corporate Communication Department, The Hague, Netherlands, 1983, 216.

193. **Wiegant, W. M., Claassen, J. A., Borghans, A. J. M. L., and Lettinga, G.,** High rate thermophilic anaerobic digestion for the generation of methane from organic wastes, in *Proc. Eur. Symp. Anaerobic Waste Water Treatment,* Van den Brink, W. J., Ed., TNO Corporate Communication Department, The Hague, Netherlands, 1983, 392.

194. **Scherer, P., Lippert, H., and Wolff, G.,** Composition of the major elements and trace elements of 10 methanogenic bacteria determined by inductively coupled plasma emission spectrometry, *Biol. Trace Element Res.,* 5, 149, 1983.

195. **Thauer, R. K.,** Nickel tetrapyrroles in methanogenic bacteria: structure, function and biosynthesis, in *Anaerobic Digestion 1981,* Hughes, D. E., Stafford, D. A., Wheatley, B. I., Baader, W., Lettinga, G., Nyns, E. J., Verstraete, W., and Wentworth, R. L., Eds., Elsevier, Amsterdam, 1982, 37.

196. **Diekert, G., Konheiser, U., Piechulla, K., and Thauer, R. K.,** Nickel requirement and factor F_{430} content of methanogenic bacteria, *J. Bacteriol.,* 148, 459, 1981.

197. **Scherer, P. and Sahm, H.,** Influence of sulphur-containing compounds on the growth of *Methanosarcina barkeri* in a defined medium, *Eur. J. Appl. Microbiol., Biotechnol.,* 12, 28, 1981.

198. **Lohani, B. N. and De Dios, R. E.,** A dynamic model for anaerobic digestion of a nutrient deficient waste, *Int. J. Dev. Technol.,* 2, 27, 1984.

199. **Lettinga, G., Van Velsen, A. F. M., Hobma, S. W., De Zeeuw, W., and Klapwijk, A.,** Use of the upflow sludge blanket (USB) reactor concept for biological wastewater treatment, especially for anaerobic treatment, *Biotechnol., Bioeng.,* 22, 699, 1980.

200. **Lettinga, G., Hulshoff, Pol, L. W., Wiegant, W., De Zeeuw, W., Hobma, S. W., Grin, P., Roersma, R., Sayed, S., and Van Velsen, A. F. M.,** Upflow sludge blanket processes, in *Proc. 3rd Int. Symp. on Anaerobic Digestion,* Wentworth, R. L., Ed., Third International Symposium on Anaerobic Digestion, Cambridge, Mass., 1983, 139.

201. **Speece, R. E., Parkin, G. F., and Gallagher, D.,** Nickel stimulation of anaerobic digestion, *Water Res.,* 17, 677, 1983.

202. **Scherer. P. and Sahm, H.,** Effect of trace elements and vitamins on the growth of *Methanosarcina barkeri,* *Acta Biotechnol.,* 1, 57, 1981.

203. **McCarty, P. L.,** Anaerobic waste treatment fundamentals. II. Environmental requirements and control, *Public Works,* 95(10), 123, 1964.

204. **Kelly, C. R. and Switzenbaum, M. S.,** Temperature and nutrient effects on the anaerobic expanded bed treating a high strength waste, in *Proc. 38th Purdue Industrial Waste Conference,* Bell, J. M., Ed., Ann Arbor Science, Ann Arbor, Mich., 1983, 591.

205. **Murray, W. D. and Van den Berg, L.,** Effects of nickel, cobalt and molybdenum on performance of methanogenic fixed film reactors, *Appl. Environ. Microbiol.,* 42, 502, 1981.

206. **Wate, S. R., Chakrabarti, T., and Subrahmanyam, P. V.,** Effect of cobalt on biogas production from cattle dung, *Indian J. Environ. Health,* 25, 179, 1983.

207. **Callander, I. J. and Barford, J. P.,** Precipitation, chelation and the availability of metals as nutrients in anaerobic digestion. I. Methodology, *Biotechnol. Bioeng.,* 25, 1947, 1983.

208. **Callander, I. J. and Barford, J. P.,** Precipitation, chelation and the availability of metals as nutrients in anaerobic digestion. II. Applications, *Biotechnol. Bioeng.,* 25, 1959, 1983.

209. **Pfeffer, J. T. and White, J. E.,** The role of iron in anaerobic digestion, in Proc. 19th Purdue Industrial Waste Conference, Engineering Extension Series, No. 117, Purdue University, West Lafayette, Ind., 1964, 887.

210. **McCarty, P. L.,** Anaerobic waste treatment fundamentals. III. Toxic materials and their control, *Public Works,* 95(11), 91, 1964.
211. **Koster, I. W. and Rinzema, A.,** Toxicity and adaptation in anaerobic processes (in Dutch), *H₂O,* 17, 575, 1984.
212. **Braun, R., Huber, P., and Meyrath, J.,** Ammonia toxicity in liquid piggery manure digestion, *Biotechnol. Lett.,* 3, 159, 1981.
213. **Lawrence, A. W. and McCarty, P. L.,** The role of sulfide in preventing heavy metal toxicity in anaerobic treatment, *J. Water Pollut. Control Fed.,* 37, 392, 1965.
214. **Mosey, F. E. and Hughes, D. A.,** The toxicity of heavy metal ions to anaerobic digestion, *Water Pollut. Control,* 74, 18, 1975.
215. **Mosey, F. E.,** Assessment of the maximum concentration of heavy metals in crude sewage which will not inhibit the anaerobic digestion of sludge, *Water Pollut. Control.,* 75, 10, 1976.
216. **Hayes, T. D. and Theis, T. L.,** The distribution of heavy metals in anaerobic digestion, *J. Water Pollut. Control Fed.,* 50, 61, 1978.
217. **Kugelman, I. J. and McCarty, P. L.,** Cation toxicity and stimulation in anaerobic waste water treatment. I. Slug feed studies, *J. Water Pollut. Control Fed.,* 37, 97, 1965.
218. **Kugelman, I. J. and McCarty, P. L.,** Cation toxicity and stimulation in anaerobic waste water treatment. II. Daily feed studies, in Proc. 19th Purdue Industrial Waste Conference, Engineering Extension Series, No. 117, Purdue University, West Lafayette, Ind., 1965, 667.
219. **Kugelman, I. J. and Chin, K. K.,** Toxicity, synergism and antagonism in anaerobic waste treatment processes, in *Anaerobic Biological Treatment Processes,* (Advances in Chemistry Series, Vol. 105), Gould, R. F., Ed., American Chemical Society, Washington, D.C., 1971, chap. 5.
220. **Melbinger, N. R. and Donnellon, J.,** Toxic effects of ammonia-nitrogen in high-rate digestion, *J. Water Pollut. Control Fed.,* 43, 1658, 1971.
221. **Van Velsen, A. F. M.,** Adaptation of methanogenic sludge to high ammonia-nitrogen concentrations, *Water Res.,* 13, 995, 1979.
222. **Parkin, G. F. and Speece, R. E.,** Modelling toxicity in methane fermentation systems, *J. Environ. Eng. Div., Am. Soc. Civ. Eng.,* 108, 515, 1982.
223. **Maaskant, W. and Hobma, S. W.,** The influence of sulphite on anaerobic waste water treatment (in Dutch), *H₂O,* 14, 596, 1981.
224. **Salkinoja-Salonen, M. S., Hakulinen, R., Valo, R., and Apajalahti, J.,** Biodegradation of recalcitrant organochlorine compounds in fixed film reactors, *Water Sci. Technol.,* 15(8/9), 309, 1983.
225. **Guthrie, M. A., Kirsch, E. J., Wukasch, R. F., and Grady, C. P. L., Jr.,** Pentachlorophenol biodegradation. II. Anaerobic, *Water Res.,* 18, 451, 1984.
226. **Parkin, G. F. and Miller, S. W.,** Response of methane fermentation to continuous addition of selected industrial toxicants, in *Proc. 37th Purdue Industrial Waste Conference,* Bell, J. M., Ed., Ann Arbor Science, Ann Arbor, Mich., 1983, 729.
227. **Speece, R. E. and Parkin, G. F.,** The response of methane bacteria to toxicity, in *Proc. 3rd Int. Symp. on Anaerobic Digestion,* Wentworth, R. L., Ed., Cambridge, Mass., 1983, 23.
228. **Parkin, G. F. and Speece, R. E.,** Attached versus suspended growth anaerobic reactors: response to toxic substances, *Water Sci. Technol.,* 15(8/9), 261, 1983.
229. **Van den Berg, L., and Kennedy, K. J.,** Comparison of advanced anaerobic reactors, in *Proc. 3rd Int. Symp. on Anaerobic Digestion,* Wentworth, R. L., Ed., Cambridge, Mass., 1983, 71.
230. **Blotevogel, K.-H., Fischer, U., and Lüpkes, K. H.,** *Methanococcus frisius* sp. nov., a new methylotrophic marine methanogen, *Can. J. Microbiol.,* 32, 127, 1986.
231. **Olivier, B. M., Mah, R. A., Garcia, J. L., and Boone, D. R.,** Isolation and characterization of *Methanogenium bourgense* sp. nov., *Int. J. Syst. Bacteriol.,* 36, 297, 1986.
232. **Worakit, S., Boone, D. R., Mah, R. A., Abdel-Samie, M.-E., and El-Halwagi, M. M.,** *Methanobacterium alcaliphilum* sp. nov., an H₂-utilizing methanogen that grows at high pH-values, *Int. J. Syst. Bacteriol.,* 36, 380, 1986.
233. **Olivier, B. M., Mah, R. A., Garcia, J. L., and Robinson, R.,** Isolation and characterization of *Methanogenium aggregans* sp. nov., *Int. J. Syst. Bacteriol.,* 35, 127, 1985.
234. **Zinder, S. H., Sowers, K. R., and Ferry, J. G.,** *Methanosarcina thermophilia* sp. nov., a thermophilic, acetotrophic, methane-producing bacterium, *Int. J. Syst. Bacteriol.,* 35, 522, 1985.
235. **van Bruggen, J. J. A., Zwart, K. B., Hermans, J. G. F., van Hove, E. M., Stumm, C. K., and Vogels, G. D.,** Isolation and characterization of *Methanoplanus endosymbiosus* sp. nov., an endosymbiont of the marine sapropelic ciliate *Metopus contortus* quennerstedt, *Arch. Microbiol.,* 144, 367, 1986.
236. **Sleat, R., Mah, R. A., and Robinson, R.,** *Acetoanaerobium notarae* gen. nov., sp. nov.: an anaerobic bacterium that forms acetate from H₂ and CO₂, *Int. J. Syst. Bacteriol.,* 35, 10, 1985.
237. **Wiegant, W. M. and Zeeman, G.,** The mechanism of ammonia inhibition in thermophilic digestion of livestock wastes, *Agric. Wastes,* 16, 243, 1986.

238. **Zeeman, G., Wiegant, W. M., Koster-Treffers, M. E., and Lettinga, G.,** The influence of the total ammonia concentration on the thermophilic digestion of cow manure, *Agric. Wastes,* 14, 19, 1985.

239. **Wiegant, W. M.,** Thermophilic Anaerobic Digestion for Waste and Wastewater Treatment, Ph.D. thesis, Agricultural University, Wageningen, Netherlands, 1986.

240. **Wiegant, W. M. and Lettinga, G.,** Thermophilic anaerobic digestion of sugars in upflow anaerobic sludge blanket reactors, *Biotechnol. Bioeng.,* 27, 1603, 1985.

241. **Wiegant, W. M., Claassen, J. A. and Lettinga, G.,** Thermophilic anaerobic digestion of high strength wastewaters, *Biotechnol. Bioeng.,* 27, 1374, 1985.

242. **Wiegant, W. M., Hennink, M., and Lettinga, G.,** Separation of the propionate degradation to improve the efficiency in the thermophilic treatment of acidified wastewaters, *Water Res.,* 20, 517, 1986.

243. **Wiegant, W. M. and de Man, A. W. A.,** Granulation of biomass in thermophilic upflow anaerobic sludge blanket reactors treating acidified wastewaters, *Biotechnol. Bioeng.,* 28, 718, 1986.

244. **Koster, I. W.,** Characteristics of the pH-influenced adaptation of methanogenic sludge to ammonium toxicity, *J. Chem. Technol. Biotechnol.,* 36, 445, 1986.

INDEX

A

ABS, see Aryl benzene sulfonates
Acclimation of microorganisms, 262
Acclimatization, 305, 306
Acenitobacter sp., 257
Acetagenesis, 292
Acetic acid, 294—295
Acetoanaerobium notrae, 294
Acetobacterium woodii, 294
Acetogenesis, 288
Acetogenium kivui, 294
Achromobacter sp., 257
Acidity, 86, 297—299
Acidogenesis, 288
Acid-producing microorganisms, 88
Activated carbon, 158, 166, 211, 212, 215, 216
Activated sludge, 22, 41, 84, 159, 161, 235—237, 273
Actual free-energy change, 292
Adaptation of bacteria, 304—306
Adapted-culture technology, 270
Adipic acid, 171
Adsorption of phenols, 14
Aeration, 23, 236—237
Aerobic degradation, 170
Air, 80
Aircraft maintenance, 206
Alcaligens euthropus, 105
Alcaligens faecalis, 272
Alcaligens sp., 104
Aldrin, 75, 90, 115—116
Aliphatic compounds, 112—116, see also specific types
  biodegradation of, 112, 113
  cyclic, 84
  metabolism of, 115
Aliphatic hydrocarbons, 83, 89, see also specific types
Alkalegens sp., 256
Alkalinity, 292, 298
Alkali salts, 244
Alkanes, 83, 89, see also specific types
n-Alkanes, 101
Alkylbenzenes, 97, see also specific types
Alkylbenzene sulfonates, 83, 255
Alkylphenols, 196—197, see also specific types
Amines, 27, see also specific types
Amino acids, see also specific types
  aromatic, 205
  fermentation of, 291—292
  in sludge, 38, 40
4-Aminoantipyrine colorimetry, 206
o-Aminophenol, 171
p-Aminophenol, 171
Aminophenols, 204, see also specific types
Ammonia, 207, 292

in coal conversion waste, 264
toxicity of, 248
Anaerobic degradation, 110
  of aromatic compounds, 185—193
    fermentation in, 190—193
    nitrate respiration in, 186—190
    photometabolism in, 185—186
    sulfate reduction in, 190
  of complex organic materials, 287—297
    fermentation and, 291—292
    hydrolysis and, 290
    substrate flows in, 287—289
  of phenolics, 171—184, 193—195, 205
Anaerobic digestion, 286
  monitoring of, 297
  nutrients and, 302
  substrate flows in, 287—289
  thermophilic, 300
Anaerobic oxidation, 292—294
Anaerobiosis, 86
Analogue metabolism, see Cometabolism
m-Anisate, 187
m-Anisic acid, 171
Antagonism, 304
Antibiotics, 251, see also specific types
Archaebacteria, 287, see also specific types
Aromatic amines, 27, see also specific types
Aromatic amino acids, 205, see also specific types
Aromatic compounds, 72, 97—112, see also specific types
  anaerobic degradation of, 185—193
    fermentation in, 190—193
    nitrate respiration in, 186—190
    photometabolism in, 185—186
    sulfate reduction in, 190
  biodegradation of, 100
  catabolism of, 255
  dechlorination of, 193
  metabolism of, 101—103
  microbiology and, 100—101
  nonhalogenated, see Nonhalogenated aromatics
  oxidative catabolism of, 106
  polycyclic, 75
Aromatic hydrocarbons, 97—103, 261, see also specific types
  catabolism of, 89, 252—255
  chlorinated, 83
  degradability of, 89
  polycyclic, 84, 97
Aromatic solvents, 70, see also specific types
Arthrobacter sp., 104, 105, 107
Aryl benzene sulfonates (ABS), 255
ASP, see Activated sludge
Aspergillus sp., 256
Asphalt-concrete layers, 132
Asphaltic fraction, 89
Authentic industrial waste waters, 212—218

Autotrophic nitrification, 256—257
Azotobacter sp., 256

B

Bacillus polymisea, 105
Bacillus sp., 256, 257
Bacteria, see also Microbiology; specific types
  adaptation of, 304—306
  aeration and, 23
  genetically altered, 84
  genetically engineered, 269
  growth of, 301
  hydrogenotrophic acetogenic, 294
  kinetics of growth of, 263
  mesophilic, 299
  methanogenic, see Methanogenic bacteria
  nitrifying, 35
  temperature and, 23
  thermophilic, 21, 299
  thermotolerant, 21
Bacterial antibiotics, 251, see also specific types
Barrier installation, 92
Batch cultures, 271
Beltsville system, 125
Bench-scale studies of SBRs, 160—166
Bentonite-sand layers, 132
Benzenes, 71, 75, 97, see also specific types
  biodegradation of, 100
  chlorinated, 90
  degradability of, 89
  dihydroxy, see Dihydroxy benzene
  hydroxy, see Hydroxy benzenes
  metabolism of, 101, 103
  monoalkyl, 25
  monohydroxy, see Monohydroxy benzenes
  polyhydroxy, see Polyhydroxy benzenes
Benzoate dioxygenase, 106
Benzoates, 89
  chlorinated, 90
  degradation of, 185, 255
  metabolism of, 101
  nitrate respiration and, 187
  sulfate reduction and, 190
Benzoic acid, 170, 186
  fermentation and, 191
  metabolism of, 101
Biochemical augmentation, 268—272
Biochemistry
  of cyanide metabolism, 258—259
  of denitrification, 257—258
  of hexachlorocyclohexanes, 115
  of nitrification, 256—257
  of nitrogen metabolism, 256—258
  of PCBs, 111—113
Biocides, 68, 72, 74, see also specific types
Biodegradability, 194
  of contaminants, 68
  of feed waste water, 164
  of pollutants, 262—272
Biodegradation, see also Degradation; specific types

  of aromatic compounds, 100
  of chlorobenzenes, 104—105
  environmental factors affecting, 85—86
  of halogenated hydrocarbons, 112
  of hexachlorocyclohexanes, 113
  in situ stimulation of, 87
  of PCBs, 110—111
  of pollutants, 82—83
  in soil, 87
  stimulated, 86—88
  of surfactants, 255—256
  in underground waters, 87
Bioextraction, 91, 117, 119—120, see also
    Extraction; specific types
  costs of, 120
  implementation of, 120
  principle of, 119—120
  in soil decontamination, 120
Biological filters, 234—235
Biological oxygen demand (BOD), 19—20, 47, 214
Biological purification, 41—47
Biological treatments, see also specific types
  full-scale plants for, 272—274
  high-performance, see High-performance
    biological treatment
  of industrial wastes, 259—272, see also specific
    types
  parameters for, 237—238
  of sulfur-containing compounds, 258
  of toxic wastes, 262—272
  of waste water, 233—238
Biomass, 47, 299
Biotank reactors, 45, 46, 49
Biphenyls, 104, see also specific types
BOD, see Biological oxygen demand
Bored piles wall, 132
Bromobenzoates, 110
Bromodichloromethane, 112
Brown coal
  carbonization of, 10
  classification of, 3
  combustion water and, 6
  formation of, 3—4
  gasification of, 10
  hydrogenation of, 10
  inherent water and, 6
  liquefaction of, 8
  origins of, 9
  processing of, 3—9
  reaction water and, 8
  refinement of, 5—8, see also specific types
  reserves of, 3, 5, 6
  utilization of, 2
Buffering capacity, 297—299
Butyribacterium methylotrophicum, 295
Butyric acid, 171

C

Caffeic acid, 171
Calcium carbonate precipitation, 299

Candida tropicalis, 100, 263, 271
n-Caproic acid, 171
Carbon
    activated, 158, 166, 211, 212, 215, 216
    filtered organic, 217
    nitrogen ratio to, 49
    total organic (TOC), 162
Carbon dioxide, 294
Carbonization, 5—6, 10, 241
Carbon tetrachloride, 113
Catabolic enzymes, 104, see also specific types
Catabolism
    of aliphaic hydrocarbons, 89
    of aromatic compounds, 89, 106, 252—256
    oxidative, 106
    of xenobiotics, 85
Catechol-1,2-dioxygenase, 102
Catechol-2,3-dioxygenase, 102
Catechol-1,2-oxygenase, 108
Catechol-2,3-oxygenase, 106
Catechols, 19, 211, see also specific types
    metabolism of, 101, 103, 106
    phenolic degradation and, 171
    photometabolism of, 186
    toluate conversion to, 109
Centrifugal aerators, 43, 48
Chemical oxygen demand (COD), 211, 214, 215,
        238, 302
Chemical treatments, 10, 14—17, 146—150, see also
        specific types
Chlorophenol, 193
Chloridazon, 254
Chlorinated aliphatic hydrocarbons, 83
Chlorinated aromatic hydrocarbons, 83
Chlorinated benzenes, 90, see also specific types
Chlorinated benzoates, 90
Chlorinated biphenyls, 104
Chlorinated ethylenes, 90
Chlorinated hydrocarbons, 84
Chlorinated phenolics, 218
Chlorinated phenols, 90, 103, 211, see also specific
        types
Chlorinated phenoxy herbicides, 105
Chlorine substitution, 90
Chlorobenzenes, 89, 90, 103—110, see also specific
        types
    biodegradation of, 104—105
    dihalogenated compounds and, 107
    enzymes and, 106
    metabolism of, 105—106
3-Chlorobenzoate, 106, 108—110
4-Chlorobenzoate, 109
Chlorobenzoates, 90, 104, see also specific types
3-Chlorocatechol, 105
4-Chlorocatechol, 106
Chlorocatechols, 106, 172, see also specific types
Chloroform, 90, 112, 113
Chlorohydroxybenzoic acid, 172
5-Chloro-2-hydroxymuconic semialdehyde, 106
4-Chloro-3-methylphenol, 105
2-Chlorophenol, 195

m-Chlorophenol, 172
o-Chlorophenol, 172
p-Chlorophenol, 172
Chlorophenols, 89, 90, 202—204, see also specific
        types
Chlororesorcinol, 173
Chlorosulfonated polyethylene, 135
Chromatography, 206, see also specific types
CLG biotank reactors, 49
Climatology, 81
Closed decontamination systems, 91—92, 125, see
        also specific types
Clostridium aceticum, 294
Clostridium formicoaceticum, 295
Clostridium sp., 113
CMAS, see Completely mixed activated sludge
Coal, see also specific types
    brown, see Brown coal
    conversion of, 212
        to heating gas, 214
        waste from, 264
        waste waters in, 215—216, 241—243
    gasification of, 6—8
    hard, 3
    hydrogenation of, 8
    synthetic, 213
Coal coking operations, 205
Coalification, 3
Coal-tar products, 97
Cocultures, 28, 170
COD, see Chemical oxygen demand
Coke ovens
    effluents from, 241
    in steel mills, 261
    wastes from, 97
    waste water of, 215
Coking, see High-temperature carbonization
Colorimetry, 206
Combustion water, 6
Cometabolism, 26, 83, 86, 267
Commensalism, 267
Competitive inhibition, 86
Completely mixed activated sludge (CMAS), 235
Composting, 69, 70, 87, 96, 122—128
    costs of, 127
    after excavation, 91
    implementation of, 125
    inoculation and, 124
    moisture and, 122—124
    nutrients and, 122
    oxygen and, 122
    porosity and, 124
    principle of, 122—124
    temperature and, 124
Compressibility of soil, 78
Condensation waters, 8, 17
Coniferyl alcohol, 173
Contaminants, 80, see also Pollutants; specific types
    biodegradability of, 68
    categories of, 72—74
    evaporation of, 144—145

at gasworks sites, 72
  thermal destruction of, 145—146
Continuous cultures, 273
Cooxidation, 26, 83
Corynebacterium sp., 105
CPA, 105
m-Cresol, 173, 212
  microbiology and, 101
  nitrate respiration and, 187
o-Cresol, 173, 187, 205, 212
p-Cresol, 173, 205, 212
  fermentation and, 191
  metabolism of, 103
  microbiology and, 101
  nitrate respiration and, 187, 189
  tyrosine and, 205
Cresols, 97 100, 101
Crude oil, 84
Crude petroleum, 89
Cultures, see also specific types
  batch, 271
  continuous, 263
  mixed enrichment, 170
  pure, 170, 263
Cyanides, 72, 143, 207, see also specific types
  biological treatment of, 267
  in coal conversion waste, 264
  detoxification of, 271
  hot gas treatments and, 145
  metabolism of, 258—259
  microorganisms in degradation of, 259
  secondary toxins and, 272
  toxicity of, 249
β-Cyanoalanine, 259
Cyclic aliphatics, 84, see also specific types
Cycloalkanes, 89, see also specific types
Cyclohexane-carboxylic acid, 173
Cyclohexanol, 173
Cyclohexanone, 174
Cyclohex-1-ene-1-carboxylic acid, 173
Cycloisomerase, 108
Cycloisomerase II, 106
Cytophaga sp., 257

D

2,4-D, 105, 109
Dano system, 91, 125
DDT, 246, 251
  dechlorination of, 254
  degradation of, 255, 267
Deacidification, 12
Dechlorination, 193, 205, 254
Decontamination of soil
  air and, 80
  bioextraction in, 120
  choice of methods in, 75—82
  climatology and, 81
  closed systems of, 91—92
  contaminants and, 80
  after excavation, 69, 91—92, 121—131, see also

Excavation
  geometry and, 80
  ground water and, 80
  in situ, 69, 90, 91, 94—96
  legislation on, 82
  microbial techniques in, 68, 90—92, see also
      specific types
  microbiological techniques in, 82—90
    comparison of, 94—96
    emission control and, 92—93, 131—138
    nonbiological vs., 96—99
    nonmicrobiological techniques in, 93, 139—150,
        see also specific types
    excavation and, 142—148
    microbiological vs., 98—99
  open systems for, 92, see also specific types
  soil type and, 75—80
  technical limitations on, 81
  topography and, 81
  urgency of, 82
  volume and, 80
Deep-jet aeration systems, 43
Degradation, see also under Biodegradation; specific
    substances; specific types
  aerobic, 170
  anaerobic, see Anaerobic degradation
  of chlorocatechols, 106
  efficiency of, 41, 42, 46, 52
  of long-chain fatty acids, 292
  of pollutants, 89—90, 251—259
  rate of, 42, 43, 46
  semi-aerobic, 87
Demethylate, 187
Demethylation, 188, 193
Demulsification, 12
Denitrification, 34, 36—37, 49
  biochemistry of, 257—258
  coking waste waters and, 216—217
  inhibitory effects on, 37
  nitrate respiration and, 186
  of sewage, 46
Dephenolizing, 12
Detergents, 83, 244, see also specific types
Dewatering, 12
Diaphragm wall, 132
Diauxic degradation, 22
Diauxy leading, 86
Dibromochloromethane, 112
Dibutylphthalate, 100
1,1-Dichloroethane, 112
1,2-Dichloroethane, 90, 112, 113
Dichlorobenzenes, 105, see also specific types
2,4-Dichlorobenzoate, 110
2,5-Dichlorobenzoate, 110
2,6-Dichlorobenzoate, 110
3,4-Dichlorobenzoate, 110
3,5-Dichlorobenzoate, 109, 110
Dichloromethane, 90, 112, 113
2,4-Dichlorophenol, 105, 195
Dichlorophenols, 90, 174, 175, see also specific
    types

Dieldrin, 75, 90, 115—116
Diethylphthalate, 100
Digesters, 287
Digestion, 87
    anaerobic
        monitoring of, 297
        nutrients and, 302
        substrate flows in, 287—289
        thermophilic, 300
Dihalogenated compounds, 107, see also specific
    types
Dihydric phenols, 197—202, see also specific types
Dihydrophloroglucinol, 175
Dihydroxy benzenes, 18, 20, see also specific types
3,5-Dihydroxybenzoate, 186
Dihydroxylated compounds, 101, see also specific
    types
Dihydroxymethylbenzenes, 175, see also specific
    types
2,4-Dimethylphenol, 195
3,4-Dimethylphenol, 195
3,5-Dimethylphenol, 195
Dimethylphenols, 175, 176, see also specific types
Dioxines, 84, see also specific types
Dioxygenase reactions, 101
Dioxygenases, 86
1,2-Dioxygenation, 106
5,6-Dioxygenation, 106
Dissolved oxygen, 41
1,2-Dioxygenation, 106
Drins, 90, see also specific types
Dye carriers, 104, see also specific types
Dyes, 97, 206, see also specific types

E

Ecosystems, 287
Effluents, 160, 241, see also specific types
Elements, 79, 301, see also specific types
Emission control, 92—93, 131—138, see also
    specific types
Endrin, 75
Energy, see also specific types
    demand for, 47, 240
    input of, 41
    yield of, 46
Environment, 85—86, 265
Enzymes, 86, see also specific types
    catabolic, 104
    chlorobenzenes and, 106
    hydrolase, 108
    partially purified, 170
    in phenol removal, 17, 26
    plasmid-coded, 28, 102
Essential trace elements, 301
Ethylbenzene, 89, 97
Ethylenes, 90, see also specific types
Ethylphenols, 97, 176, 177, see also specific types
Eubacterium limosum, 295
Eugenol, 177
Evaporation of contaminants, 144—145

Excavated soil, 142—150, see also Excavation
Excavation
    nonmicrobiological soil decontamination and, 93
    process of, 121
    soil decontamination after, 69, 91—93
Exoenzymes, 290, see also specific types
Extraction, 142—143, 148, 149, see also Bioextrac-
    tion; Leaching; specific types
    applications of, 143
    phenol, 14
    principle of, 142
Extractive dephenolizing, 12

F

Fatty acids, see also specific types
    anaerobic oxidation of, 292
    formation of, 294
    long-chain, 292
    profile of, 18
    synthesis of, 294
    volatile, 17, 294
Fermentation, 170, 185, 190—193
    of amino acids, 291—292
    of methane, 297—306
        nutrient requirements for, 301—303
        temperature and, 299—301
        toxicity and, 303—306
    pH and, 291
    of sugars, 291—292
Ferricyanides, 72, 84, see also specific types
Fertilizers, 118, 246, see also specific types
Ferulic acid, 177
Filtered organic carbon (FOC), 217
Filters, see also specific types
    biological, 234—235
    trickling, 22, 24, 41—42
Final purification, 42, 46
Flavobacterium sp., 105, 257
Flocculation, 22, 32
2-Fluorbenzoate, 106
4-Fluorbenzoate, 106
Foaming, 41
FOC, see Filtered organic carbon
Formalin, 19
Fossil fuel processing industries, 239
Free-energy change, 292
Frictional properties of soil, 80
Full-scale biological treatment plants, 272—274
Fungi, 105, see also specific types

G

Gallic acid, 177
Gas chromatography, 206
Gases, see also specific types
    barriers to, 135—136
    collection of, 92
    monitoring of emissions of, 138
    removal of, 132—137
    treatment of, 92, 137

Gasification, 6—8, 10, 20, 205, 213
Gas plant wastes, 97
Gasworks site contamination, 72
Genetically altered bacteria, 84, see also specific
types
Genetic engineering, 85, 269, 270
Genetics, 109, 111—113
Gentisate, 103
Gentisic acid, 177, 189
Geohydrological properties of soil, 78
Gloeocercospora sorghi, 259
Gluconobacter sp., 257
Glucose chemical oxygen demand, 211
Gouderak, 75
Ground water, 77, 80
level of, 78
monitoring of, 137—138
Guaiacols, 178, 211

H

Haloaromatic compounds, 104, see also specific
types
conversion of to halocatechols, 108
halogen elimination from, 106
metabolism of, 104
Halobacterium sp., 257
Halocatechols, 108, see also specific types
3-Halocatechols, 106
Halocitrate, 105
3-Haloderivatives, 106
Halogenated compounds, 70, 89, 90, see also specific
types
Halogenated hydrocarbons, 90, 112—113, see also
specific types
biodegradation of, 112
chemical treatments and, 147
extraction and, 143
hot gas treatments and, 145
volatile, 89, 90
Halogenated methanes, 112, see also specific types
Hard coal, 3
Hazardous wastes, 157—167, see also specific types
HCH, 90
γ-HCH, see Hexachlorocyclohexanes
Heat exchangers, 145
Heavy metals, 74, 86, see also specific types
chemical treatments and, 147
extraction and, 143
hot gas treatment of, 144
leaching of, 88
microorganisms and, 83
removal of, 144
toxicity of, 250
in waste waters, 207
n-Heptanoic acid, 178
Herbicides, 104, see also specific types
chlorinated phenoxy, 105
phenoxy, 84, 105
phenoxyacetate, 105
toxicity of, 250

Heterotrophic nitrification, 257
Hexachlorocyclohexanes, 89, 113—115, see also
specific types
High-performance biological treatment, 41—46, 48,
52
High performance liquid chromatography, 206
High-temperature carbonization (HTC), 5, 216—217
Homogentisate, 103
Homopyrocatechol, 178
Hot gas treatments, 144—145
HTC, see High-temperature carbonization
Hydrocarbons, see also specific types
aromatic, see Aromatic hydrocarbons
chemical treatments and, 147
chlorinated, 84
extraction and, 143
halogenated, see Halogenated hydrocarbons
hot gas treatments and, 145
hydrophilic, 143
microbial degradation of, 267
petroleum, 83
polycyclic aromatic, 84
toxicity of, 249—250
volatile halogenated, 89
Hydrogen
nonmethanogenic conversions of, 294—295
synthesis of volatile fatty acids from, 294
transfer of, 190, 291
Hydrogenation, 8, 10
Hydrogenotrophic acetogenic bacteria, 294
Hydrogen partial pressure, 293
Hydrogen transfer, 293
Hydrolase enzymes, 108, see also specific types
Hydrolysis, 288
of lignin, 207
optimum pH for, 290
as rate-limiting step, 290
of suspended solids, 290
Hydrophilic hydrocarbons, 143
Hydroquinone, 178, 189
Hydroxy benzenes, 17, 25, see also specific types
n-Hydroxybenzoate, 187
p-Hydroxybenzoate, 187
Hydroxybenzoates, 191, see also specific types
Hydroxybenzoic acids, 178, 179
p-Hydroxycinnamic acid, 179
2-Hydroxycyclohexane-carboxylic acid, 179
2-Hydroxycyclohexanone, 179
α-Hydroxymuconic semialdehyde, 179
2-Hydroxymuconic semialdehyde, 102
p-Hydroxyphenylacetate, 186
Hyohomicrobium sp., 112
Hypalon, 135
Hypohomicrobium sp., 257

I

Industrial processing systems, 129—131, see also
specific types
Industrial Revolution, 229—230
Industrial waste waters, 205—211

authentic, 212—218
biological treatment of, 259—272, see also specific
    types
classification of, 239
Inherent water, 6
Inhibition, 195
Inoculation, 124
Insecticides, 84, 246, 250, see also specific types
In situ decontamination of soil, 69, 90, 91, 94, 117—
    120, see also specific types
In situ stimulation of biodegradation, 87
In situ treatment, 93, 148—150
Intermediary acidogenesis, 288
Intermediary products, 292—294, see also specific
    types
Interspecies hydrogen transfer, 190, 291, 293
Iodobenzoates, 110, see also specific types
Isobutyl acetate, 14
iso-propylbenzene, 97
Isopropyl ether, 14

J

Jet grouting, 132

K

Ketone oil, 12
Ketones, 12—14, see also specific types
Kinetics
    of aerobic degradation, 30—35
    of bacterial growth, 263
    of batch cultures, 271
    of degradation, 30—35, 41
Klebsiella pneumoniae, 268
Kneer system, 91, 92, 125
Koppers recirculation, 14
Krimpen, 75

L

Lactobacillus plantarum, 266
Lagoons, 87
Lakes, 261, 262
Landfarming, 69, 70, 87, 96, 117—119
    costs of, 118—119, 122
    excavation and, 91, 122
    implementation of, 118, 122
    in situ decontamination and, 91
    principle of, 117—118, 122
    state of development of, 118
LAS, see Linear alkylbenzen sulfonates
Leachate, 158
Leaching, 88, 148, see also Extraction
Legislation, 82, 230—233
Lignin hydrolysis, 207
Lignite, see Brown coal
Lindane, see Hexachlorocyclohexanes
Linear alkylbenzene sulfonates (LAS), 255
Liquefaction, 8, 205—206, 288
Lithosphere, 79

Long-chain fatty acids, 292, see also specific types
Low-temperature carbonization (LTC), 5, 241
LTC, see Low-temperature carbonization
Lytic processes, 34

M

Magdeburg P-process, 41
Maintenance coefficient, 37
Maintenance metabolism, 32
Maleylacetate, 106
Mass spectrometry, 206
MCPA, 105, 106
Mercaptans, 258
Mesophilic methanogenic bacteria, 299
Metabolism, see also Cometabolism
    adaptation and, 305
    of alkylphenols, 196—197
    of aminophenols, 204
    analogue, see Cometabolism
    of aromatic compounds, 101—103
    of chlorobenzenes, 105—106
    of chlorophenols, 202—204
    of cyanide, 258—259
    of dihydric phenols, 197—202
    of haloaromatic compounds, 104
    maintenance, 32
    of naphthols, 195
    nitrogen, 256—258
    of nitrophenols, 204
    of phenols, 195
    stages of, 288
Metallo-organic compounds, 261, see also specific
    types
Metals, see also specific types
Methanes, see also specific types
    fermentation of, 297—306
        acidity and, 297—299
        buffering capacity and, 297—299
        nutrient requirements for, 301—303
        temperature and, 299—301
        toxicity and, 303—306
    halogenated, 112
Methanogenesis, 23, 89, 288, 295—297, 303
Methanogenic bacteria, 110, 287, 296, see also
    specific types
    acclimization potential of, 306
    elements in, 301
    mesophilic, 299
    occurrence of, 286
    susceptibility of to toxic compounds, 303
    thermophilic, 299
Methanogenic cocultures, 28
Methanol, 294—295
Methanosarcina sp., 297
Methanothrix sp., 297
Methoxyphenols, 180
Methyl branching, 83
3-Methylcatechol, 103
4-Methylcatechol, 103
Methylcyclohexane, 101

3-Methylgentisate, 103
Microbial decontamination of soil, 68
Microbial degradation, of pollutants
    biochemical pathways for, 252—255
    surfactants and, 255—256
Microbial interactions, 265—268, see also specific
    types
Microbiology, see also Bacteria; Microorganisms;
    specific aspects
  aromatic compounds and, 100—101
  brown coal refinement and, 17—37
  fundamentals of, 21—23
  soil decontamination and, 82—90
  of toxic waste treatment, 264—272
  waste water composition and, 17—19
Micrococcus sp., 100, 257
Microorganisms, see also Bacteria; Microbiology;
    specific types
  acid-producing, 88
  acclimation of, 262
  chemical compound interactions with, 82, 83
  cyanide-degrading, 259
  landfarming and, 118
  in xenobiotic degradation, 88—89
Mineral acids, 244
Mineralization, 110
Mineral oil, 70, 72
Minerals, 76, see also specific types
Mixed enrichment cultures, 170
Mixed liquor suspended solids (MLSS), 162, 265
Mixed liquor volatile suspended solids (MLVSS),
    162
MLSS, see Mixed liquor suspended solids
MLVSS, see Mixed liquor volatile suspended solids
Moisture, 118, 122—124
Monitoring, 92—93, 137—138, see also specific
    types
  of anaerobic digestion, 297
  for gaseous emissions, 138
  for ground water, 137—138
  of lakes, 261, 262
  process, 137
  of rivers, 261, 262
  of streams, 261, 262
Monoalkyl benzenes, 25
Monochlorobenzene, 105
Monochlorophenols, 90
Monohydroxy benzenes, 17, 25, 27, see also specific
    types
Moraxella sp., 255, 257, 258
cis,cis-Muconic acid, 180
Mutant-culture technology, 270
Mycobacterium sp., 256

N

Naphthalene, 89, 102
Naphthols, 180, 195, see also specific types
Natural ecosystems, 287
Neoprene, 135
Nitrate respiration, 89, 185—190, 205
Nitrification, 35—37, 49, see also specific types

autotrophic, 256—257
  heterotrophic, 257
  types of, 256—257
Nitrifying bacteria, 35, see also specific types
Nitriles, 89
Nitroaromatics, 89
Nitrobacter sp., 256
Nitrococcus sp., 256
Nitrogen
  carbon ratio to, 49
  COD ratio to, 302
  lack of, 86
  metabolism of, 256—258
  reduction of, 47
  in sewage, 46
Nitrogen-containing compounds, 89, see also specific
    types
p-Nitrophenol, 254
2-Nitrophenol, 195
4-Nitrophenol, 195
Nitrophenols, 181, 204, see also specific types
Nitrosococcus sp., 256
Nitrosolobus sp., 256
Nitrosomonas sp., 256
Nitrosopira sp., 256
Nitrospira sp., 256
Nocardia method, 21
Nocardia sp., 100, 103, 256
Nonhalogenated aromatics, 89—90, see also specific
    types
Nonhalogenated aromatic solvents, 70, see also
    specific types
Nonionic detergents, 244, see also specific types
Nonmethanogenic conversions, 294—295
Nontoxic waste waters, 238—247
Nutrients, see also specific types
  anaerobic digestion and, 302
  COD ratio to, 302
  composting and, 122
  methane fermentation and, 301—303
  supply of, 41

O

Oil-like products, 68, see also specific types
Oil products, 72, 84, see also specific types
Oil refinery products, 97, see also specific types
Oils, 68, 89, see also specific types
Olefin plants, 244
Open decontamination systems, 91, 125—127, see
    also specific types
Orcinol, 181
Organic compounds, see also specific types
  anaerobic degradation of, 287—297
    fermentation and, 291—292
    hydrolysis and, 290
    substrate flows in, 287—289
  removal of, 162
  in soil, 76
Organochlorine pesticides, 246, see also specific
    types
Organohalogenated pesticides, 246, see also specific

types
Organophosphates, 246, see also specific types
Organophosphorus pesticides, 258
Oxidation
  anaerobic, see Anaerobic oxidation
  of meleylacetate, 106
  of phenols, 14, 19—20
β-Oxidation, 292
Oxidative catabolism, 106
2-Oxocyclohexane, 181
Oxygen
  absence of (anaerobiosis), 86
  biological demand for, see Biological oxygen
      demand
  chemical demand for, see Chemical oxygen
      demand
  composting and, 122
  dissolved, 41
  influence of, 19, 22, 26, 31
  landfarming and, 118
  theoretical demand for, 215
  transfer of, 41
  yield of, 43
Oxygen uptake-rate studies, 162—165

P

Panel sheet piling, 132
Paper mills, 206, 207, 217—218
Paracoccus sp., 257
Partially purified enzymes, 170
PCBs, see Polychlorinated biphenyls
Peat, 3
Pentachlorophenol, 104, 181
Perchlorethylene, 112
Permeability of soil, 78
Peroxidase, 14
Peroxydases, 26
Pesticides, 83, 89, 104, see also specific types
  classification of, 247
  degradation of, 90
  microbial degradation of, 267
  organochlorine, 246
  organohalogenated, 246
  organophosphorus, 258
  toxicity of, 250
  waste water and, 246
Petrochemical industries, 205, 244
Petroleum, 89
Petroleum hydrocarbons, 83, see also specific types
Petroleum industries, 205, 208—210
pH
  control of, 298
  fermentation and, 291
  hydrolysis and, 290
  landfarming and, 118
Pharmaceuticals, 97, 261
Phenolics, 170, see also specific types
  anaerobic degradation of, 171—184, 193—195,
      205
  chlorinated, 218
  identification of, 206

in petroleum industry waste water, 208—210
  quantitation of, 206
  toxicity of, 248
  treatment in waste waters, 211—218
  in waste waters, 8, 10, 205—211
Phenols, 96—103, 262, see also specific types
  adsorption of, 14
  air oxidation of, 19—20
  biological treatment of, 267
  chemical oxidation of, 14
  chlorinated, 90, 103, 211
  degradation of, 90, 100
  detoxification of, 271
  dihydric, see Dihydric phenols
  enzymes in removal of, 17, 26
  extraction of, 14, 143
  fermentation and, 191, 195
  in industrial waste waters, 214, 215
  metabolism of, 101, 103, 185, 195
  microbiology and, 101
  nitrate respiration and, 186, 189
  photometabolism of, 186
  SBRs and, 162
  toxicity of, 248
  trihydric, see Trihydric phenols
  tyrosine and, 205
  yeast and, 22
Phenosolvan, 12, 20, 24
Phenoxyacetate herbicides, 105
Phenoxy herbicides, 84, see also specific types
Phenylalanine, 205
Phenylureas, 84
Phloroglucinol, 181, 182, 191
Phloroglucinolcarboxylic acid, 189
Phosphorus, 41, 165
  COD ratio to, 302
  lack of, 86
Photometabolism, 185—186
Phthalate esters, 90, 100, see also specific types
Phthalates, 89, 97—103, see also specific types
p-Hydroxyphenylacetic acid, 179
Physical methods of waste water treatment, 10—14,
      see also specific types
Pimelic acid, 182
PKM process, 49
Plasmid-coded enzymes, 28, 102, see also specific
      types
Plastic film, 132
Plasticizers, 89, 97
Plastics, 83, 97
Plastic sheet piling, 132
Plating industry waste water, 247
Pollutants, see also Contaminants; specific types
  degradability of, 89—90, 262—272
  degradation of, 82—83
  microbial degradation of, 251—259
    biochemical pathways for, 252—255
    surfactants and, 255—256
  toxicity of, 247—250
Pollution, 230—233, see also specific topics
Polyauxic degradation, 31, 32
Polychlorinated biphenyls (PCBs), 75, 84, 90, 104,

110—113, see also specific types
biochemistry of, 111—113
degradation of, 105, 110—111
genetics of, 111—113
uses of, 110
Polycyclic aromatic compounds, 75, see also specific types
Polycyclic aromatic hydrocarbons, 84, 97, see also specific types
Polyhydroxy benzenes, 17—19, see also specific types
Polymers, 83, see also specific types
Ponds, 87
Porosity, 124
Predation, 267
Preliminary purification, 12
Process industrial systems, 92
Process monitoring, 137
Propionebacterium sp., 257
Propionic acid, 182
Proteins, 40, see also specific types
Protocatechuate, 186
Protocatechuic acid, 182
Pseudomonas acidovorans, 100, 272
Pseudomonas aeruginosa, 254, 255, 267
Pseudomonas denitrificans, 257
Pseudomonas putida, 255, 268
  aromatic compounds and, 100
  chlorobenzenes and, 105
  dioxygenase system of, 25
  metabolism and, 101
Pseudomonas sp.
  aromatic hydrocarbons and, 254
  chlorobenzenes and, 104—106
  3-chlorobenzoate and, 106, 109
  denitrification and, 257
  dihalogenated compounds and, 107
  halogenated hydrocarbons and, 112
  improvement in, 108, 109
  metabolism and, 101, 102
  nitrate respiration and, 187
  nitrification of, 256
Pseudomonas cepacia, 105
Pulp mills, 206, 207, 211, 217—218
Pure cultures, 170, 263
Purification
  biological, 41—47, 49
  degree of, 46
  final, 42, 46
  preliminary, 12
  of sewage, 46—47
Pyridine, 12—14
Pyrocatechase II, 106
Pyrogallol, 182, 191
Pyruvate, 258

R

RDX, see Trinitrotrimethylenetriamine
Reaction water, 8
Recalcitrant compounds, 83—85, see also specific types

Recirculation, 14
Recycling of sludge, 32
Refined oils, 89
Refineries, 243—244, see also specific types
Resin manufacturing, 206
Resorcinol, 189, 191
α-Resorcylic acid, 182
β-Resorcylic acid, 182
Rhodococcus sp., 100
Rhodopseudomonas pulustris, 186
Rhodopseudomonas sp., 257
Rivers, 26, 262

S

Salicylate, 102
Salicylic acid, 183, 189
Salt, 86
Santhomonas sp., 257
Saturated hydrocarbon fractions, 89, see also specific types
SBR, see Sequencing batch reactors
Schnorr system, 91, 125
Secondary toxicants, 264, 272, see also specific types
Semi-aerobic degradation, 87
Sequencing batch reactors (SBR), 157—167
  bench-scale studies of, 160—166
  full-scale demonstration facility of, 166
  operating strategies for, 161—162
Sewage
  biological purification of, 46—47
  denitrification of, 46
  nitrogen in, 46
  purification of, 46—47
  two-stage treatment of, 32
Sinapic acid, 183
Sludge, see also specific types
  activated, see Activated sludge
  age of, 46
  amino acids in, 38, 40
  applications of, 123
  disposal of, 37—38
  as feed protein, 38
  flocculation of, 22
  formation of, 38
  proteins in, 40
  recycling of, 32
  treatment of, 38
  vitamins in, 40
Sludge volume index (SVI), 162
Slurry trench technique, 132
Soil
  biodegradation in, 87
  composition of, 77
  compressibility of, 78
  decontamination of, see Decontamination of soil
  elements in, 79
  excavated, 142—150, see also Excavation
  frictional properties of, 80
  geohydrological properties of, 78
  in situ treatment of, 148—150
  mechanical properties of, 78—80

minerals in, 76
organic compounds in, 76
permeability of, 78
physicochemical properties of, 75—77
stratification of, 75
structure of, 76
types of, 75—80
xenobiotic turnover in, 82
Soil pollution, 71—75, see also specific types
Solids, see also specific types
detoxification of, 87—88
mixed liquor suspended, 162, 265
mixed liquor volatile suspended, 162
suspended, see Suspended solids
volatile suspended, 265
Solvents, 68, 72, 74, 104, see also specific types
Sour digesters, 298, 303
Spirillum sp., 257
SS, see Suspended solids
Standard free-energy change, 292
Steam stripping, 146, 150
Steel mills, 261
Steel sheet piling, 131
Stemphylium loti, 259
Stimulated biodegradation, 86—88
Stockholm Conference, 230
Stratification of soil, 75
Streams, 261, 262
Streptococcus faecalis, 266
Substrate flows, 287—289
Substrate-specific demands, 40—41, see also specific
    types
Succinic acid, 183
Sugars, 291—292, see also specific types
Sulfate reduction, 185, 190
Sulfides, 207, 258, 264
Sulfonates, 255, see also specific types
Sulfur-containing compounds, 89, 258, see also
    specific types
Surfactants, 255—256, 267, see also specific types
Surrogate indicators, 238
Suspended solids (SS), 162, 265, 290
SVI, see Sludge volume index
Synergism, 304
Synthetic coal, 213
Synthetic polymers, 83
Synthetic waste waters, 211—212
Syringaldehyde, 183
Syringic acid, 183
Syringol, 183

T

Tar hydrogenation, 10
Tar products, 72, see also specific types
TCC, see Tricarboxylic acid cycle
TCDD, 84
Telodrin, 90
Temperature
  bacteria and, 23
  composting and, 124
  extreme, 86

landfarming and, 118
methane fermentation and, 299—301
short-term variations in, 299
Tetrachloroethylene, 112, 113
Theoretical oxygen demand (TOD), 215
Thermal methods of brown coal refinement, 5—8,
    see also specific types
Thermal treatments, 143—146, 150, see also specific
    types
Thermophilic anaerobic digestion, 300
Thermophilic bacteria, 21, see also specific types
Thermophilic methanogenic bacteria, 299
Thermotolerant bacteria, 21, see also specific types
Thiobacillus sp., 88, 257
Thiocyanates, 258, see also specific types
  biological treatment of, 267
  detoxification of, 271
  toxicity of, 249
Thiols, 258, see also specific types
Thiopenes, 258, see also specific types
Thiosulfates, 258, see also specific types
TNT, see Trinitrotoluene
TOC, see Total organic carbon
TOD, see Theoretical oxygen demand
Toluate, 109
Toluate oxidase, 109
Toluene, 97
  biodegradation of, 100
  degradability of, 89
  metabolism of, 101—103
  microbiology and, 100, 101
Topography, 81
Total organic carbon (TOC), 162
Tower trickling filter method, 41—42
Toxicants, 264, 272, 303, see also specific types
Toxicity, see also specific types
  of ammonia, 248
  of cyanide, 249
  of heavy metals, 250
  of hydrocarbons, 249—250
  methane fermentation and, 303—306
  of pesticides, 250
  of phenolics, 248
  of pollutants, 247—250
  of thiocyanates, 249
Toxic wastes
  biological treatment of, 262—272
  full-scale biological waste treatment plants for,
      272—274
  microbiology in treatment of, 264—272
Toxic waste waters, 238—247, see also specific
    types
Trace elements, 301, see also specific types
Trenches, 134—135
Triazines, 84, see also specific types
Tricarboxylic acid cycle (TCC), 25
1,1,1-Trichloroethane, 112
Trichloroethylene, 112
2,4,6-Trichlorophenol, 195
Trichosporon cutaneum, 263
Trickling filters, 22, 24, 41—42
Triga system, 92, 125

Trihydric phenols, 197—202, see also specific types
3,4,5-Trimethoxybenzoate, 188
Trimethoxybenzoic acids, 183
Trimethylbenzenes, 89, 97, see also specific types
Trinitrotoluene (TNT), 127
Trinitrotrimethylenetriamine (RDX), 127
Tryptophan, 205
Two-stage sewage treatment, 32
Tyrosine, 184, 205

U

Underground water, 87

V

Vanillic acid, 184
Vanillins, 184, 211
Vanillyl alcohol, 184
Ventilation, 133—135
Veratic acid, 184
Vitamins, 40, see also specific types
Volatile fatty acids, 17, 294, see also specific types
Volatile halogenated hydrocarbons, 89, 90, see also
    specific types
Volatile suspended solids (VSS), 265
Volgermeerpolder, 75
VSS, see Volatile suspended solids

W

Wastes, see also specific types
  application of, 87
  coke oven, 97
  gas plant, 97
  hazardous, see Hazardous wastes
  solid, see Solids
Waste waters, see also specific types
  biodegradability of in SBRs, 164
  biological treatments of, 233—238
  from brown coal refinement, 9
    aerobic degradation and, 30—35
    chemical methods of treatment of, 10, 14—17
    in Germany, 48—49
    microbiology in treatment of, see Microbiology
    physical methods of treatment of, 10—14
    sludge disposal in, 37—38
    treatment of, see specific types
  characteristics of, 160, 238—247, 251, see also
      specific types
    nontoxic, 238—247
    from refineries, 243—244
    toxic, 238—247

classification of, 239
coal conversion, 215—216, 241—243
coke oven, 215
coking, 216—218
composition of, 17—19
digestion and, 87
koppers recirculation and, 14
in pesticide industry, 246
petrochemical industry, 244
from petroleum industries, 208—210
phenolics in, 205—211
plating industry, 247
preliminary purification of, 12
refinery, 243—244
synthetic, 211—212
treatment of, 211—218
Water
  combustion, 6
  condensation, 8, 17
  ground, see Ground water
  inherent, 6
  lack of, 86
  phenolic, 8, 10
  reaction, 8
  underground, 87
  waste, see Waste waters
  xenobiotic turnover in, 82
Well drilling, 92
Windrow system, 125
Wood preserving plants, 206

X

Xanthobacter autotropicus, 112
Xenobiotics, see also specific types
  catabolism of, 85
  degradation of, 88—89
  microorganisms and, 88—89
  molecular structure of, 83
  persistent, 84
  turnover of, 82
o-Xylene, 103
p-Xylene, 103
Xylenes, 97, see also specific types
  biodegradation of, 100
  degradability of, 89
  metabolism of, 101, 102
  microbiology and, 100, 101
Xylenols, 97, see also specific types

Y

Yeast, 22, 24, see also specific types